Boundary Elements: Implementation and Analysis of Advanced Algorithms

E
V
C

Notes on Numerical Fluid Mechanics (NNFM) Volume 54

Volume 54 Boundary Elements: Implementation and Analysis of Advanced Algorithms. Proceedings of the Twelfth GAMM-Seminar, Kiel, January 19–21, 1996 (W. Hackbusch / G. Wittum, Eds.)

Volume 53 Computation of Three-Dimensional Complex Flows. Proceedings of the IMACS-COST Conference on Computational Fluid Dynamics, Lausanne, September 13–15, 1995 (M. Deville / S. Gavrilakis / I. L. Ryhming, Eds.)

Volume 52 Flow Simulation with High-Performance Computers II. DFG Priority Research Programme Results 1993–1995 (E. H. Hirschel, Ed.)

Volume 51 Numerical Treatment of Coupled Systems. Proceedings of the Eleventh GAMM-Seminar, Kiel, January 20–22, 1995 (W. Hackbusch / G. Wittum, Eds.)

Volume 50 Computational Fluid Dynamics on Parallel Systems. Proceedings of a CNRS-DFG Symposium in Stuttgart, December 9 and 10, 1993 (S. Wagner, Ed.)

Volume 49 Fast Solvers for Flow Problems. Proceedings of the Tenth GAMM-Seminar, Kiel, January 14–16, 1994 (W. Hackbusch / G. Wittum, Eds.)

Volume 48 Numerical Simulation in Science and Engineering. Proceedings of the FORTWIHR Symposium on High Performance Scientific Computing, München, June 17–18, 1993 (M. Griebel / Ch. Zenger, Eds.)

Volume 47 Numerical Methods for the Navier-Stokes Equations (F.-K. Hebeker, R. Rannacher, G. Wittum, Eds.)

Volume 46 Adaptive Methods – Algorithms, Theory and Applications. Proceedings of the Ninth GAMM-Seminar, Kiel, January 22–24, 1993 (W. Hackbusch / G. Wittum, Eds.)

Volume 45 Numerical Methods for Advection – Diffusion Problems (C. B. Vreugdenhil / B. Koren, Eds.)

Volume 44 Multiblock Grid Generation – Results of the EC/BRITE-EURAM Project EUROMESH, 1990–1992 (N. P. Weatherill / M. J. Marchant / D. A. King, Eds.)

Volume 43 Nonlinear Hyperbolic Problems: Theoretical, Applied, and Computational Aspects Proceedings of the Fourth International Conference on Hyperbolic Problems, Taormina, Italy, April 3 to 8, 1992 (A. Donato / F. Oliveri, Eds.)

Volume 42 EUROVAL – A European Initiative on Validation of CFD Codes (W. Haase / F. Brandsma / E. Elsholz / M. Leschziner / D. Schwamborn, Eds.)

Volume 41 Incomplete Decompositions (ILU) – Algorithms, Theory and Applications (W. Hackbusch / G. Wittum, Eds.)

Volume 40 Physics of Separated Flow – Numerical, Experimental, and Theoretical Aspects (K. Gersten, Ed.)

Volume 39 3-D Computation of Incompressible Internal Flows (G. Sottas / I. L. Ryhming, Eds.)

Volume 38 Flow Simulation on High-Performance Computers I (E. H. Hirschel, Ed.)

Volume 37 Supercomputers and their Performance in Computational Fluid Mechanics (K. Fujii, Ed.)

Volume 36 Numerical Simulation of 3-D Incompressible Unsteady Viscous Laminar Flows (M. Deville / T.-H. Lê / Y. Morchoisne, Eds.)

Volumes 1 to 29, 45 are out of print.
The addresses of the Editors and further titles of the series are listed at the end of the book.

Boundary Elements: Implementation and Analysis of Advanced Algorithms

Proceedings of the Twelfth GAMM-Seminar
Kiel, January 19–21, 1996

Edited by
Wolfgang Hackbusch and
Gabriel Wittum

vieweg

Vieweg ist a subsidiary company of Bertelsmann Professional Information.

Produced by Langelüddecke, Braunschweig
Printed on acid-free paper
ISBN-13: 978-3-322-89943-9 e-ISBN-13: 978-3-322-89941-5
DOI: 10.1007/978-3-322-89941-5

FOREWORD

The GAMM Committee for "Efficient Numerical Methods for Partial Differential Equations" organizes seminars and workshops on subjects concerning the algorithmic treatment of partial differential equations. The topics are discretisation methods like the finite element and the boundary element method for various type of applications in structural and fluid mechanics. Particular attention is devoted to advanced solution methods.

The latest seminar in this series was the 12th Kiel-Seminar which took place on January 19–21, 1996 at Christian-Albrechts-University of Kiel and focussed on the topic

Boundary Elements:
Implementation and Analysis of Advanced Algorithms.

The seminar was attended by 65 scientists from 10 countries. 23 lectures were given, including two survey lectures.

In the last years, a remarkable progress in the numerical treatment of boundary element methods (BEM) has been obtained in Germany. This is, in particular, a result of a Schwerpunktverfahren supported by the DFG. Many aspects of the BEM are not only analysed but also implemented. Therefore, these proceedings present a number of papers on implementational aspects besides the analysis of advanced techniques.

Kiel, July 1996

W. Hackbusch
G. Wittum

Advanced information on the GAMM-seminars Kiel is electronically available via the WWW-site

`http://www.numerik.uni-kiel.de/gamm.html`

CONTENTS

MULTISCALE METHODS FOR BOUNDARY INTEGRAL EQUATIONS AND THEIR APPLICATION TO BOUNDARY VALUE PROBLEMS IN SCATTERING THEORY AND GEODESY

Bernd H. Kleemann
Berliner Institut für Optik, GmbH
Rudower Chaussee 6
D–12484 Berlin, Germany

Andreas Rathsfeld
Weierstraß-Institut für
Angewandte Analysis und Stochastik
Mohrenstr. 39
D–10117 Berlin, Germany

Reinhold Schneider
Fachbereich Mathematik
Technische Hochschule Darmstadt
Schloßgartenstr. 7
D–64289 Darmstadt, Germany

SUMMARY

In the present paper we give an overview on multiscale algorithms for the solution of boundary integral equations which are based on the use of wavelets. These methods have been introduced first by Beylkin, Coifman, and Rokhlin [5]. They have been developed and thoroughly investigated in the work of Alpert [1], Dahmen, Proessdorf, Schneider [16-19], Harten, Yad-Shalom [25], v.Petersdorff, Schwab [33-35], and Rathsfeld [39-40]. We describe the wavelet algorithm and the theoretical results on its stability, convergence, and complexity. Moreover, we discuss the application of the method to the solution of a two-dimensional scattering problem of acoustic or electromagnetic waves and to the solution of a fixed geodetic boundary value problem for the gravity field of the earth. The computational tests confirm the high compression rates and the saving of computation time predicted by the theory.

1. INTRODUCTION

Boundary element methods (BEM) are known to be a good tool for the solution of some classes of boundary value problems for partial differential equations. Frequently, however, the fact that the arising stiffness matrices are dense causes a lot of problems in the implementation. For example, the solution of a boundary element system with about 100 000 unknowns requires a storage capacity of about 80 Gigabytes. Clearly, also the computation time for the processing of such an amount of data is considerably long. To reduce storage and computation time, there have been introduced several methods. We only mention the method of multipole expansion by Rokhlin and Greengard [42,23], the panel clustering of Hackbusch and Nowak [24,46], and the splitting methods of Amosov [3], Saranen and Vainikko [44]. Note that *h-p* schemes can also be considered as a method to improve computation time and to reduce storage.

In the present paper, we will discuss algorithms which are called *multiscale methods* and which are well-known from their application to finite element methods. For finite element methods, however, the matrices of the linear equations are sparse, and the multiscale approach is used only to construct preconditioners for the iterative solution of the matrix equations. This feature is important for the BEMs, too. Namely, in case of integral operators of order different from zero, the condition numbers of the traditional BEM matrices increases rapidly if the mesh size tends to zero whereas the condition numbers of the BEM matrices corresponding to wavelet bases multiplied by diagonal preconditioners remain bounded. In addition to this preconditioning, the multiscale setting also provides a good tool to develop fast algorithms for the multiplication of a vector times the matrices of the BEM systems, i.e., to develop fast iterative solvers. This facility has been used without a wavelet frame by Brandt and Lubrecht [6] and together with a wavelet scheme in the pioneer paper by Beylkin, Coifman, and Rokhlin [5]. Note that throughout the present paper we will use the notion of wavelets in a generalized sense.

The *wavelet approach* has been thoroughly investigated and further developed by several authors. A new basis of multi wavelets for the discretization of integral equations has been introduced by Alpert [1] and a wavelet basis for the space of test functionals in case of collocation has been treated by Dahmen, Proessdorf, Schneider [16-19]., Harten, and Yad-Shalom [25]. Moreover, Dahmen, Proessdorf, and Schneider [16-19] were probably the first who have matched the compression error to the discretization error of the Galerkin and the collocation method, respectively. The same authors [19,47] have stressed the importance of higher orders for the dual wavelets in biorthogonal settings, and they have defined preconditioners for the BEM matrices corresponding to the wavelet basis [18] (cf. also [14,26,48]). A further improvement of the estimates for the error due to the compression as well as special aspects for special BEM applications have been investigated by v.Petersdorff and Schwab [34-35]. The final step in the reduction of the number of arithmetic operations has been achieved by Schneider [47]. He has proved that, for the computation of a numerical solution with N degrees of freedom approximating the exact solution with an error less than the usual discretization error, $O(N)$ matrix entries are sufficient. This result is obtained if the compression is extended to matrix entries corresponding to test and trial wavelets with overlapping support. Furthermore, a suitable quadrature algorithm for the computation of the stiffness matrix with respect to the wavelet bases has been developed simultaneously by Schneider [47], v.Petersdorff and Schwab [34] for the case of the Galerkin method and by Rathsfeld [39-40] for collocation.

The first results on wavelet methods over non-uniform grids are due to Rathsfeld [39], v. Petersdorff and Schwab [33]. The issue of adaptive wavelet methods has been discussed by Dahlke, Dahmen, Hochmuth, and Schneider [11]. Finally, we mention that the first steps to handle unstructured grids have been performed by Carnicer, Dahmen, Peña [7], and Sweldens [49] (cf. also [15]), and numerical tests have been reported e.g. in the papers by Dorobantu [21] and by Dahmen, Kleemann, Proessdorf, and Schneider [12,13].

The *plan of the paper* is as follows. In Section 2 we will make a few remarks on boundary integral operators and introduce the integral operators for the direct time-harmonic scattering problem and for the geodetic boundary value problem. The trial space and its multiscale decomposition will be introduced in Section 3. We will define the wavelet functions in a general setting. In particular, we will describe the piecewise linear wavelets with smallest support which are employed in the numerical experiments. Section 4 is devoted to the wavelet algorithm for the fast solution of the Galerkin BEM equations. Thus, we will present compression strategies, error estimates, a remark on the quadrature approximation, and theorems on the diagonal preconditioning. In Section 5 we will consider the collocation method in an analogous manner. We will introduce wavelet test functionals and define the wavelet algorithm for the collocation. Finally, we present the results of numerical tests.

2. BOUNDARY INTEGRAL OPERATORS

2.1. GENERAL BOUNDARY INTEGRAL OPERATORS

The first step of BEM consists in the reduction of boundary value problems to equivalent integral operators. For example, suppose a boundary value problem for an elliptic partial differential equation over a domain D in the $n+1$ dimensional Euclidean space $\mathbb{R}^{n+1}$ is to be solved. Then we can seek a solution in the form of an integral representation, e.g., in form of the representation formula for the solution of the partial differential equation. Clearly, the representation includes one or more yet unknown density functions defined over the boundary Γ of D. Applying the operator of the boundary value condition to this representation, we end up with one or more integral equations over Γ for the unknown density functions. With the BEM these integral equations are solved numerically. Finally, we get the solution to the boundary value problem by substituting the approximate density functions into the integral representation.

Now we suppose that the boundary Γ is piecewise smooth, i.e., we suppose Γ to be the union of the closed bounded surface pieces $\overline{\Gamma}_m$, $m = 1, \ldots, m_\Gamma$ such that, for every m, there exists a coordinate mapping $\kappa_m : \square \to \Gamma_m$ from a reference domain $\square \subseteq \mathbb{R}^n$ to Γ_m. The domain $\square$ is something like a unit square or a simple triangle, and the mapping κ_m is supposed to be continuously differentiable up to a certain degree. Moreover, we suppose that this mapping extends to a mapping over a small neighbourhood of $\square$. Note that, frequently, the parametric surface representation is provided by CAD, CAGD (cf. e.g. freeform surfaces, NURBS, and the standards for format: IGES).

Usually, the boundary integral operator over Γ is a matrix of integral operators of different types. In the simplest case, the boundary integral equation takes the form

$$Au(x) := au(x) + \int_\Gamma K_A(x,y)u(y)\mathrm{d}_y\Gamma = v(x), \quad x \in \Gamma, \tag{1}$$

where a is a constant and $K_A(x,y)$ is the so called Schwartz kernel of A. Frequently, the integral on the left-hand side of (1) is to be understood in a generalized sense. For a certain fixed $\mathbf{r}$, the kernel $K_A(x,y)$ satisfies the Calderon Zygmund estimate

$$|\partial_x^\alpha \partial_y^\beta K_A(x,y)| \leq C(\alpha,\beta,A,\Gamma)\, |x-y|^{-(n+\mathbf{r}+|\alpha|+|\beta|)}\,, \tag{2}$$

provided $n+\mathbf{r}+|\alpha|+|\beta| > 0$. Note that $\mathbf{r}$ is called the order of A and that, for sufficiently smooth Γ and classical pseudodifferential operators A, the operator A maps the Sobolev space $H^s(\Gamma)$ into $H^{s-\mathbf{r}}(\Gamma)$. Usually, we have $a=1$, $\mathbf{r}=0$ or $a=0$, $\mathbf{r}\neq 0$. Finally, we remark that, in local coordinates, (1) takes the form

$$\begin{aligned} Au(\kappa_k(t)) &= au(\kappa_k(t)) + \sum_{m=1}^{m_\Gamma} \int_\square K_A(\kappa_k(t),\kappa_m(s))u(\kappa_m(s))|\kappa_m'(s)|\mathrm{d}s = v(\kappa_k(t)), \\ & \quad t \in \square,\ k=1,\ldots,m_\Gamma, \end{aligned} \tag{3}$$

where $|\kappa_m'(s)|$ denotes the absolute value of the derivative $\kappa_m'(s)$ for $n=1$ and the absolute value of the vector product $\partial_{s_1}\kappa_m(s) \times \partial_{s_2}\kappa_m(s)$ for $n=2$.

For the stability of the numerical methods, the concept of strong ellipticity plays a crucial role. We call A strongly elliptic if A satisfies the so called Gårding inequality, i.e.,

$$\mathrm{Re}\langle Au,u\rangle_{L^2(\Gamma)} \geq \gamma\|u\|_{H^{\mathbf{r}/2}(\Gamma)} - |\langle Tu,u\rangle_{L^2(\Gamma)}| \tag{4}$$

for any $u \in H^{\mathbf{r}/2}(\Gamma)$. In (4) the operator $T \in \mathcal{L}(H^{\mathbf{r}/2}(\Gamma), H^{-\mathbf{r}/2}(\Gamma))$ is supposed to be compact and γ stands for a positive constant independent of u.

2.2. AN INTEGRAL OPERATOR IN SCATTERING THEORY

Now we turn to a boundary integral equation for a scattering problem (for details cf. [10]). We consider the scattering of time–harmonic acoustic and electromagnetic waves by an infinitely long cylindrical obstacle with a simply connected bounded cross section $D_- \subset \mathbb{R}^2$. We suppose the boundary Γ of D_- to be analytic and set $D := \mathbb{R}^2 \setminus D_-$. The wave corresponding to a wave number $\bar{k}$ and defined in the outer domain D is the sum of a given incoming wave w^{in} and the unknown scattered wave w. For a sound-soft obstacle, this w is the solution of the following exterior boundary value problem for the Helmholtz equation:

$$\Delta w(x) + \bar{k}^2 w(x) = 0 \quad \text{in } D, \tag{5}$$

$$w(x) = -w^{\mathrm{in}}(x) \quad \text{on } \Gamma, \tag{6}$$

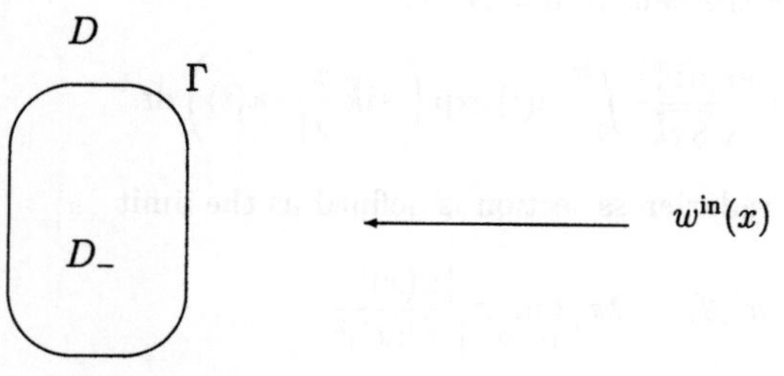

Figure 1: Scattering problem.

$$\frac{\partial w(x)}{\partial r} - \mathrm{i}\bar{k}w(x) \;=\; o(r^{-1/2}),\ r = |x| \to \infty\,. \tag{7}$$

Here (7) is supposed to hold uniformly in all directions. This condition is called Sommerfeld radiation condition. We seek w in form of a single-layer potential

$$w(x) = \frac{1}{2\mathrm{i}}\int_\Gamma H_0^{(1)}(\bar{k}|x-y|)\tilde{u}(y)\mathrm{d}_y\Gamma \tag{8}$$

with the Hankel function $H_0^{(1)}$ of order zero and of the first kind. Using the global parametrization $\kappa = (x_1, x_2) : \square \longrightarrow \Gamma$ over the 2π periodic interval $\square$, we arrive at the boundary integral equation

$$(Au)(t) := \frac{1}{2\pi}\int_0^{2\pi} K(t,\tau)u(\tau)\mathrm{d}\tau = v(t)\,,\ 0 \le t \le 2\pi, \tag{9}$$

where

$$\begin{aligned} u(t) &:= \tilde{u}(\kappa(t))\sqrt{[x_1'(t)]^2 + [x_2'(t)]^2}, \quad v(t) := -w^{\mathrm{in}}(\kappa(t))\,,\\ K(t,\tau) &:= \frac{\pi}{\mathrm{i}}H_0^{(1)}(\bar{k}r(t,\tau))\,,\ t \ne \tau\,,\\ r(t,\tau) &:= \sqrt{[x_1(t)-x_1(\tau)]^2 + [x_2(t)-x_2(\tau)]^2}\,. \end{aligned}$$

Note that the integral operator A on the left-hand side of (9) is strongly elliptic, its order **r** is -1, and the dimension of the boundary manifold is $n = 1$. In addition to (2), we have

$$|K(t,\tau)| \le C(0,0,A,\Gamma)\log\left(\max\{1, |t-\tau|^{-1}\}\right), \tag{10}$$

i.e., the integral equation has a logarithmic kernel.

Knowing the solution u of (9), several interesting quantities can be derived. The scattered wave w is known to have an asymptotics of the form

$$w(x) = \frac{\exp(\mathrm{i}\bar{k}|x|)}{\sqrt{|x|}}\left\{u^\infty\left(\frac{x}{|x|}\right) + O\left(\frac{1}{|x|}\right)\right\},\ |x| \longrightarrow \infty. \tag{11}$$

Here, the function u^∞ is called the far field pattern or scattering amplitude of w. It can be expressed as a functional of the density $u = A^{-1}v$.

$$u^\infty\left(\frac{x}{|x|}\right) = \frac{\exp(i\frac{\pi}{4})}{\sqrt{8\pi \bar{k}}} \int_0^{2\pi} u(t) \exp\left(-i\bar{k}\frac{x}{|x|}\cdot\kappa(t)\right) dt. \tag{12}$$

For $x/|x| = (\cos\theta, \sin\theta)$, the radar cross section is defined as the limit

$$\sigma^c(\theta) := 2\pi \lim_{|x|\to\infty} x \frac{|w(x)|^2}{|w^{in}(x)|^2}.$$

It is not hard to see that $\sigma^c(\theta) = 2\pi|u^\infty(x/|x|)|^2$.

2.3. AN INTEGRAL OPERATOR IN GEODESY

One fundamental problem in geodesy consists in the computation of the gravity field around the earth from the gravity data measured over the known surface of the earth (for more details cf. [27,22]). In other words, if D is the exterior of the earth, Γ the earth's surface, ω the angular velocity of the rotation, g the gravity, and w the unknown potential of the gravity field, then w is the solution of

$$\begin{aligned} \Delta w(x) &= 2\omega^2, \quad x \in D, \\ |\text{grad}\, w(x)| &= g(x), \quad x \in \Gamma. \end{aligned}$$

Introducing a well approximating reference potential $w_0 \approx w$ and neglecting higher order non-linear terms, we arrive at the following linear oblique derivative boundary value problem for the unknown difference potential $\delta w := w - w_0$:

$$\begin{aligned} \Delta[\delta w](x) &= 0, \qquad x \in D, \\ \partial_{l(x)}[\delta w](x) &= \delta g(x), \quad x \in \Gamma, \end{aligned}$$

where $\delta g := (g^2 - g_0^2)/2|g_0|$, $g_0 :=$ grad w_0, and $l := g/|g_0|$. For the vector field l, we suppose that, at any $x \in \Gamma$, $l(x)$ is not tangential. We seek δw in form of a single layer potential with unknown density function $\tilde{u}$ over the earth's surface Γ.

$$\delta w(x) = \frac{1}{4\pi}\int_\Gamma \frac{\tilde{u}(y)}{|x-y|} d_y\Gamma. \tag{13}$$

Substituting this potential into the oblique derivative boundary condition and applying the jump relation for the gradient of the single layer potential, we obtain the following singular integral equation:

$$-2\pi\cdot\cos[\mathbf{n}(x), l(x)]\cdot\tilde{u}(x) + p.v.\int_\Gamma \frac{\cos[l(x), y-x]}{|y-x|^2}\cdot\tilde{u}(y)\cdot d_y\Gamma \;=\; 4\pi\cdot\delta g(x). \tag{14}$$

Here $\mathbf{n}$ denotes the direction of the outward pointing normal on the earth's surface Γ and the integral is defined in the principal value sense (cf. [29]). Of course, we are not able to discretize the whole surface of the earth. This would require too much storage

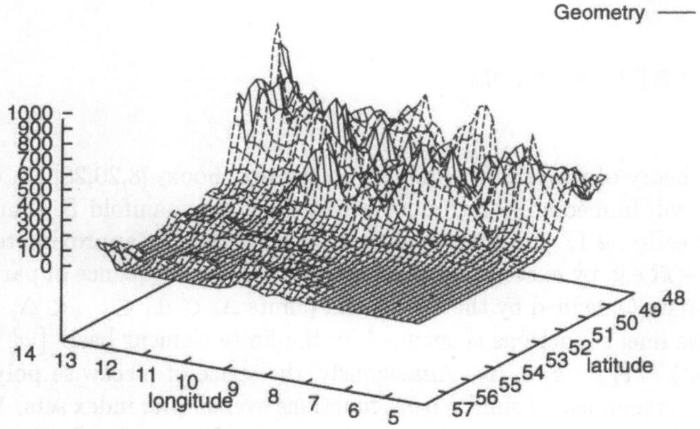

Figure 2: Height depending on latitude and longitude.

and computation time. Thus, we restrict the surface Γ to a quadratic surface piece of points with latitude between 48.60^0 and 56.65^0 and longitude between 5.35^0 and 13.40^0 (cf. Figure 2). Using a parametrization $\kappa : \square := [0,1] \times [0,1] \longrightarrow \Gamma$ based on Overhauser interpolation (cf. [32]) of uniformly distributed data, we transform (14) into the boundary integral equation $Au = v$ over $\square$ of the form

$$-2\pi \cos[\mathbf{n}(\kappa(t)), l(\kappa(t))]u(t) + p.v. \int_\square \frac{\cos[l(\kappa(t)), \kappa(s) - \kappa(t)]}{|\kappa(s) - \kappa(t)|^2} |\kappa(s)| u(s) ds \;=\; v(t), \quad (15)$$

where $u(t) := \tilde{u}(\kappa(t))$ and $v(t) := 4\pi \cdot \delta g(\kappa(t))$. Note that the singular integral operator A is a strongly elliptic boundary integral operator defined over a "manifold" of dimension $n = 2$ and its order $\mathbf{r}$ is equal to zero.

3. TRIAL FUNCTIONS AND MULTISCALE DECOMPOSITION

3.1. THE WAVELET FUNCTIONS

The general theory of wavelets is exhibited in the text books [8,20,28]. In contrary to these books, we will immediately introduce wavelets on the manifold Γ. Thus, suppose that Γ is as in Section 2.1. Using the parametrizations, we can approximate boundary functions $u : \Gamma \to I\!R$ e.g. by curved finite elements. We chose a sequence of partitions over the reference domain $\square$ defined by the sets of grid points $\Delta_0 \subset \Delta_1 \subset \ldots \subset \Delta_j \subset \ldots$. The space of piecewise linear functions is spanned by the finite element basis $\{\varphi_k^{j,*} : k \in \Delta_j\}$ defined by $\varphi_k^{j,*}(k') = \delta_{k,k'}$, $k' \in \Delta_j$. Analogously, the space of piecewise polynomials of degree less than d is spanned by similar basis functions over similar index sets. We suppose that these basis functions are normalized such that $\|\varphi_k^{j,*}\|_{L^2} \sim 1$. Setting $\varphi_{m,k}^j(x) = \varphi_k^{j,*}(\kappa_m^{-1}(x))$, $x \in \Gamma_m$, we obtain the finite element space $\text{span}\{\varphi_{m,k}^j : k \in \Delta_j\}$. We denote by S_j the intersection of the span of all these spaces defined for $m = 1, \ldots, m_\Gamma$ with the space $C^\sigma(\Gamma)$ of σ times continuously differentiable functions. In particular, we choose $\sigma = 0$ for the piecewise linear elements and $\sigma = -1$, i.e., no intersection with $C^\sigma(\Gamma)$, for the piecewise constant elements. We note here that a choice of higher values of σ can lead to a lot of difficulties, especially, for the subsequent constructions of wavelet bases. Even the case $\sigma = 0$ is not trivial. Nevertheless, we arrive at a multiscale analysis $S_0 \subset S_1 \subset \ldots \subset S_j \subset \ldots$ of order d ($d = 2$ for piecewise linear or bilinear functions and $d = 1$ for piecewise constants) and regularity $\gamma = \sigma + 3/2$. In other words, the spaces S_j are contained in the Soblolev spaces $H^s(\Gamma)$ of order s less than γ and the S_j contain polynomials of degree less than d. For these spaces S_j, we assume the approximation property

$$\inf_{u_j \in S_j} \|u - u_j\|_{H^s} \leq C 2^{-j(t-s)} \|u\|_{H^t}, \qquad s \leq t \leq d,\ s < \gamma \tag{16}$$

and the inverse property

$$\|u_j\|_{H^t} \leq C 2^{j(t-s)} \|u_j\|_{H^s}, \qquad s \leq t < \gamma. \tag{17}$$

Now we set $W_{-1} = S_0$ and choose complement spaces W_j of the subspaces S_j in S_{j+1}. We arrive at the multiscale decomposition $S_j = \sum_{l=-1}^{j-1} W_l$ and call the elements of these complement spaces wavelets. By $\{\psi_{l,k}^m : m = 1, \ldots, m_\Gamma,\ k \in \nabla_l := \Delta_{l+1} \backslash \Delta_l\}$ we denote a basis of W_l which is supposed to satisfy $\|\psi_{l,k}^m \circ \kappa_m\|_{L^2} \sim 1$. The properties of the new basis and the basis transforms depend strongly on the dual multiscale decomposition. To introduce this, we define the duality pairing for functions over Γ by

$$\langle f, g \rangle := \sum_{m=1}^{m_\Gamma} \int_\square f(\kappa_m(s)) \overline{g(\kappa_m(s))} \mathrm{d}s, \tag{18}$$

and consider a dual multiscale analysis $\tilde{S}_0 \subset \tilde{S}_1 \subset \ldots \subset \tilde{S}_j \subset \ldots$ together with the dual multiscale decomposition $\tilde{S}_j = \sum_{l=-1}^{j-1} \tilde{W}_l$ and the basis functions $\tilde{\varphi}_{m,k}^j$ and $\tilde{\psi}_{l,k}^m$. Note that

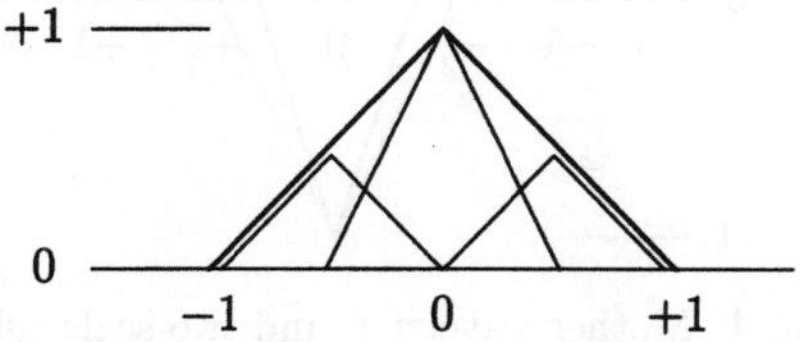

Figure 3: Scaling function φ and refinement equation.

the multiscale decomposition is called dual to $\{\psi_{l,k}^m\}$ if

$$\langle \varphi_{m,k}^j, \tilde{\varphi}_{m',k'}^j \rangle = \delta_{k,k'}\delta_{m,m'}, \quad \langle \psi_{l,k}^m, \tilde{\psi}_{l',k'}^{m'} \rangle = \delta_{l,l'}\delta_{k,k'}\delta_{m,m'}. \tag{19}$$

If such a dual system exists for $\{\psi_{l,k}^m\}$, then the wavelets $\psi_{l,k}^m$ are called biorthogonal wavelets (cf. [9]). We denote the order of the dual system by d^* and the parameter of regularity by γ^*.

For the wavelet algorithm applied to the BEM, the following two properties will be essential:

i) The wavelet functions $\psi_{l,k}^m$ are supposed to have a possibly small support. Typically, their support is the union of a small number of subdomains in the partition corresponding to the grid Δ_l.

ii) The wavelet functions $\psi_{l,k}^m$ with $l \geq 0$ fulfill the moment conditions of order d^*, i.e.,

$$\int_\square \psi_{l,k}^m(\kappa_m(s))s^\alpha ds = 0, \quad |\alpha| < d^*. \tag{20}$$

Clearly, condition ii) is a consequence of (19). If the properties i) and ii) are satisfied, then with the multiscale algorithm for the numerical solution of the boundary integral equation (1) over Γ we seek an approximate solution for the exact solution u in the form

$$u_j = \sum_{m=1}^{m_\Gamma} \sum_{l=-1}^{j-1} \sum_{k \in \nabla_l} u_{l,k}^m \psi_{l,k}^m. \tag{21}$$

Thus, the number N of degrees of freedom is of order $O(2^{jn})$. To simplify the subsequent formulae, we will assume $m_\Gamma = 1$ and we write $\kappa = \kappa_1$, $\varphi_k^j = \varphi_{1,k}^j$, and $\psi_{l,m} = \psi_{l,m}^1$. For the applications from Sections 2.2 and 2.3, the assumption $m_\Gamma = 1$ is satisfied.

3.2. PIECEWISE LINEAR AND BILINEAR WAVELETS

Now we consider the boundary curve Γ of Section 2.2 given via the global parametrization κ over $\square = [0, 2\pi]$. Since we solve the parameterized integral equation, we define the multiscale functions over the periodic interval $[0, 2\pi]$. This means we define the functions

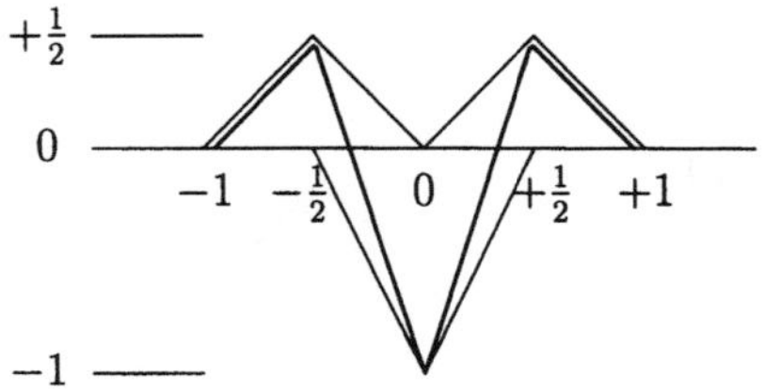

Figure 4: Mother wavelet ψ and two-scale relation.

as in the Section 3.1 but with Γ replaced by $[0, 2\pi]$ and κ replaced by the identity mapping. For $[0, 2\pi]$, the nested grids can be chosen as $\Delta^j := \{t_k^j : t_k^j := kh_j, k = 0, ..., N-1\}$ with $h_j := 2^{-j} \cdot 2\pi$ and $N = N_j := 2^j$. The finite element basis function φ_k^j is given over $[t_{k-1}^j, t_{k+1}^j]$ by $\varphi_k^j(s) := \varphi(s/h_l - k)$ with

$$\varphi(s) = \begin{cases} 1 - |s| & \text{if } |s| < 1 \\ 0 & \text{else,} \end{cases} \tag{22}$$

(cf. Figure 3) and over $[0, 2\pi] \setminus [t_{k-1}^j, t_{k+1}^j]$ we set φ_k^j to be zero. Note that the so called scaling function φ is a refinable function, i.e., we have (cf. Figure 3)

$$\varphi(x) = \frac{1}{2}\varphi(2x+1) + \varphi(2x) + \frac{1}{2}\varphi(2x-1). \tag{23}$$

In order to define the wavelet basis functions, we introduce the shape function ψ by (cf. Figure 4)

$$\psi(x) := \frac{1}{2}\varphi(2x+1) - \varphi(2x) + \frac{1}{2}\varphi(2x-1). \tag{24}$$

Note that ψ is called mother wavelet. Now the basis function $\psi_{l,k}$ on the grid of level l is the shift of the dilated function $\psi(\cdot/h_l)$ to the point t_k^{l+1} of the difference grid ∇_l, i.e., $\psi_{l,k}(t) := \psi((t - t_k^{l+1})/h_l)$ if $t \in [t_{k-2}^{l+1}, t_{k+2}^{l+1}]$ and $\psi_{l,k}(t) := 0$ else. It is not hard to see (cf. [39], Lemma 3.5) that these wavelets are biorthogonal in the sense of [9], with the exception that the dual wavelets have no finite support but decay exponentially. From [9] we infer that these functions represent a Riesz basis. Moreover, we get $d = 2$, $d^* = 2$, $\gamma = 3/2$.

Now we turn to the definition of bilinear wavelets for the boundary integral operator of Section 2.3. The boundary is given by the global parametrization κ over $\square = [0,1] \times [0,1]$. We define the functions as in Section 3.1 but with Γ replaced by $[0,1] \times [0,1]$ and κ replaced by the identity. Since the functions will be defined by tensor products, we begin with univariate functions over $[0,1]$. Analogously to the functions over the periodic interval $[0, 2\pi]$, we introduce the one-dimensional grids Δ_j^B, ∇_l^B and the univariate functions $\varphi_k^{B,j}$ and $\psi_{l,k}^B$ over $[0,1]$. More exactly, we set $N_l := 3 \cdot 2^l$, $\Delta_l^B := \{k/N_l,\ k = 1, \ldots, N_l - 1\}$ and $\nabla_l^B := \Delta_{l+1} \setminus \Delta_l$ and take only the hat functions $\varphi_k^{B,j}(s) := \varphi(N_j s - k)$, $k = 1, \ldots, N_j - 1$ vanishing at the end points of the interval. We define the wavelets $\psi_{l,k}^B$ analogously to the case of $[0, 2\pi]$. However, we modify the $\psi_{l,k}^B$ for $k = 1$ and $k = N_{l+1} - 1$ such that they vanish at the end points, too. The two-dimensional wavelets are now the tensor products of these univariate functions. In other words, we set $\Delta_j := \Delta_j^B \times \Delta_j^B$ and $\nabla_l := \Delta_{l+1} \setminus \Delta_l = (\nabla_l^B \times \Delta_l^B) \cup (\Delta_l^B \times \nabla_l^B) \cup (\nabla_l^B \times \nabla_l^B)$ and define the bilinear multiscale basis function by

$$\psi_{-1,(k,m)}(s,t) = \varphi_k^{B,0}(s)\cdot\varphi_m^{B,0}(t), \tag{25}$$

$$\psi_{l,(k,m)}(s,t) = \begin{cases} \psi_{l,k}^B(s)\cdot\varphi_m^{B,l}(t) & \text{if } k\in\nabla_l^B \text{ and } m\in\Delta_l^B \\ \varphi_k^{B,l}(s)\cdot\psi_{l,m}^B(t) & \text{if } m\in\nabla_l^B \text{ and } k\in\Delta_l^B \\ \psi_{l,k}^B(s)\cdot\psi_{l,m}^B(t) & \text{if } k,m\in\nabla_l^B, \end{cases} \quad l=0,\dots j-1.$$

For these wavelets, we again get $d = 2$, $d^* = 2$, and $\gamma = 3/2$.

3.3. WAVELET TRANSFORMS AND GENERAL CONSTRUCTION OF WAVELETS

Now we return to the general case of Section 3.1. In the multiscale setting, we have the two basis $\{\varphi_k^j\}$ and $\{\psi_{l,k}\}$ of the trial space S_j. The corresponding basis transform T_j is called wavelet transform. More exactly, T_j is defined by $T_j(d_{l,k})_{l,k} = (c_k^j)_k$ if $\sum_{l=-1}^{j-1}\sum_{k\in\nabla_l} d_{l,k}\psi_{l,k} = \sum_{k\in\Delta_j} c_k^j\varphi_k^j$. These transforms T_j and their inverses are uniformly bounded with respect to the l^2 matrix norm if the wavelet functions represent a Riesz basis, i.e, if the L^2 norm of a function is equivalent to the l^2 norm of its coefficients with respect to the wavelet basis. For the biorthogonal bases of [9] and the representation $u_j = \sum_{l=-1}^{j-1}\sum_{k\in\nabla_l} d_{l,k}\psi_{l,k}$, we even have the more general norm equivalence valid for Sobolev spaces (cf. e.g. [31,14,47])

$$c_1\,\|u_j\|_{H^s}^2 \le \sum_{l=-1}^{j-1}\sum_{k\in\nabla_l}|2^{ls}d_{l,k}|^2 \le c_2\,\|u_j\|_{H^s}^2, \tag{26}$$

where $\gamma^* < s < \gamma$ and c_1, c_2 denote non-negative constants. Note that this equivalence is fundamental for the preconditioning.

The application of the transform T_j can be realized by a fast pyramid scheme. To this end, we observe that the embeddings $S_j \subset S_{j+1}$ and $W_j \subset S_{j+1}$ imply the so called two-scale relations

$$\varphi_k^j = \sum_{q\in\Delta_{j+1}} m_{q,k}^{0,j}\varphi_q^{j+1}, \quad k\in\Delta_j, \tag{27}$$

$$\psi_{j,k} = \sum_{q\in\Delta_{j+1}} m_{q,k}^{1,j}\varphi_q^{j+1}, \quad k\in\nabla_j := \Delta_{j+1}\backslash\Delta_j. \tag{28}$$

We introduce the corresponding matrices by $\mathbf{M}_{j,0} := (m_{q,k}^{0,j})_{q\in\Delta_{j+1},k\in\Delta_j}$ and $\mathbf{M}_{j,1} := (m_{q,k}^{1,j})_{q\in\Delta_{j+1},k\in\nabla_j}$. These matrices are sparse and, for the linear wavelets of Section 3.2, even band matrices since the two-scale relations (27) and (28) on arbitrary levels are similar two (23) and (24), respectively. Using the matrices $\mathbf{M}_{j,0}$ and $\mathbf{M}_{j,1}$, the transition from the representation $u^{j+1} = \sum_k c_k^j\varphi_k^j + \sum_k d_{j,k}\psi_{j,k}$ to the representation $u^{j+1} = \sum_k c_k^{j+1}\varphi_k^{j+1}$ is realized by the matrix multiplications $(c_k^{j+1})_k = \mathbf{M}_{j,0}(c_k^j)_k + \mathbf{M}_{j,1}(d_{j,k})_k$. Moreover, the transition from $u^j = \sum_{l=-1}^{j-1}\sum_k d_{l,k}\psi_{l,k}$ to $u^j = \sum_k c_k^j\varphi_k^j$ can be performed by stepping from level $l = 0$ to level $l = j-1$ and multiplying by $\mathbf{M}_l := (\mathbf{M}_{l,0}, \mathbf{M}_{l,1})$ in each step. Indeed, if we set $d^l := (d_{l,k})_k$ and $c^l := (c_k^l)_k$ for $\sum_k c_k^l\varphi_k^l = \sum_{l'=-1}^{l-1}\sum_k d_{l',k}\psi_{l',k}$, then $c^j = T_j(d^{l'})_l$

can be computed by the multiplications corresponding to the pyramid scheme

$$\begin{array}{ccccccccccc} & d^0 & & d^1 & & & \cdots & d^{j-1} & & \\ & \searrow & \mathbf{M}_{0,1} & \searrow & \mathbf{M}_{1,1} & & & \searrow & \mathbf{M}_{j-1,1} & \\ d^{-1} & \rightarrow & c^1 & \rightarrow & c^2 & \cdots & c^{j-1} & \rightarrow & c^j & . \\ & \mathbf{M}_{0,0} & & \mathbf{M}_{1,0} & & & & \mathbf{M}_{j-1,0} & & \end{array} \tag{29}$$

Since the matrices are sparse, the application of T_j requires only $O(N_j)$ operations, where N_j is the dimension of the space S_j.

For the inverse transform T_j^{-1}, we remark that, analogously to (27) and (28), we get the two-scale relation

$$\varphi_q^{j+1} = \sum_{k\in\Delta_j} g_{q,k}^{0,j}\varphi_k^j + \sum_{k\in\nabla_j} g_{q,k}^{1,j}\psi_{j,k}, \; q \in \Delta_{j+1}. \tag{30}$$

We introduce the matrices $\mathbf{G}_{j,0} := (g_{q,k}^{0,j})_{q\in\Delta_{j+1},k\in\Delta_j}$, $\mathbf{G}_{j,1} := (g_{q,k}^{1,j})_{q\in\Delta_{j+1},k\in\nabla_j}$, $\mathbf{M}_j := (\mathbf{M}_{j,0}, \mathbf{M}_{j,1})$ and $\mathbf{G}_j := (\mathbf{G}_{j,0}, \mathbf{G}_{j,1})$. Clearly, $\mathbf{G}_j^* = \mathbf{M}_j^{-1}$ and the wavelet transform T_j^{-1} can be computed by running backward through (29). On each level we have to multiply by $\mathbf{G}_j^*$, which is a band matrix for the biorthogonal wavelets in [9]. In case of the linear wavelets of Section 3.2, $\mathbf{G}_j^*$ is not sparse anymore. However, the multiplication by $\mathbf{G}_j^*$ can be replaced by a fast solver for band matrices applied to the matrix equation including matrix $\mathbf{M}_j$.

Now we conclude this section with a remark on the construction of new wavelets. In many application, a simple wavelet basis with stable wavelet transforms T_j (i.e., the T_j and T_j^{-1} are uniformly bounded with respect to j) is available, however, these functions are not smooth enough or they do not satisfy the moment conditions of the desired order. Then the simple wavelets can be modified by changing the coefficients in the two-scale relations.

Lemma 1 [7,15,19,49] *Let $\breve{\mathbf{M}}_j = (\mathbf{M}_{j,0}, \breve{\mathbf{M}}_{j,1})$ and $\breve{\mathbf{G}}_j = (\breve{\mathbf{G}}_{j,0}, \breve{\mathbf{G}}_{j,1}) = (\breve{\mathbf{M}}_j^{-1})^*$ stand for the matrices of the simple and stable wavelets. Furthermore, suppose $\mathbf{K}_j = (k_{q,k}^j)_{q,k\in\nabla_j}$ is invertible and $\mathbf{L}_j = (l_{q,k}^j)_{q\in\Delta_j,k\in\nabla_j}$ is arbitrary. Then the matrix $\mathbf{M}_{j,1} := \mathbf{M}_{j,0}\mathbf{L}_j + \breve{\mathbf{M}}_{j,1}\mathbf{K}_j$ together with $\mathbf{M}_{j,0}$ defines another stable wavelet system, where $\mathbf{G}_{j,0} := \breve{\mathbf{G}}_{j,0} - \breve{\mathbf{G}}_{j,1}(\mathbf{K}_j^*)^{-1}\mathbf{L}_j^*$, $\mathbf{G}_{j,1} := \breve{\mathbf{G}}_{j,1}(\mathbf{K}_j^*)^{-1}$.*

In particular, one can choose $\mathbf{K}_j = \mathbf{I}$ to get $\psi_{j,k} = \sum_{q\in\Delta_j} l_{q,k}^j\varphi_q^j + \breve{\psi}_{j,k}, \quad k \in \nabla_j$. Thus, for any given stable completion, it is possible to construct new wavelets with a desired order of vanishing moments. Note that a dual construction is also possible. Moreover, for sparse matrices $\breve{\mathbf{M}}_j$ and $\breve{\mathbf{G}}_j$ and suitable $\mathbf{K}_j$ and $\mathbf{L}_j$, the resulting matrices $\mathbf{M}_j$ and $\mathbf{G}_j$ are sparse again. As starting points for these constructions we can choose the discontinuous piecewise polynomials or hierarchical (nodal) basis functions for Lagrangian finite elements. Note that the values of the dual scaling functions can be obtained by the subdivision scheme

$$\lim_{J\to\infty} \sum_{\nu\in\Delta_J} \tilde{f}_{k,\nu}^{j,J}\varphi_\nu^J \;=\; \tilde{\varphi}_k^j, \quad (\tilde{f}_{k,\nu}^{j,J})_{\nu\in\Delta_J} := \prod_{l=j}^{J} \mathbf{G}_{l,0}\mathbf{e}_k^j, \tag{31}$$

$$\mathbf{e}_k^j \;:=\; (0, \cdots, \overset{k}{\breve{1}}, \cdots, 0)^{\mathrm{T}}.$$

4. THE WAVELET-GALERKIN METHOD

4.1. CONVENTIONAL GALERKIN SCHEMES

For the solution of the boundary integral equation (1), Galerkin schemes are very popular. With this scheme we seek an approximate solution $u_j \in S_j$ for the exact solution u of $Au = v$ by solving

$$\langle Au_j, w\rangle = \langle v, w\rangle, \quad w \in S_j. \tag{32}$$

These Galerkin equations are equivalent to the equation $Q_j^* A u_j = Q_j^* v$, where Q_j^* denotes the $\langle \cdot, \cdot \rangle$- adjoint operator of Q_j and Q_j stands for the orthogonal projection of $H^{\mathbf{r}/2}(\Gamma)$ onto S_j. Recall that $\mathbf{r}$ is the order of operator A. The Galerkin method is called stable in the Sobolev space H^s if, for sufficiently large j, the Galerkin operators $\mathcal{A}_j := Q_j^* A|_{S_j} \in \mathcal{L}(S_j)$ are invertible and the $(H^{s-\mathbf{r}}, H^s)$- norm of the inverse operators are uniformly bounded. The following result is well known:

Theorem 1 *If A is a bounded, strongly elliptic, and invertible operator mapping H^s into $H^{s-\mathbf{r}}$ for $\mathbf{r} - d \le s \le d$, then the Galerkin method is stable in H^s for $\mathbf{r} - \gamma < s < \gamma$. The approximate solution u_j of (32) converges in H^s to the exact solution u of $Au = v$ for any $v \in H^{s-\mathbf{r}}$ with $s > \mathbf{r} - \gamma$ and the Galerkin error $\|u - u_j\|_{H^s}$ is less than $C2^{-j(t-s)}\|u\|_{H^t}$ for $\mathbf{r} - d \le s \le t \le d$, $s < \gamma$, $\mathbf{r} - \gamma < t$.*

The conventional BEM algorithm determines the Galerkin solution u_j with respect to the finite element basis. In view of (32), the coefficients u_k^j of $u_j = \sum_{k \in \Delta_j} u_k^j \varphi_k^j$ are obtained by solving the linear system

$$\sum_{k \in \Delta_j} \langle A\varphi_k^j, \varphi_{k'}^j \rangle u_k^j = \langle v, \varphi_{k'}^j \rangle \ , \quad k' \in \Delta_j \ . \tag{33}$$

The condition number $\|A_j\|_{\mathcal{L}(l^2)} \cdot \|A_j^{-1}\|_{\mathcal{L}(l^2)}$ of the $N \times N$ matrix $A_j := (\langle A\varphi_k^j, \varphi_{k'}^j \rangle)_{k',k \in \Delta_j}$ is of order $O(2^{j|\mathbf{r}|})$. Thus, to find efficient iterative algorithms for (33) with $\mathbf{r} \neq 0$ is a difficult task. Diagonal preconditioners are not suitable.

4.2. THE WAVELET ALGORITHM

The wavelet method determines the Galerkin solution u_j in form of the wavelet representation $u_j = \sum_l \sum_k u_{l,k}\psi_{l,k}$. The unknown coefficients $u_{l,k}$ are the solutions of the system

$$\sum_{l=-1}^{j-1} \sum_{k \in \nabla_l} \langle A\psi_{l,k}, \psi_{l',k'} \rangle u_{l,k} = \langle v, \psi_{l',k'} \rangle, \quad k' \in \nabla_{l'}, \ -1 \le l' \le j-1 \ . \tag{34}$$

For the preconditioning of the corresponding matrix $\tilde{A}_j := (\langle A\psi_{l,k}, \psi_{l',k'}\rangle)_{(l',k'),(l,k)}$, we introduce the diagonal matrix $D_j^\sigma := (2^{\sigma l}\delta_{l,l'}\delta_{k,k'})_{(l',k'),(l,k)}$.

Theorem 2 [18,14,47] *Suppose $\gamma^* > -\frac{\mathbf{r}}{2}$. The condition numbers of the preconditioned matrices $D_j^{-\mathbf{r}/2}\tilde{A}_j D_j^{-\mathbf{r}/2}$ are uniformly bounded with respect to j.*

Consequently, the system (34) can be solved iteratively with a number of iterations not depending on the mesh size 2^{-j}.

Another advantage of the wavelet basis consists in the pseudo locality of the operator representation, i.e., if the supports of the trial basis function $\psi_{l,k}$ and the test wavelet $\psi_{l',k'}$ are far from each other, then the corresponding matrix entry is small. Even if the support of one of these two wavelets is contained in the support of the other wavelet and if the distance of the first support to the discontinuity points of the other wavelet is relatively large, then the entry is small again.

Lemma 2 *a)* [5,18,35,47] *If $n + 2d^* + \mathbf{r} > 0$ and if* $\mathrm{dist}(\mathrm{supp}\,\psi_{l,k}, \mathrm{supp}\,\psi_{l',k'}) > 0$ *then, for the corresponding entry of $\tilde{A}_j$, we get*

$$|\langle A\psi_{l,k}, \psi_{l',k'}\rangle| \le C\,\frac{2^{l(-\frac{n}{2}-d^*)}2^{l'(-\frac{n}{2}-d^*)}}{[\mathrm{dist}(\mathrm{supp}\,\psi_{l,k}, \mathrm{supp}\,\psi_{l',k'})]^{n+\mathbf{r}+2d^*}}. \tag{35}$$

b) [47] *If* $\mathrm{supp}\,\psi_{l,k}$ *is contained in* $\mathrm{supp}\,\psi_{l',k'}$ *such that* $\mathrm{supp}\,\psi_{l,k}$ *does not intersect the set* $\mathrm{sing\,supp}\,\psi_{l',k'}$ *of discontinuity points for $\psi_{l',k'}$ and the derivatives of $\psi_{l',k'}$, then we get*

$$|\langle A\psi_{l,k}, \psi_{l',k'}\rangle| \le C\,\frac{2^{l(-\frac{n}{2}-d^*)}2^{l'(-\frac{n}{2})}}{[\mathrm{dist}(\mathrm{supp}\,\psi_{l,k}, \mathrm{sing\,supp}\,\psi_{l',k'})]^{n+\mathbf{r}+d^*}}. \tag{36}$$

A similar estimate holds if the roles of $\psi_{l,k}$ and $\psi_{l',k'}$ are interchanged.

Knowing these decay properties, we can set up a compressed matrix such that the additional error due to the neglect of small matrix entries is less than the discretization error in Theorem 1. We choose constants $a_i > 0$, $i = 0,\ldots,3$, $\alpha \ge 0$, β_i, $i = 1,\ldots,6$ and set

$$\begin{aligned}
\mathcal{B}_{l,l'} &:= \max\{a_0 2^{-l}, a_0 2^{-l'}, a_1 j^{\alpha} 2^{\beta_1 j - \beta_2 l - \beta_3 l'}\}, \\
\mathcal{B}^S_{l,l'} &:= \max\{a_2 2^{-\max\{l,l'\}}, a_3 2^{\beta_4 j - \beta_5 \max\{l,l'\} - \beta_6(l+l')}\}.
\end{aligned} \tag{37}$$

With these constants, the compressed matrix $A_j^\varepsilon = (a^\varepsilon_{(l',k'),(l,k)})_{(l',k'),(l,k)}$ is defined by

$$a^\varepsilon_{(l',k'),(l,k)} := \begin{cases}
\langle A\psi_{l,k}, \psi_{l',k'}\rangle\,, & \text{if } \mathrm{dist}(\mathrm{supp}\,\psi_{l,k}, \mathrm{supp}\,\psi_{l',k'}) \le \mathcal{B}_{l,l'} \\
& \text{and if} \\
& \quad \text{either } \mathrm{dist}(\mathrm{supp}\,\psi_{l,k}, \mathrm{sing\,supp}\,\psi_{l',k'}) \le \mathcal{B}^S_{l,l'} \\
& \qquad \text{for } l \ge l' \\
& \quad \text{or } \mathrm{dist}(\mathrm{supp}\,\psi_{l',k'}, \mathrm{sing\,supp}\,\psi_{l,k}) \le \mathcal{B}^S_{l,l'} \\
& \qquad \text{for } l' \ge l \\
0\,, & \text{otherwise.}
\end{cases} \tag{38}$$

By $\mathcal{A}_j^\varepsilon \in \mathcal{L}(S_j)$ we denote the operator whose matrix $(\langle \mathcal{A}_j^\varepsilon\psi_{l,k}, \psi_{l',k'}\rangle)_{(l',k'),(l,k)}$ is A_j^ε. Further, we denote by $u_j^\varepsilon = \sum_l \sum_k u^\varepsilon_{l,k}\psi_{l,k}$ the solution of $\mathcal{A}_j^\varepsilon u_j^\varepsilon = Q_j^* v$, i.e., of

$$\sum_{l=-1}^{j-1} \sum_{k\in\nabla_j} a^\varepsilon_{(l',k'),(l,k)} u^\varepsilon_{k,l} = \langle v, \psi_{k',l'}\rangle, \quad l' = -1,\ldots,j-1,\ k' \in \nabla_{l'}. \tag{39}$$

Theorem 3 *Suppose that the assumptions of Theorem 1 are satisfied and that* $d^* > d - \mathbf{r}$. *Moreover, with* $d' \in (d, d^* + \mathbf{r})$ *we choose* $\alpha = 0$, $\beta_1 = \beta_4 = (2d' - \mathbf{r})/(2d^* + \mathbf{r})$, $\beta_2 = \beta_3 = (d' + d^*)/(2d^* + \mathbf{r})$, $\beta_5 = d^*/(2d^* + \mathbf{r})$, *and* $\beta_6 = d'/(2d^* + \mathbf{r})$. *If the constants* a_i, $i = 0, \ldots, 3$ *are sufficiently large and if* A_j^ε *is given by (38), then:*

i) The sequence $\{A_j^\varepsilon\}$ *is stable in* H^s *for* $\mathbf{r} - \gamma < s < \gamma$.

ii) The number of non-zero entries in A_j^ε *is less than* $O(N) = O(2^{nj})$.

iii) The error $\|u - u_j^\varepsilon\|_{H^s}$ *is bounded by* $O(2^{-j(t-s)}\|u\|_{H^t})$ *for* $\mathbf{r} - d \leq s \leq t \leq d$, $s < \gamma$, $\mathbf{r}/2 \leq t$.

4.3. THE ASSEMBLING OF THE MATRIX VIA QUADRATURE ALGORITHM

Normally, the major part of the computation time for the BEM is spent for the assembling of the stiffness matrix. The subsequent solution of the matrix equation is much faster. This ratio of computation times does not change even for the wavelet method. In view of the fact that we have to compute a compressed sparse matrix instead of the full matrix (cf. Theorem 3 ii)), we expect the same reduction of the computation time for the assembling of the matrix as for the iterative solution of the linear system. If analytic formulae for the computation of the matrix entries are available, then the development of an efficient algorithm for the matrix assembling should be easy. However, if the entries are to be computed by quadrature, then it is not so easy to set up a quadrature algorithm such that the numerical error due to the quadrature does not exceed the discretization error (cf. the Galerkin error in Theorem 1) and that the number of quadrature knots is essentially less than that for the conventional BEM. At least for piecewise analytic boundaries Γ and piecewise analytic kernel functions $K_A(x, y)$, the compressed matrix can be computed by quadrature with no more than $O(N)$ times a power of $\log N$ operations. The corresponding quadrature algorithms are due to v.Petersdorff, Schwab [34], and Schneider [47] (cf. also [15]). We will mention four fundamental principles of the algorithm, only. Note that, for the matrix entry $\langle A\psi_{l,k}, \psi_{l',k'}\rangle$, we have to compute

$$\int_{\tau'}\int_{\tau} \psi_{l',k'}(\kappa(x))K_A(\kappa(x), \kappa(y))\psi_{l,k}(\kappa(y))|\kappa'(y)|\,|\kappa'(x)|\mathrm{d}y\mathrm{d}x, \tag{40}$$

where $\tau \subset \operatorname{supp}\psi_{l,k}$ and $\tau' \subset \operatorname{supp}\psi_{l',k'}$ denote closed patches of the partitions on level l and l', respectively. The wavelets $\psi_{l,k} \circ \kappa$ and $\psi_{l',k'} \circ \kappa$ restricted to τ and τ', respectively, are polynomials of degree less than d. The quadrature algorithm for the assembling of the stiffness matrix A_j^ε should be based on the following:

i) If the order $\mathbf{r}$ of A is non-negative, then the integrals in (40) for $\tau \cap \tau' \neq \emptyset$ are singular or hypersingular. These integrals should be regularized before quadrature rules are applied (cf. [30,45,34]).

ii) For $\tau = \tau'$ as well as for τ and τ' sharing a common edge or vertex, the integrals in (40) are to be treated by transformation techniques (cf. [45,34]).

iii) If $\tau \cap \tau' = \emptyset$, then the domains τ and τ' should be divided in a further step. The final quadrature partitions $\tau = \cup_i \tau_i$ and $\tau' = \cup_m \tau'_m$ should satisfy

$$\operatorname{dist}(\tau'_m, \tau_i) \geq Q \max\{\operatorname{diam} \tau'_m, \operatorname{diam} \tau_j\}, \tag{41}$$

where Q is a fixed positive constant.

iv) The integrals of (40) restricted to $\tau'_m \times \tau_i$ should be approximated by tensor product Gauß rules. The orders of the Gauß rules should be chosen using quadrature estimates for analytic functions. Following this idea, the Gauß orders increase logarithmically with decreasing distance $\operatorname{dist}(\tau'_m, \tau_i)$.

5. THE COLLOCATION METHOD

5.1. THE CONVENTIONAL COLLOCATION

Suppose we seek an approximate solution $u_j \in S_j$ for the solution of the boundary integral equation $Au = v$. If u_j is determined by the Galerkin equation (32) or (34), then we have to compute stiffness matrices the entries of which contain double integrals (cf. (40)). In order to reduce the amount of work for the quadrature, the Galerkin equations are replaced by the simpler collocation equations. For this, we suppose that S_j is the space of piecewise linear or n- linear continuous functions defined over the partition on the j-th level, i.e., S_j is of order $d = 2$ and γ is equal to 3/2. The collocation solution $u_j \in S_j$ is the solution of

$$(Au_j)(\kappa(k)) = v(\kappa(k)), \quad k \in \Delta_j. \tag{42}$$

Recall that Δ_j is the set of grid points of the level j partition on which S_j is defined. We introduce the interpolation projection P_j onto S_j by

$$P_j f \in S_j, \quad P_j f(\kappa(k)) = f(\kappa(k)), \quad k \in \Delta_j. \tag{43}$$

Clearly, the collocation system (42) is equivalent to $P_j A u_j = P_j v$. Analogously to the Galerkin method, the collocation is called stable in H^s if, for sufficiently large j, the collocation operators $\mathcal{A}_j := P_j A|_{S_j} \in \mathcal{L}(S_j)$ are invertible and the (H^{s-r}, H^s)- norms of the inverse operators are uniformly bounded.

Theorem 4 *i)* [36,4,38] *Suppose that Γ is at least three times continuously differentiable and homeomorphic to the n-dimensional torus and that $\kappa : \square = [0,1]^n \longrightarrow \Gamma$ is a global parametrization. We assume that S_j is the set of n- linear splines defined over the uniform partition of $\square$ into 2^{nj} n-cubes (cf. Section 3). Moreover, we assume A to be a strongly*

elliptic classical pseudo-differential operator of order $\mathbf{r} < 1$. *Then the collocation method is stable in* H^s *for* $\mathbf{r} \le s < \frac{3}{2}$. *The collocation solution* u_j *defined by (42) converges in* H^s *to the exact solution* u *of* $Au = v$ *for any* $v \in H^{s-\mathbf{r}}$ *with* $s > \mathbf{r} + \frac{1}{2}$, *and the collocation error* $\|u_j - u\|_{H^s}$ *is bounded by* $C2^{-j(t-s)}\|u\|_{H^t}$ *for* $\mathbf{r} \le s \le t \le 2$, $s < \frac{3}{2}$, $\mathbf{r} + \frac{n}{2} < t$.
ii) [37] *Suppose that* $\Gamma = [0,1] \times [0,1]$, *that* S_j *and* Δ_j *are defined over* Γ *as in Section 3.2, and that* A *is a strongly elliptic pseudo-differential operator of order* $\mathbf{r} = 0$. *Then the collocation method is stable in* L^2. *The collocation solution* u_j *defined by (42) converges in* L^2 *to the exact solution* u *of* $Au = v$ *for any* $v \in L^2$ *such that* $\|P_j v - v\|_{L^2} \longrightarrow 0$. *If* u *is in* H^2 *and vanishes over the boundary of* $[0,1] \times [0,1]$, *then* $\|u_j - u\|_{L^2}$ *is less than* $C2^{-2j}\|u\|_{H^2}$.

Choosing the conventional finite element basis $\{\varphi_k^j\}_{k\in\Delta_j}$, the collocation equation (42) is equivalent to the system

$$\sum_{k\in\Delta_j} (A\varphi_k^j)(\kappa(k'))u_k^j = f(\kappa(k')), \quad k' \in \Delta_j \tag{44}$$

for the coefficients u_k^j of $u_j := \sum_{k\in\Delta_j} u_k^j \varphi_k^j$. Thus, the stiffness matrix of the collocation is $A_j := ((A\varphi_k^j)(\kappa(k')))_{k',k\in\Delta_j}$. The condition number of A_j is of order $O(2^{j|\mathbf{r}|})$ and, for $\mathbf{r} \neq 0$, the construction of efficient preconditioners is difficult.

5.2. WAVELETS IN THE SPACE OF TEST FUNCTIONALS

The space of test functionals in case of the collocation method (42) is spanned by the Dirac delta functionals $\{\delta_{\kappa(k)},\ k \in \Delta_j\}$, i.e., by the functionals of function evaluation at the collocation points. Following Brandt, Lubrecht [6], Harten and Yad-Shalom [25], we can introduce a basis of wavelet functionals in this space. For simplicity, we take wavelets with one or two vanishing moments and restrict the consideration first to the case that $\Gamma = \square$ is the periodic interval $[0, 2\pi]$ and later to $\square = [0,1] \times [0,1]$. In other words, we define the wavelets for the test functionals employed in the numerical tests for the solution of the boundary integral equations from Sections 2.2 and 2.3.

For $\Gamma = \square = [0, 2\pi]$, we retain the notation of Section 3.2 and introduce the wavelet functionals

$$\begin{aligned} \eta_{-1,k}(f) &:= f(t_k^0), & k \in \Delta_0, \\ \eta_{l,k}(f) &:= f(t_k^{l+1}) - \frac{1}{2}\{f(t_{k-1}^{l+1}) + f(t_{k+1}^{l+1})\}, & k \in \nabla_l, l = 0, \ldots, j-1. \end{aligned} \tag{45}$$

It is not hard to verify that the basis functionals $\eta_{l,k}$ span the space span$\{\delta_{\kappa(k)},\ k \in \Delta_j\}$. They satisfy the second order moment conditions, i.e., they vanish at constant and linear functions. Furthermore, it is easy to see that the "classical" hierarchical basis $\{\varphi_k^{l+1},\ l = -1, \ldots, j-1,\ k \in \nabla_l\}$ is dual to $\{\eta_{l,k}\}$, i.e., $\eta_{l,k}(\varphi_{k'}^{l'+1}) = \delta_{l,l'}\delta_{k,k'}$ for $l, l' = -1, \ldots, j-1$ and $k \in \nabla_l$, $k' \in \nabla_{l'}$. Consequently, the interpolation $P_j f$ can be written as $\sum_l \sum_k \eta_{l,k}(f)\, \varphi_k^{l+1}$ and, analogously to (26), we arrive at

$$c_1 \|u_j\|_{H^s}^2 \le \sum_{l=-1}^{j-1} \sum_{k\in\nabla_l} |2^{ls}\eta_{l,k}(u_j)|^2 \le c_2 \|u_j\|_{H^s}^2, \tag{46}$$

where $\frac{1}{2} < s < \frac{3}{2}$. Now, the wavelet transform with respect to the dual basis maps the vector $(f(t_k^j))_{k\in\Delta_j}$ into the vector $(\eta_{l,k}(f))_{l=-1,\dots,j-1,\,k\in\nabla_l}$, and the inverse transform vice versa. If $(f(t_k^j))_{k\in\Delta_j}$ is known, than $(\eta_{l,k}(f))_{l=-1,\dots,j-1,\,k\in\nabla_l}$ can be computed directly by (45). For the inverse transform, a pyramid scheme should be applied. The transition from level l to level $l+1$ is accomplished by using the two-scale relation (45). Note that, for $t_k^{l+1} \in \nabla_l \subset \Delta_{l+1}$ the neighbours $t_{k\pm1}^{l+1}$ belong to Δ_l.

Now consider $\Gamma = \square = [0,1] \times [0,1]$. Retaining the definitions of $\Delta_l := \Delta_l^B \times \Delta_l^B$ and $\nabla_l = \Delta_{l+1} \setminus \Delta_l$ from Section 3.2, we define the test wavelets

$$\eta_{-1,(k,m)}(f) = f\left(\frac{k}{N_0}, \frac{m}{N_0}\right), \tag{47}$$

$$\eta_{l,(k,m)}(f) = \begin{cases} f\left(\frac{k}{N_{l+1}}, \frac{m}{N_{l+1}}\right) - f\left(\frac{k+1}{N_{l+1}}, \frac{m}{N_{l+1}}\right) & \text{if } k \in \nabla_l^B \text{ and } m \in \Delta_l^B \\ f\left(\frac{k}{N_{l+1}}, \frac{m}{N_{l+1}}\right) - f\left(\frac{k}{N_{l+1}}, \frac{m+1}{N_{l+1}}\right) & \text{if } m \in \nabla_l^B \text{ and } k \in \Delta_l^B \\ f\left(\frac{k}{N_{l+1}}, \frac{m}{N_{l+1}}\right) - f\left(\frac{k}{N_{l+1}}, \frac{m+1}{N_{l+1}}\right) & \text{if } k, m \in \nabla_l^B, \end{cases}$$
$$l = 0, \dots j-1.$$

It is not hard to verify that the basis functionals $\eta_{l,k}$, $l = -1, \dots, j-1$, $k \in \nabla_l$ span the space span$\{\delta_{\kappa(k)},\ k \in \Delta_j\}$. They satisfy the first order moment condition, i.e., they vanish at constant functions. The dual basis is a hierarchical basis of piecewise constant functions, and the wavelet transforms can be computed like those for the univariate case. Unfortunately, there is no analogue to (46).

5.3. THE WAVELET COLLOCATION METHOD

The wavelet method determines the solution u_j in form of the wavelet representation $u_j = \sum_{l=-1}^{j-1} \sum_{k\in\nabla_l} u_{l,k}\psi_{l,k}$. The unknown coefficients $u_{l,k}$ are the solutions of the system

$$\sum_{l=-1}^{j-1} \sum_{k\in\nabla_l} \eta_{l',k'}(A\psi_{l,k})\, u_{l,k} = \eta_{l',k'}(v), \quad k' \in \nabla_{l'}, \quad -1 \le l' \le j-1 \,. \tag{48}$$

For the preconditioning of the corresponding stiffness matrix $\tilde{A}_j := (\eta_{l',k'}(A\psi_{l,k}))_{(l',k'),(l,k)}$, we recall the definition $D_j^\sigma := (2^{\sigma l}\delta_{l,l'}\delta_{k,k'})_{(l',k'),(l,k)}$.

Theorem 5 [47] *Suppose $n = 1$ and that the assumptions of Theorem 4 i) are satisfied. Then, for $\frac{1}{2} < t < \frac{3}{2}$, the condition numbers of the matrices $D_j^{-\mathbf{r}+t}\tilde{A}_j D_j^{-t}$ are uniformly bounded with respect to j.*

Consequently, the system (48) can be solved iteratively with a number of iterations not depending on the mesh size 2^{-j}. For example, we consider the preconditioning for the boundary integral operator of Section 2.2, where the original cross section is an ellipse. Observing $\mathbf{r} = -1$, we arrive at the preconditioned matrices $D_j^{1+t}\tilde{A}_j D_j^{-t}$. However, taking the dependences on the wave number $\bar{k}$ into account, the preconditioner could be slightly improved. We replace D_j^t by $D_{\bar{k},j}^t := (\{\bar{k}^2 + 2^{2l}\}^{t/2}\delta_{l,l'}\delta_{k,k'})_{(l',k'),(l,k)}$. In Table 1 we present the numbers of GMRes iterations (cf. [43,50]) and the computation times for several types of preconditioning applied to the compressed stiffness matrix (cf. the subsequent

Table 1: Number of GMRes iterations and computation time for the iterative solver.

Preconditioning												
j N_j	2 4	3 8	4 16	5 32	6 64	7 128	8 256	9 512	10 1024	11 2048	12 4096	12 4096
without	3	5	9	14	18	21	29	33	37	40	44	160.3s
$D_j^1 A_j^\varepsilon$	3	5	9	15	18	20	25	27	28	28	29	104.6s
$D_j^{\frac{1}{2}} A_j^\varepsilon D_j^{\frac{1}{2}}$	3	5	9	15	17	19	24	27	26	26	27	96.4s
$A_j^\varepsilon D_{\bar{k},j}^1$	3	5	9	14	**15**	**16**	**21**	**23**	**23**	**22**	**22**	81.5s

Theorem 6). The computations have been performed on a DEC 3000 AXP 600 α-processor workstation for the case of an ellipse Γ and a wave number $\bar{k} = 10$. The results show that the number of iterations are independent of the mesh size and that they are essentially reduced by the preconditioning.

Now it is not hard to see that a lemma analogously to Lemma 2 is valid for the collocation, too. Hence, we can set up a compressed matrix $A_j^\varepsilon = (a^\varepsilon_{(l',k'),(l,k)})_{(l',k'),(l,k)}$ by (cf. (37))

$$a^\varepsilon_{(l',k'),(l,k)} := \begin{cases} \eta_{l',k'}(A\psi_{l,k}) & \text{if } \operatorname{dist}(\operatorname{supp}\psi_{(l,k)}, \operatorname{supp}\eta_{(l',k')}) \le \mathcal{B}_{l,l'} \\ 0\,, & \text{otherwise.} \end{cases} \tag{49}$$

Introducing the dual basis $\{\tilde{\psi}_{l,k}\}_{l,k}$ of $\{\eta_{l,k}\}_{l,k}$, we define by $\mathcal{A}_j^\varepsilon$ the operator mapping S_j into $\operatorname{span}\{\tilde{\psi}_{l,k}\}_{l,k}$ whose matrix with respect to the bases $\{\psi_{l,k}\}_{l,k}$ and $\{\tilde{\psi}_{l,k}\}_{l,k}$ is A_j^ε. Further, we denote by $u_j^\varepsilon = \sum_l \sum_k u^\varepsilon_{l,k}\psi_{l,k}$ the solution of

$$\sum_{l=-1}^{j-1} \sum_{k\in\nabla_l} a^\varepsilon_{(l',k'),(l,k)} u^\varepsilon_{l,k} = \eta_{k',l'}(v), \quad l' = -1,\ldots,j-1,\ k' \in \nabla_{l'}, \tag{50}$$

and let $\tilde{d}^*$ stand for the number of vanishing moments of the test wavelets $\eta_{l,k}$, $l \ge 0$. We have $\tilde{d}^* = 2$ for the univariate wavelets of Section 5.2 and $\tilde{d}^* = 1$ for the two-dimensional wavelets.

Theorem 6 [17,18,47] *Suppose that the assumptions of Theorem 4 i) or ii) are satisfied and that the number d^* of vanishing moments for the trial wavelets is greater than $-\mathbf{r}$. Choosing $d' \in (d, \tilde{d}^* + \mathbf{r})$, we set $\alpha = 5/[2(2+\mathbf{r}+\tilde{d}^*)]$, $\beta_1 = (d' - \mathbf{r})/(2+\mathbf{r}+\tilde{d}^*)$, $\beta_2 = \tilde{d}^*/(2+\mathbf{r}+\tilde{d}^*)$, and $\beta_3 = (d'+2)/(2+\tilde{d}^*)$. If the a_i, $i = 0,1$ are sufficiently large and if $\mathcal{A}_j^\varepsilon$ is given by (49), then:*

i) The sequence $\{\mathcal{A}_j^\varepsilon\}$ is stable in H^s for $\mathbf{r} \le s < \frac{3}{2}$.

ii) The number of non-zero entries in A_j^ε is less than $O(N \log N) = O(j2^{nj})$.

iii) The numerical error $\|u - u_j^\varepsilon\|_{H^s}$ for the approximate solution u_j^ε of (50) is less than $O(2^{-j(t-s)}\|u\|_{H^t})$ for $\mathbf{r} \le s \le t \le 2$, $s < \frac{3}{2}$, $\mathbf{r} + \frac{n}{2} < t$.

In Table 2 we compare the collocation method with the Galerkin method. We give the convergence orders, the regularity parameters γ, γ^*, $\tilde{\gamma}^*$, and the numbers of vanishing moments d^*, $\tilde{d}^*$. These regularity parameters and vanishing moment conditions are required for the applicability of the discretization schemes (conformity), for the optimal compression rates, and for the preconditioning. Note that γ, γ^*, d^*, and d have been introduced in Section 3.1. The number $\tilde{\gamma}^*$ is the regularity parameter for the system $\{\tilde{\psi}_{l,k}\}$ dual to $\{\eta_{l,k}\}$, i.e., $\tilde{\psi}_{l,k} \in H^s$ for $s < \tilde{\gamma}^*$.

Table 2: Comparison of Galerkin and collocation method.

Galerkin:		
Regularity	$\gamma > \frac{\mathbf{r}}{2}$ for conformity	$\gamma^* > -\frac{\mathbf{r}}{2}$ for preconditioning
Van. Moments	$d^* > d - \mathbf{r}$	
Max. Conv. Rate		in $H^{\mathbf{r}-d}$: $2^{-j(2d-\mathbf{r})}$
Collocation:		
Regularity	$\gamma > \mathbf{r}$ for conformity	$\tilde{\gamma}^* > \frac{n}{2}$, $\gamma > \frac{n}{2} + \mathbf{r}$, $\tilde{\gamma}^* + \gamma^* > -\mathbf{r}$ for preconditioning
Van. Moments	$\tilde{d}^* > d - \mathbf{r}$	$d^* > -\mathbf{r}$
Max. Conv. Rate		in $H^{\mathbf{r}}$: $2^{-j(d-\mathbf{r})}$

For the boundary integral operator in Section 2.2, the assumptions of Theorem 6 are not satisfied. However, similarly to Theorem 6, we can prove that the compressed matrix A_j^ϵ defined with $a_0 = a_1 = a$, $\alpha = 5/6$, $\beta_1 = 2/3$, $\beta_2 = 4/3$, and $\beta_3 = 2/3$ contains no more than $O(N)$ times a power of $\log N$ non-zero entries. Moreover, the wavelet collocation including the compression is stable in L^2 and converges with order two. In Figure 5 we present the numerically obtained numbers of non-zero entries and compare these with the quadratic behaviour $O(N^2)$ for the fully populated matrices and with the linear behaviour $O(N)$. For a number $N = N_j = 4096$ and for the parameters $\bar{k} = 10$, $a = 0.3$, $\bar{k} = 10$, $a = 1.0$, and $\bar{k} = 1$, $a = 0.3$, respectively, the relative L^2 error of the single layer density function u is less than $3 \cdot 10^{-5}$, $3 \cdot 10^{-6}$, and $1 \cdot 10^{-7}$, respectively. Figure 6 shows the computation times for the iterative solution of the matrix equation with and without compression. The linear systems are solved by GMRes on a DEC 3000 AXP 500 α-processor workstation.

Now we turn to the boundary integral operator of Section 2.3 and consider the collocation method based on the bilinear splines of Section 3.2 together with the wavelet algorithm based on the wavelets from (25) and (47). Again, the assumptions of Theorem 6 are not satisfied. However, similarly to Theorem 6, we can prove that the compressed matrix A_j^ϵ defined with $a_0 = 1$, $a_1 = a$, $\alpha = 0$, $\beta_1 = 4/3$, $\beta_2 = 1$, and $\beta_3 = 1$ contains no more then $O(N^{4/3})$ entries. Moreover, the wavelet collocation including the compression is stable in L^2 and converges with order two. Though the compression is asymptotically not optimal, we believe that the presented choice of wavelets and compression parameters leads to faster computation times for $N \leq 10\,000$. In Table 3 we show the compression rates (compression rate = number N^2 of entries in the full matrix divided by the number of entries in the compressed matrix), the number of quadrature knots, and the computation times (including the time for the set up of the stiffness matrix). The time t_C and the number of knots k_C are given for the BEM without wavelets. The time t_D, the number

Table 3: Compression rates, number of quadrature knots, and computation times.

j	1	2	3	4	5
N_j	25	121	529	2209	9025
r_C	1	1	1	1	
k_C	$8 \cdot 10^3$	$1.25 \cdot 10^5$	$1.95 \cdot 10^6$	$3.1 \cdot 10^7$	
t_C	2.92s	6.64s	65.49s	1 005.81s	
r_D	1.19	2.69	6.83	12.64	13.99
k_D	$8 \cdot 10^3$	$1.0 \cdot 10^5$	$1.05 \cdot 10^6$	$1.07 \cdot 10^7$	$1.3 \cdot 10^8$
t_D	3.08s	6.06s	39.55s	405.17s	5 580.07s
r_P	1.19	2.69	6.83	19.44	36.90
k_P	$8 \cdot 10^3$	$1.0 \cdot 10^5$	$1.05 \cdot 10^6$	$9.4 \cdot 10^6$	$9.1 \cdot 10^7$
t_P	3.09s	6.06s	39.55s	351.63s	3 641.05s

of knots k_D, and the rate r_D correspond to the wavelet algorithm, where the compression parameter a is chosen such that the relative compression error for the unknown density u taken at four different points of $[0,1] \times [0,1]$ is less than 10^{-5}. For t_P, k_P, and r_P, the parameter a of the wavelet algorithm is chosen such that the relative compression error for a derivative of the single layer potential is less than 10^{-5}. The last error is taken over eight points in the exterior of the earth close to $\kappa(0.5, 0.5)$. Note that in the computation with the wavelet algorithms presented in Table 3 we solve the linear system (44) iteratively. The multiplication of a vector by A_j, however, is realized by applying the wavelet transform to the vector, by multiplying with the compressed and wavelet transformed matrix A_j^c, and by applying the inverse wavelet transform.

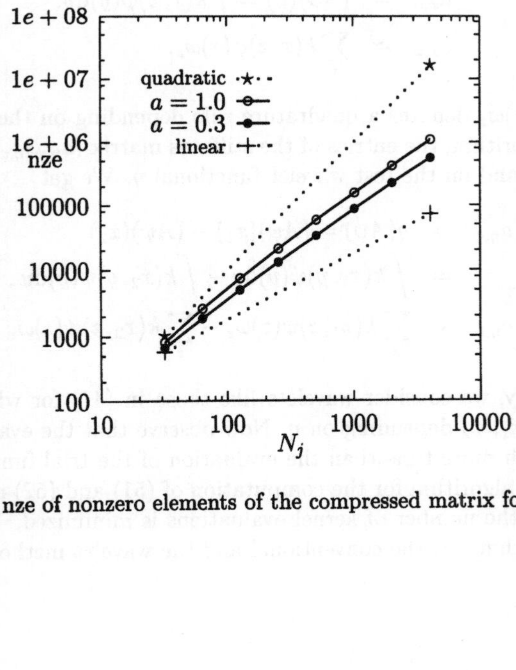

Figure 5: Number nze of nonzero elements of the compressed matrix for two constants a.

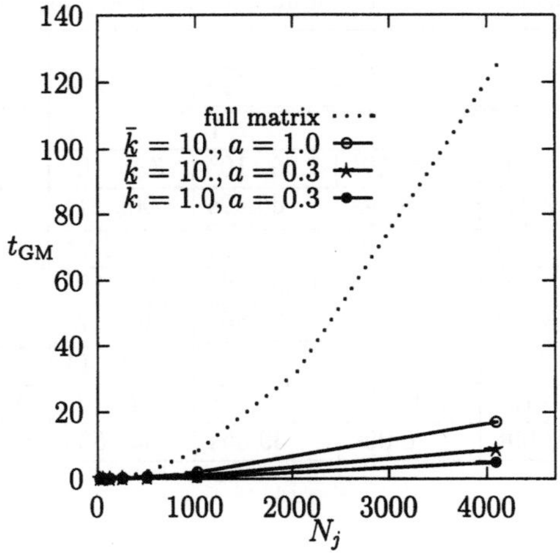

Figure 6: CPU-time t_{GM} in seconds for different wave numbers $\bar{k}$ and constants a.

5.4. THE ASSEMBLING OF THE MATRIX VIA QUADRATURE ALGORITHM

In order to concentrate on the essential issues of the matrix assembling we simplify formulae and notation. The conventional collocation algorithm is based on a stiffness matrix $(a_{x,\varphi})_{x,\varphi}$, where x runs through the set of collocation points and φ through the set of finite element basis functions. The entries take the form

$$\begin{aligned} a_{x,\varphi} &= (A\varphi)(x) = \int k(x,y)\varphi(y)\mathrm{d}y, \\ a_{x,\varphi} &\approx \sum_z k(x,z)\varphi(z)\omega_z, \end{aligned} \tag{51}$$

where $\int f \sim \sum_z f(z)\omega_z$ denotes a quadrature rule depending on the collocation point x. For the wavelet algorithm, the entries of the stiffness matrix $(a_{\eta,\psi})_{\eta,\psi}$ depend on the basis wavelet function ψ and on the test wavelet functional η. We get

$$\begin{aligned} a_{\eta,\psi} &= \eta(A\psi) = (A\psi)(x_1) - (A\psi)(x_2) \\ &= \int k(x_1,y)\psi(y)\mathrm{d}y - \int k(x_2,y)\psi(y)\mathrm{d}y\,, \\ a_{\eta,\psi} &\approx \sum_z k(x_1,z)\psi(z)\omega_z - \sum_z k(x_2,z)\psi(z)\omega_z\,, \end{aligned} \tag{52}$$

where, for simplicity, we consider wavelets like those in (47) for which $\eta(f) = f(x_1) - f(x_2)$ with certain x_1, x_2 depending on η. Now observe that the evaluation of the kernel functions costs much more time than the evaluation of the trial functions. Thus, the do loops of an efficient algorithm for the computation of (51) and (52) are to be arranged in such an order that the number of kernel evaluations is minimized. We, finally, arrive at the following algorithms for the conventional and the wavelet method, respectively:

```
do For all collocation points x.

    Determine the quadrature
    rule with knots z and
    weights ω_z depending on x.

    do For all quadrature knots z.

        Compute U = [k(x,z)ω_z].

        Determine all basis trial
        functions φ with φ(z) ≠ 0.

        do For all φ with φ(z) ≠ 0.

            Compute T = U · φ(z).

            Add T to a_{x,φ}.

        enddo
    enddo
enddo
```

```
do For all collocation points x.

    Determine all η s.t. x ∈ supp η.

    Determine the quadrature
    rule with knots z and
    weights ω_z depending on x.

    do For all quadrature knots z.

        Compute U = [k(x,z)ω_z].

        Determine all basis trial
        functions ψ with ψ(z) ≠ 0.

        do For all ψ with ψ(z) ≠ 0.

            Compute S = U · ψ(z).

            do For all η s.t. x ∈ supp η.
                Add ±S to a_{η,ψ}.
            enddo

        enddo
    enddo
enddo
```

Comparing these algorithms, we observe that to the do loop over φ with $\varphi(z) \neq 0$ there corresponds the longer do loop over all ψ with $\psi(z) \neq 0$ and with $a_{\eta,\psi}$ not neglected in the compression step (though this is not indicated in the algorithm). To the addition of T to $a_{x,\varphi}$ there corresponds the do loop over all η such that $x \in \operatorname{supp} \eta$. We have chosen the wavelets in (25) and (47) so that the longer do loop over the ψ is not too long and that the additional do loop over the η is as short as possible. The wavelet algorithm can be faster than the conventional collocation only if the number of quadrature knots z in the corresponding do loop over z is less for the wavelet algorithm. Hence, the quadrature algorithm is essential. Before we turn to the quadrature algorithm we mention two important aspects. First, in order to reduce the time for input/output operations between CPU and main memory, we recommend to introduce an extra working field. For a fixed x, the entries $a_{\eta,\psi}$ with η such that $x \in \operatorname{supp} \eta$ and with arbitrary ψ should be stored in this two-dimensional field. Second, for the fast evaluation of the trial functions, the "tree" structure of the wavelet basis can be used. In fact, the values at a fixed point z for the wavelets and finite element functions of a certain level l can easily be obtained from the values for the functions of level $l+1$.

The quadrature algorithm for the assembling of the stiffness matrix A_j^ϵ should be based on the following (cf. [40]):

i) If the order $\mathbf{r}$ of A is non-negative, then the singular or hypersingular integrals in (52) should be regularized (cf. [30,45,34]) before quadrature rules are applied. E.g., for the singular kernel k of the boundary integral equation in (15), we split the kernel into the sum of a singular main part k_S and a weakly singular kernel. Then

Table 4: Quadrature errors and number of knots for the integral of a logarithmic function.

j	accuracy 2^{-3j}	knots dyadic	max.knots per subinterval	knots equally spaced	max.knots per subinterval
3	$2\cdot 10^{-3}$	8	2	16	2
5	$3\cdot 10^{-5}$	24	3	96	3
7	$1\cdot 10^{-6}$	48	4	384	3
10	$1\cdot 10^{-9}$	108	6	4194	4
12	$2\cdot 10^{-11}$	154	7	20864	5
14	$2\cdot 10^{-13}$	208	8	-	-

the integral in the integral operator takes the form

$$\int_{[0,1]\times[0,1]} k(x,y)u_h(y)\mathrm{d}y_1\mathrm{d}y_2 = u_h(x)\int k_S(x,y)\mathrm{d}y + \int \left[k(x,y)u_h(y) - k_S(x,y)u_h(x)\right]\mathrm{d}y. \tag{53}$$

Now the quadrature rules can be applied to the last integral which is weakly singular, only. The first integral on the right-hand side can be computed by an analytic formula.

ii) If the quadrature rule for the integral in $a_{\eta,\psi}$ is chosen in dependence not only on the collocation point $x \in \operatorname{supp}\eta$ but also on ψ (this means that a quadrature algorithm different from that presented above is used) and if the kernel is analytic, then composite Gauß rules over meshes graded geometrically towards the singularity point x can be applied. The order of the Gauß rule should increase slowly with decreasing distance to x. We have tested such a rule for $\int_0^{2\pi}\log(x-\pi)\mathrm{d}x$. In Table 4 we present the quadrature error, the number of knot points, and the maximal order of the Gauß rules. We compare a composite Gauß rule with variable Gauß order over the dyadic mesh $\{\pi \pm 2^{-l}\pi,\ l=0,\dots,j\}$ and a composite rule over the uniform partition $\{\pi \pm m\cdot 2^{1-j}\pi,\ m=0,\dots,2^{j-1}\}$ with a fixed order of the Gauß rule on each subinterval. Clearly, the rule over the graded mesh performs better. Note that in special cases quadrature rules with end point correction can also be applied (cf. [41,2]).

iii) The quadrature rule should be a composite tensor product Gauß rule and the underlying partition should be a refinement of the partition over which the trial functions are polynomials. More precisely, if the collocation point x is fixed, then we have to compute $\int k(x,y)\psi(y)\mathrm{d}y$ only for those ψ for which there is an η such that $x \in \operatorname{supp}\eta$ and such that $a_{\eta,\psi}$ is not discarded in the compression step. To this set of basis wavelets ψ there corresponds a minimal partition such that the ψ are polynomials on the subdomains. This partition is coarser than the partition of the level j, and we will call it starting partition. If we refine the starting partition over some subdomains near the singularity point x and take a composite Gauß rule, then we arrive at a suitable rule $\int f \sim \sum_z f(z)\omega_z$. Since the starting partition is coarser than the level j partition, the number of quadrature knots should be less than N^2.

iv) If the subdomain of the starting partition in iii) belongs to a level l coarser than level j, then no further partition step is needed. The order of the Gauß rule, which should be applied over the subdomain, can be fixed but should be greater or equal to the polynomial degree d of the trial functions plus the number of vanishing moments for the trial wavelets.

v) If the subdomain of the starting partition in iii) belongs to level j, then the quadrature technique for conventional collocation methods should be applied.

Note that the numbers of quadrature knots and the computation times (on a DEC 3000 AXP 400 α-processor workstation) presented in Table 3 are obtained with a quadrature algorithm based on i), iii)-v). The reduction in computation time is much less than the reduction in storage. This is due to the mesh grading over the subdomains of v). Using optimal rules for these subdomains, the reduction factor for the computation time should be similar to the compression rate.

REFERENCES

[1] ALPERT, B.K.: "A class of basis in L^2 for the sparse representation of integral operators", SIAM J. Math. Anal. 24, pp. 246–262, 1993.

[2] ALPERT, B.K.: "Rapidly-convergent quadratures for integral operators with singular kernels", Technical Report LBL 30092, Lawrence Berkeley Laboratory, University of California, Berkeley, CA, 1990.

[3] AMOSOV, B.A.: "On the approximate solution of elliptic pseudodifferential equations over smooth and closed curves", in Russian, Z. Anal. Anwendungen 9, pp. 545–563, 1990.

[4] ARNOLD, D.N. and WENDLAND, W.L: "The convergence of spline collocation for strongly elliptic equations on curves", Numer. Math. 47, pp. 317–431, 1985.

[5] BEYLKIN, G., COIFMAN, R., and ROKHLIN, V.: "Fast wavelet transforms and numerical algorithms I", Comm. Pure Appl. Math. 44, pp. 141–183, 1991.

[6] BRANDT, A. and LUBRECHT, A.A.: "Multilevel matrix multiplication and fast solution of integral equation", J. Comput. Phys. 90, pp. 348–370, 1991.

[7] CARNICER, J.M., DAHMEN, W., and PEÑA, J.M.: "Local decomposition of refinable spaces", to appear in Applied and Computational Harmonic Analysis, IGPM Report 112, RWTH Aachen, 1994.

[8] CHUI, C.K.: "An introduction to wavelets", Academic Press, Boston, 1992.

[9] COHEN, A., DAUBECHIES, I., and FEAUVEAU, J.-C.: "Biorthogonal bases of compactly supported wavelets", Comm. Pure and Appl. Math. 45, pp. 485–560, 1992.

[10] COLTON, D. and KRESS, R.: "Integral equation methods in scattering theory", Springer-Verlag, Berlin, Heidelberg, New York, 1983.

[11] DAHLKE, S., DAHMEN, W., HOCHMUTH, R., and SCHNEIDER, R.: "Stable multiscale bases and local error estimation for elliptic problems", IGPM Report 124, RWTH Aachen, 1996.

[12] DAHMEN, W., KLEEMANN, B., PROESSDORF, S., and SCHNEIDER, R.: "A multiscale method for the double layer potential equation on a polyhedron", in Dikshit, H.P. and Micchelli, C.A.: "Advances of Computational Mathematics", World Scientific Publ. Co., Inc., New Delhi, pp. 15–57, 1994.

[13] DAHMEN, W., KLEEMANN, B., PROESSDORF, S., and SCHNEIDER, R.: "Multiscale methods for the solution of the Helmholtz and Laplace equations", in Wendland, W.: "Boundary Element Methods", Reports from the Final Conference of the Priority Research Programme 1989-1995 of the German Research Foundation, Oct. 2-4, 1995, Stuttgart, Springer-Verlag, 1996.

[14] DAHMEN, W. and KUNOTH, A.: "Multilevel preconditioning", Numer. Math. 63, pp. 315–344, 1992.

[15] DAHMEN, W., KUNOTH, A., and SCHNEIDER, R.: "Operator equations, multiscale concepts and complexity", to appear in Proceedings of the AMS-SIAM Summer School, Park City, 1995, Preprint No. 206, WIAS, Berlin, 1996.

[16] DAHMEN, W., PROESSDORF, S., and SCHNEIDER, R.: "Multiscale methods for pseudo-differential equations", in Schumaker, L. and Webb, G.: "Recent Advances in Wavelet Analysis", Wavelet Analysis and its Application 3, pp. 191–235, 1994.

[17] DAHMEN, W., PROESSDORF, S., and SCHNEIDER, R.: "Wavelet approximation methods for pseudo-differential equations I: Stability and convergence", Math. Zeitschr. 215, pp. 583–620, 1994.

[18] DAHMEN, W., PROESSDORF, S., and SCHNEIDER, R.: "Wavelet approximation methods for pseudo-differential equations II: Matrix compression and fast solution", Advances in Comp. Math. 1, pp. 259-335, 1993.

[19] DAHMEN, W., PROESSDORF, S., and SCHNEIDER, R.: "Multiscale methods for pseudo-differential equations on smooth manifolds", in Chui, C.K., Montefusco, L., and Puccio, L.: "Wavelets: Theory, Algorithms, and Applications", Academic Press, pp. 1–40, 1994.

[20] DAUBECHIES, I.: "Ten lectures on wavelets", CBMS Lecture Notes 61, SIAM, Philadelphia, 1992.

[21] DOROBANTU, M.: "Potential integral equations of the 2D Laplace operator in wavelet basis", Preprint TRITA-NA-9401, Royal Institute of Technology, University of Stockholm, 1994.

[22] ENGELS, J. and KLEES, R.: "Galerkin versus collocation - Comparison of two discretization methods for the boundary integral equations in $\mathbb{R}^3$", Manuscripta Geodetica 17, pp. 245–256, 1992.

[23] GREENGARD, L. and ROKHLIN, V.: "A fast algorithm for particle simulation", J. Comput. Phys. 73, pp. 325–348, 1987.

[24] HACKBUSCH, W. and NOWAK, Z.P.: "On the fast matrix multiplication in the boundary element method by panel clustering", Numer. Math. 54, pp. 463–491, 1989.

[25] HARTEN, A. and YAD-SHALOM, I.: "Fast multiresolution algorithms for matrix-vector multiplication", SIAM J. Numer. Anal. 31, pp. 1191–1218, 1994.

[26] JUNKHERR, J.: "Effiziente Lösung von Gleichungssystemen die aus der Diskretisierung von schwach singulären Integralgleichungen 1. Art herrühren", Ph. D. Thesis, Christian-Albrechts-Universität Kiel, 1994.

[27] KLEES, R.: "Lösung des geodätischen Randwertproblems mit Hilfe der Randelementmethode", German Geodetic Kommission (DGK), Series C, No.382, Munich, 1992.

[28] MEYER, Y.: "Wavelets and Operators", Cambridge University Press, Cambridge, 1992.

[29] MIKHLIN, S.G. and PROESSDORF, S.: "Singular integral operators", Springer-Verlag, Berlin, Heidelberg, New York, Tokyo, 1986.

[30] NEDELEC, J.C.: "Integral equations with non integrable kernels", Integral Equations and Operator Theory 5, pp. 562–572, 1982.

[31] OSWALD, P.: "Multilevel finite element approximation", Teubner, Stuttgart, 1994.

[32] OVERHAUSER, A.W.: "Analytic definition of curves and surfaces by parabolic blending", Techn. Report No. S168-40, Scientific Research Staff Publication, Ford Motor Company, Detroit, 1968.

[33] PETERSDORFF, T.v. and SCHWAB, C.: "Boundary element methods with wavelets and mesh refinement", to appear in Proc. of ICIAM 1995, Hamburg, Special issue of ZAMM, Research Report No. 95-10, SAM, Eidgenössische Technische Hochschule Zürich, 1995.

[34] PETERSDORFF, T.v. and SCHWAB, C.: "Fully discrete multiscale Galerkin BEM", to appear in Dahmen, W., Kurdila, A.J., and Oswald, P.: "Multiresolution Analysis and PDE", Series: Wavelet Analysis and its Applications, Academic Press 1996, Research Report No. 95-08, SAM, Eidgenössische Technische Hochschule Zürich, 1995.

[35] PETERSDORFF, T.v. and SCHWAB, C.: "Wavelet approximations for first kind boundary integral equations on polygons", submitted to Numer. Math., Technical Note BN-1157, Institute for Physical Science and Technology, University of Maryland at College Park, 1994.

[36] PROESSDORF, S. and SCHMIDT, G.: "A finite element collocation method for singular integral equations", Math. Nachr. 100, pp. 33–60, 1981.

[37] PROESSDORF, S. and SCHNEIDER, R.: "A spline collocation method for multidimensional strongly elliptic pseudodifferential operators of order zero", Integral Equations and Operator Theory 14, pp. 399–435, 1991.

[38] PROESSDORF, S. and SCHNEIDER, R.: "Spline approximation methods for multidimensional periodic pseudodifferential equations", Integral Equations and Operator Theory 15, pp. 626–672, 1992.

[39] RATHSFELD, A.: "A wavelet algorithm for the solution of the double layer potential equation over polygonal boundaries", Journal of Integral Equations and Applications 7, pp. 47–97, 1995.

[40] RATHSFELD, A.: "A wavelet algorithm for the boundary element solution of a geodetic boundary value problem", Preprint No. 225, WIAS, Berlin, 1996.

[41] ROKHLIN, V.: "End point corrected trapezoidal quadrature rules for singular functions", Computers and Mathematics with Applications 20, pp. 52-61, 1990.

[42] ROKHLIN, V.: "Rapid solution of integral equations of classical potential theory", J. Comput. Phys. 60, pp. 187–207, 1983.

[43] SAAD, Y. and SCHULTZ, M.H.: "GMRES: A generalized minimal residual algorithm for solving nonsymmetric linear systems", SIAM J. Sci. Stat. Comput. 7, No.3, pp. 856–869, 1986.

[44] SARANEN, J. and VAINIKKO, G.: "Fast solvers of integral and pseudodifferential equations on closed curves", to appear.

[45] SAUTER, S. A.: "Über die effiziente Verwendung des Galerkinverfahrens zur Lösung Fredholmscher Integralgleichungen", Ph. D. Thesis, Christian-Albrechts-Universität, Kiel, 1992.

[46] SAUTER, S. A.: "Der Aufwand der Panel-Clustering-Methode für Integralgleichungen", Bericht Nr. 9115, Inst. f. Inform. u. Prakt. Math., Christian-Albrechts-Universität, Kiel, 1991.

[47] SCHNEIDER, R.: "Multiskalen- und Waveletkompression: Analysisbasierte Methoden zur effizienten Lösung großer vollbesetzter Gleichungssysteme", Habilitationsschrift, Fachbereich Mathematik, Technische Hochschule Darmstadt, 1995.

[48] STEPHAN, E.P. and TRAN, T.: "A multi-level additive Schwarz method for hypersingular integral equations", Report AMR94/24, University of New South Wales, Sydney, 1994.

[49] SWELDENS, W.: "The lifting scheme: a custom-design construction of biorthogonal wavelets", Technical Report, Industrial Mathematics Initiative, Department of Mathematics, University of South Carolina, 1994.

[50] WALKER, H.F.: "Implementation of the GMRES method using Householder transformations", SIAM J. Sci. Stat. Comput. 9, No. 1, pp. 152–163, 1988.

CUBATURE TECHNIQUES FOR 3-D GALERKIN BEM

Stefan A. Sauter
Lehrstuhl für Praktische Mathematik
Mathematisches Seminar Bereich II
Christian-Albrechts-Universität zu Kiel
Hermann-Rodewald-Str. 3/1
24098 Kiel, Germany

SUMMARY

In this paper we present cubature methods for the approximation of surface integrals arising from Galerkin discretizations of 3-d boundary integral equations. This numerical integrator is fully implicit in the sense that the form of the kernel function, the surface parametrization, the trial and test space, and the order of the singularity of the kernel function is **not** used explicitly. Different kernels can be treated by just replacing the subroutine which evalutes the kernel function in certain surface points.

Furthermore, the implementation of the integrator is relatively easy since it can be checked on simple test kernels as, e.g., polynomials where the exact integrals are available.

We discuss the convergence of the cubature methods together with a stability and consistency analysis in order to determine the minimal cubature orders a priori.

INTRODUCTION

In this paper we will discuss Galerkin discretizations of Fredholm integral equations on two-dimensional surfaces. From the theoretical point of view as, e.g., convergence, stability, and symmetry behaviour, the Galerkin method is advantageous compared with collocation or Nyström discretizations. In practice, however, especially in the case of 2-d surfaces, the collocation or Nyström method is usually implemented. The only reason for this is that, for Galerkin discretizations, the elements of the system matrix are given by two two-dimensional integrations over pairs of surface patches which, in particular, contain a kernel function with a characteristic singular and near-singular behaviour.

In our paper we will present efficient cubature methods for the approximation of that kind of integrals. (Note that in more than one dimension the terminus *quadrature* is replaced by *cubature*). These integrators are fully automatic, which means, that they neither depend on the *explicit* form of the kernel function nor on the choice of the trial space and surface parametrization. Different kernels can be treated by just replacing the subroutine which evaluates the kernel function at certain surface points while the *numerical integrator* has not to be modified. In contrast to analytic and semi-analytic methods such kind of strategies have the advantage that they are relatively easy to implement and behave robustly with respect to cancellation errors. They can be tested isolated for polynomial test kernel where the exact solution is available.

The cubature strategy is based on the following idea. First all singular integrals are transformed into integrals with *analytic* integrands by employing regularizing coordinate transforms. These coordinate transforms were first introduced in [12] and [4] where they were combined with semi-analytic optimization techniques based on expansions of the kernel function and surface parametrization. A fully implicit treatment of such integrals for special applications as wavelets and *hp*-methods were presented in [11]. Here, we present a modified version of these coordinate transforms for the *h*-version of the Galerkin BEM together with an error analysis. Furthermore, the results of a stability and consistency analysis will be presented which allow to choose the order of integration minimally. We emphasize the importance of a proper consistency analysis for this kind of problem, since the bottleneck in any 3-d implementation of Galerkin discretizations of boundary integrals equations is the evaluation of the four dimensional integrals defining the matrix entries. Only slight over-integration usually will lead to an increase of the CPU-time by orders of magnitude. Such a stability and consistency analysis for the *regular* integrals was presented in [13]. Here, we will focus on the singular and near-singular integrals which can be regarded as Part II of the mentioned paper [13].

Due the limited space in these proceedings the proofs are only sketched. We concentrate here on the practical point of view, i.e., present the cubature techniques such that they can be implemented in a straightforward manner. The full mathematical analysis and implementation details are worked out in [9].

BOUNDARY INTEGRAL EQUATIONS

We consider the variational formulation of Fredholm integral equations. Let Γ be a piecewise analytic, orientable Lipschitz surface of a bounded domain $\Omega \subset \mathbb{R}^3$. Let $W^{k,p}(\Gamma)$ denote the usual Sobolev space as defined, e.g., in the book of Adams [1]. The norm of $W^{k,p}$ is denoted by $\|\cdot\|_{k,p}$ and the seminorm containing only the highest derivatives by $|\cdot|_{k,p}$. For $p = 2$ we write H^k instead of $W^{k,2}$ and $\|\cdot\|_k$ instead of $\|\cdot\|_{k,2}$.

For given right-hand side $f \in H^{s_1'}$, we are seeking $u \in H^{s_1}(\Gamma)$ such that

$$(v, \lambda_1 u)_0 + (v, K_1 u)_0 = (v, \lambda_2 f)_0 + (v, K_2 f)_0, \qquad \forall v \in H^{s_1}(\Gamma) \tag{1}$$

where $(\cdot,\cdot)_0$ denotes the L^2-scalar product. The associated operators $\lambda_i I + K_i$ are assumed to be of order $2s_i$ and explicitly given by

$$K_i u := p.f. \int_\Gamma k_i(x,y)\, u(y)\, dy$$

where the integral has to be understood in the finite part sense (see [16]) and k_i denotes the kernel functions. The functions $\lambda_i : \Gamma \to \mathbb{R}$ are assumed to be piecewise constant on smooth patches of the surface and may jump across corners and edges. The regularity exponent s_1' for the right-hand side should be such that $\lambda_2 f + K_2 f \in H^{-s_1}$, i.e., $s_1' = -s_1 - 2s_2$. The left-hand side of (1) defines the bilinear forms $a, a_1, a_2 : H^{s_1}(\Gamma) \times H^{s_1}(\Gamma) \to \mathbb{C}$ as follows

$$\begin{aligned} a_1(u,v) &:= (v, \lambda_1 u)_0, \\ a_2(u,v) &:= (v, K_1 u)_0, \\ a &:= a_1 + a_2. \end{aligned} \tag{2}$$

The right-hand side defines the functionals $F, F_1, F_2 : H^{s_1} \to \mathbb{C}$:

$$\begin{aligned} F_1(v) &:= (v, \lambda_2 f)_0, \\ F_2(v) &:= (v, K_2 f)_0, \\ F &:= F_1 + F_2. \end{aligned}$$

Remark 1 *The order of the operator $\lambda_i I + K_i$ can be negative which implies in particular $\lambda_i \equiv 0$.*

Throughout this paper we assume that the bilinear form a is continuous

$$|a(u,v)| \le C_a \|u\|_{s_1} \|v\|_{s_1}, \qquad \forall u, v \in H^{s_1}(\Gamma) \tag{3}$$

and elliptic

$$\|a(u,u)\| \ge C_e \|u\|_{s_1}^2 - C_c \|u\|_t^2 \qquad \forall u \in H^{s_1}(\Gamma) \tag{4}$$

with suitable $t < s_1$. For simplicity, we assume that (1) has a unique solution for every $f \in H^{s_1'}(\Gamma)$. Otherwise, one has to introduce suitable normalizations or to restrict $H^{s_i}(\Gamma)$ to some quotient space. Then, the theory presented below directly applies to these situations.

The following model problem illustrates the abstract setting.

Example 2 *Consider the Laplace equation in the exterior of the bounded domain $\Omega \subset \mathbb{R}^3$.*

$$\begin{aligned} \Delta u &= 0, && \text{in } \mathbb{R}^3 \backslash \bar{\Omega}, \\ \tfrac{\partial u}{\partial n} &= f, && \text{on } \Gamma := \partial\Omega, \\ u(x) &\to 0, && \|x\| \to \infty, \end{aligned} \tag{5}$$

where $n(x)$ denotes the normal in $x \in \Gamma$ pointing into Ω and $\partial/\partial n := \langle n, \nabla \cdot \rangle$. Let the bilinear form $a : H^{1/2} \times H^{1/2} \to \mathbb{R}$ be defined by

$$a(u,v) := \int_\Gamma v(x)\, p.f. \int_\Gamma u(y) \frac{\partial}{\partial n_x} \frac{\partial}{\partial n_y} \frac{1}{\|x-y\|} dy dx$$

and the functional of the right-hand side by

$$F(v) := \frac{1}{2}(v,f)_0 - \int_\Gamma v(x)\, p.f. \int_\Gamma f(y) \frac{\partial}{\partial n_x} \frac{1}{\|x-y\|} dy dx.$$

Let f be such that $F \in \left(H^{1/2}/\mathbb{R}\right)'$. Then the problem of finding $u \in H^{1/2}/\mathbb{R}$ such that

$$a(u,v) = F(v), \qquad \forall v \in \left(H^{1/2}/\mathbb{R}\right)'$$

has a unique solution. This function u is the trace of the solution of (5).

Let Ω be a three-dimensional domain with piecewise analytic, orientable Lipschitz surface Γ. We assume that Γ is a smooth image of an interpolating polyhedron. To be more concrete, let $\tilde{\Gamma}$ be a surface of a polyhedron consisting of plane faces $\tilde{\Gamma}_j$ such that there exists a bijective mapping $\eta : \tilde{\Gamma} \to \Gamma$ with the following properties.

1. η is bi-Lipschitz continuous, i.e.,

$$C_1 \|\tilde{x} - \tilde{y}\| \leq \|\eta(\tilde{x}) - \eta(\tilde{y})\| \leq C_2 \|\tilde{x} - \tilde{y}\|, \qquad \forall \tilde{x}, \tilde{y} \in \tilde{\Gamma},$$

2. there exists plane extensions $\tilde{\Gamma}_j^\star$ of $\tilde{\Gamma}_j$, i.e., $\tilde{\Gamma}_j \subset\subset \tilde{\Gamma}_j^\star$ where $\eta \mid_{\tilde{\Gamma}_j}$ can be extended analytically.

In this paper we restrict the presentation to triangular elements. Generalizations to the quadrilateral case are straightforward by using the techniques presented in [15] and [14]. Let $\tilde{\tau} := \left\{\tilde{K}_1, \tilde{K}_2, \ldots, \tilde{K}_N\right\}$ denote a grid of the surface $\tilde{\Gamma}$ consisting of plane triangles having the property that

$$\begin{aligned} &\tilde{\Gamma} := \overline{\bigcup_{\tilde{K} \in \tilde{\tau}} \tilde{K}}, \\ &\tilde{K} \cap \tilde{K}' = \emptyset, \qquad \forall \tilde{K}, \tilde{K}' \in \tau \text{ with } \tilde{K} \neq \tilde{K}'. \end{aligned} \tag{6}$$

This grid induces a grid on the true surface Γ by $\tau := \left\{\eta\left(\tilde{K}\right)\right\}_{\tilde{K} \in \tilde{\tau}}$. Note that the transformation η is independent of the grid $\tilde{\tau}$.

In order to define finite element spaces on Γ, we lift polynomial spaces on a reference element onto the surface elements. In this light, we define, for a domain $D \subset \mathbb{R}^d$, the space of polynomials of total degree p by

$$\Pi_p(D) := \left\{ v(x) = \sum_{|\alpha| \leq p} c_\alpha x^\alpha \text{ for some } c_\alpha \in \mathbb{C} \right\},$$

where $\alpha \in \mathbb{N}_0^d$ is a multi-index and $|\alpha|$ denotes the l^1-norm.

For triangular elements we employ the master element

$$K^0 := \left\{ \hat{x} \in \mathbb{R}_+^2 \mid 0 \leq \hat{x}_1 \leq 1, 0 \leq \hat{x}_2 \leq \hat{x}_1 \right\}$$

and the polynomials of degree p are given by $\Pi_p(K^0)$. The transformation of K^0 onto $\tilde{K}$ is affine linear and denoted by $\kappa_{\tilde{K}}$ while the mapping on the true surface element is given by $\kappa_K := \eta \circ \kappa_{\tilde{K}}$. The space $\Pi_p(K^0)$ is lifted onto the surface patch K by κ_K :

$$\Pi_p(K) := \left\{ u : K \to \mathbb{C} \mid u \circ \kappa_K \in \Pi_p\left(K^0\right)\right\}. \tag{7}$$

The global finite element space is given by

$$V_{k,p} := \left\{ u \in C^k(\Gamma) \mid \forall K \in \tau : u \mid_K \in \Pi_p(K)\right\}. \tag{8}$$

The Galerkin boundary element discretization to (1) is characterized by replacing the infinite dimensional Sobolev spaces H^{s_1} by the finite dimensional subspace $V_{k,p}$: for given $f \in H^{s_1'}$, we are seeking $u_G \in V_{k,p}$ such that

$$a\left(u_G, v\right) = F\left(v\right), \qquad \forall v \in V_{k,p}. \tag{9}$$

In this paper we are only considering conforming methods, i.e., the exponent k in (8) has to be chosen such that $V_{k,p} \subset H^{s_1}$.

In order to compute the solution to problem (9), we introduce a basis $\{\varphi_i\}_{1<i\leq N}$ of $V_{k,p}$. Any function $u \in V_{k,p}$ has a unique basis representation

$$u\left(x\right) = \sum_{i=1}^{n} \mathbf{u}_i \varphi_i\left(x\right).$$

For $1 \leq i, j \leq N$, let the system matrix $\mathbf{K}$ and the vector of the right-hand side be defined by

$$\begin{aligned} \mathbf{K}_{i,j} \; &: \; = a\left(\varphi_i, \varphi_j\right), \\ \mathbf{F}_i \; &: \; = F\left(\varphi_i\right). \end{aligned} \tag{10}$$

Then the solution of

$$\mathbf{K}\mathbf{u}_G = \mathbf{F}$$

is linked to the solution of (9) by

$$u_G\left(x\right) = \sum_{i=1}^{N} \mathbf{u}_{G,i} \varphi_i\left(x\right).$$

For the computation of the matrix $\mathbf{K}$ and right-hand side $\mathbf{F}$, integrals over (pairs of) surface patches have to be evaluated. This can be done exactly only in very special situations. In general, one has to use numerical cubature techniques to approximates the coefficients. The effect of numerical integration is considered as an approximation of the continuous (bi-)linear forms. The bilinear forms $\tilde{a}$, $\tilde{a}_1$, and $\tilde{a}_2$ are defined by replacing the integrals by numerical cubature, while the definition of $\tilde{F}, \tilde{F}_1$ and $\tilde{F}_2$ is analogously. Note that these approximate (bi-)linear forms have to be defined only on the discrete level, i.e., for functions in $V_{k,p}$ and not necessarily for functions in $H^{s_1}(\Gamma)$.

The Galerkin discretization including numerical cubature is given by seeking $\tilde{u}_G \in V_{k,p}$ such that

$$\tilde{a}\left(\tilde{u}_G, v\right) = \tilde{F}\left(v\right), \quad \forall v \in V_{k,p}.$$

In the next section we will present families of cubature methods which approximates the continuous (bi-)linear forms to any desired accuracy. Then, we will estimate the cubature error which, in combination with a stability and consistency analysis, enables us to determine the *minimal* cubature orders a priori.

KERNEL PROPERTIES OF BOUNDARY INTEGRAL EQUATIONS

Before presenting the cubature methods we have to collect the characteristic properties of the kernel functions. The kernel functions are suitable Gâteau derivatives of the fundamental solution of the underlying boundary value problem. The following properties are satisfied for all common kernels which arise by transforming elliptic boundary value problems in $\mathbb{R}^3$ into integral equations (see [16]). We formulate these conditions in a way which will turn out to be convenient for the cubature analysis. We assume that $k(x, y)$ can be written in the form

$$k(x,y) = \sum_{|\alpha|\geq t} s_\alpha(x,y) \frac{(y-x)^\alpha}{\|y-x\|^{\sigma+t}} \tag{11}$$

with functions $s_\alpha : \Gamma \times \Gamma \to \mathbb{C}$ being analytic on pairs of smooth surface patches of Γ. The integer σ denotes the *order* of the singularity and we assume that $\sigma \leq 3$ holds. For $\sigma \leq 2$, we assume that the kernel is anti-symmetric, i.e.,

$$k(x,y) + k(y,x) = \sum_{|\alpha|\geq t+1} \tilde{s}_\alpha(x,y) \frac{(y-x)^\alpha}{\|y-x\|^{\sigma+t}}. \tag{12}$$

This condition is stronger than the common Tricomi-Giraud-Mikhlin condition and implies that the kernel is Cauchy-singular, i.e.,

$$\lim_{\varepsilon\to 0} \int_{K\backslash B_\varepsilon(x)} k(x,y) f(y)\, dy$$

exists for all elements $K \in \tau$, Lipschitz-continuous functions $f \in C^{0,1}(K)$, and $x \in \Gamma\backslash\partial K$.

For $\sigma = 3$, the kernel function is called hyper-singular and an appropriate regularization has to be performed *on the continuous level.* We assume that the bilinear form a_2 of (2) can be rewritten as follows

$$a_2(u,v) = \sum_{i=0}^{m_1}\sum_{j=0}^{m_2} \int_\Gamma \int_\Gamma D_i[v](x,y)\, k_{i,j}^{reg}(x,y)\, D_j[u](x,y)\, dy dx$$

with $m_{1,2} \in \{0,1\}$. The operators D_i are of order i satisfying

$$\|D_i[w](x,y)\|_0 \leq C\|w\|_i,$$

while the kernel functions $k_{i,j}^{reg}$ satisfy (11) and (12) with $\sigma_{i,j}^{reg} \leq 2$. Furthermore, throughout the paper, we assume the relations $\sigma_{i,j}^{reg} + i + j \leq 3$ and $s_1 \leq m_1 \leq p$. An analogous representation has to be employed if the kernel function k_2 is hyper-singular. For weakly or Cauchy singular kernels we put $k_{0,0}^{reg} = k$ and $m_1 = m_2 = 0$.

Example 3 *The bilinear form $a(u,v)$ of example (2) can be rewritten as follows*

$$a(u,v) = \int_\Gamma \int_\Gamma D_1[v](x,y)\, D_1[u](x,y)\, k^{reg}(x,y)\, dydx$$

with $D_1[w](x,y) := (w(y) - w(x)) / \|y - x\|$ and

$$k^{reg}(x,y) := \|y-x\|^2 \frac{\partial^2}{\partial n_x \partial n_y} \frac{1}{\|x-y\|}.$$

The following representation is proved in [10])

$$a(u,v) := -\int_\Gamma (-n(x) \times \nabla_x)\, u(x) \int_\Gamma \frac{1}{\|x-y\|} (n(y) \times \nabla_y)\, v(y)\, dydx.$$

Using this kind of regularization it is natural to assume in the following that the bilinear form has a representation with a Cauchy-singular kernel function ($\sigma \leq 2$) while, possibly, the functions u and v have to replaced by $D_j[\cdot]$. We state that for much more general kernel functions such representation are available, too (see [10], [6], [7], [5], [3]). Another approach is presented in [8] which can be treated by a slight modification of the cubature techniques below.

We are now ready for the presentation of the cubature techniques.

CUBATURE TECHNIQUES FOR GALERKIN DISCRETIZATIONS

In view of the previous section, the (bi-) linear forms can be decomposed as follows

$$a_1(u,v) = \sum_{K \in \mathcal{T}} a_K(u,v) \qquad a_2(u,v) = \sum_{K_x \times K_y \in \mathcal{T} \times \mathcal{T}} a_{K_x \times K_y}(u,v)$$

$$F_1(v) = \sum_{K \in \mathcal{T}} F_K(v) \qquad F_2(v) = \sum_{K_x \times K_y \in \mathcal{T} \times \mathcal{T}} F_{K_x \times K_y}(v)$$

with

$$a_K(u,v) := \int_K \lambda_1 uv dx, \qquad a_{K_x \times K_y}(u,v) := \lim_{\varepsilon \to 0} \int_{\substack{K_x \times K_y \\ \|x-y\| \geq \varepsilon}} A(x,y)\, k_1(x,y)\, dydx,$$

$$F_K(v) := \int_K \lambda_2 f v dx, \qquad F_2(v) := \lim_{\varepsilon \to 0} \int_{\substack{K_x \times K_y \\ \|x-y\| \geq \varepsilon}} B(x,y)\, k_2(x,y)\, dydx.$$

The functions A and B are given by $A(x,y) = D_1[u](x,y)\, D_2[v](x,y)$ and $B(x,y) := D_3[f](x,y)\, D_4[v](x,y)$ with the operators D_j of order m_j defined in the previous section. The kernel functions k_i are assumed to satisfy the assumptions of the previous section with $\sigma \leq 2$.

In the first part we will present coordinate transforms which render the integrands above analytic such that they can be treated by standard cubature methods. We emphasize that these transformations does not contain any approximation process.

The integrands of the integrals corresponding to a_K and F_K are analytic and can be approximated by standard techniques. In the following we will concentrate on the integrals a_2 and F_2. Here, the integrands have a singular, near-singular, or regular behaviour which requires different transformations. The following four basis cases have to be distinguished.

1. "Identical Panels": $K_x = K_y$.

2. "Common Edge": K_x and K_y share exactly an edge.

3. "Common Vertex": K_x and K_y share exactly a vertex.

4. "Regular Case": $\overline{K}_x$ and $\overline{K}_y$ are disjoint.

Condition (6) does not guarantee that every pair K_x, $K_y \in \tau$ satisfies one of the four cases listed above. However, it is clear that one of the panels K_x, K_y can be subdivided such that the arising sub-panels satisfy one of the conditions above. The methods presented below then have to be applied to the sub-panels. Hence, we assume for the following that any pair of panels $K_x \times K_y$ satisfies one and only one of the conditions above.

In the following we will discuss these four cases in detail.

THE CASE OF IDENTICAL PANELS

Here, we consider integrals of the type

$$I_i := \lim_{\varepsilon \to 0} \int_{\substack{K \times K \\ \|x-y\| \geq \varepsilon}} H(x,y)\, k(x,y)\, dy dx$$

with H is either the function A or B defined above. The kernel function k is assumed to satisfy the assumptions on the kernel functions with $\sigma \leq 2$. We transform this integral onto the scaled master element $\Delta^h := h\Delta^0$ by the transformation $\kappa_K^h(\hat{x}) := \kappa_K(\hat{x}/h)$ with κ_K defined by (7) and $h := \operatorname{diam} K$. In local coordinates, the integrand is given by

$$\begin{aligned} k_{loc}(\hat{x},\hat{y}) &:= k\left(\kappa_K^h(\hat{x}), \kappa_K^h(\hat{y})\right) \\ H_{loc}(\hat{x},\hat{y}) &:= g_K(\hat{x})\, g_K(\hat{y})\, H\left(\kappa_K^h(\hat{x}), \kappa_K^h(\hat{y})\right) \end{aligned}$$

with

$$g_K(\hat{x}) := \left| \left\{ \left\langle \frac{\partial \kappa_K^h(\hat{x})}{\partial \hat{x}_i}, \frac{\partial \kappa_K^h(\hat{x})}{\partial \hat{x}_j} \right\rangle \right\}_{1 \leq i,j \leq 2} \right|^{1/2}.$$

In local coordinates the integral I_i takes the form

$$I_i := \lim_{\varepsilon \to 0} \int_0^h \int_0^{\hat{x}_1} \underset{\|\hat{x}-\hat{y}\| \geq \varepsilon}{\int_0^h \int_0^{\hat{y}_1}} H_{loc}(\hat{x},\hat{y})\, k_{loc}(\hat{x},\hat{y})\, d\hat{y} d\hat{x}.$$

The problem with this integral is that the location of the singularity is *not* fixed and cannot be cancelled by degenerate coordinate transforms as, e.g., polar coordinates. The purpose of the following coordinate transforms is to fix the singularity by introducing *relative coordinates* as proposed in [12] and [4]. The variable $\hat{y}$ is replaced by $\hat{y} = z + \hat{x}$ and hence, the singularity of the kernel function is fixed at the origin $z = \hat{y} - \hat{x} = 0$.

The domain of the $(\hat{x}, z)$-integration can be described by the union of the following six domains

$$D_1 = \left\{ \begin{array}{c} -h \le z_1 \le 0, \\ -h \le z_2 \le z_1, \\ -z_2 \le \hat{x}_1 \le h, \\ -z_2 \le \hat{x}_2 \le \hat{x}_1. \end{array} \right\} \qquad D_2 = \left\{ \begin{array}{c} 0 \le z_1 \le h, \\ z_1 - h \le z_2 \le 0, \\ -z_2 \le \hat{x}_1 \le h - z_1, \\ -z_2 \le \hat{x}_2 \le \hat{x}_1. \end{array} \right\}$$

$$D_3 = \left\{ \begin{array}{c} -h \le z_1 \le 0, \\ z_1 \le z_2 \le 0, \\ -z_1 \le \hat{x}_1 \le h, \\ -z_2 \le \hat{x}_2 \le z_1 + \hat{x}_1 - z_2. \end{array} \right\} \qquad D_4 = \left\{ \begin{array}{c} 0 \le z_1 \le h, \\ z_1 \le z_2 \le h, \\ z_2 - z_1 \le \hat{x}_1 \le h - z_1, \\ 0 \le \hat{x}_2 \le z_1 + \hat{x}_1 - z_2. \end{array} \right\}$$

$$D_5 = \left\{ \begin{array}{c} -h \le z_1 \le 0, \\ 0 \le z_2 \le z_1 + h, \\ z_2 - z_1 \le \hat{x}_1 \le h, \\ 0 \le \hat{x}_2 \le z_1 + \hat{x}_1 - z_2. \end{array} \right\} \qquad D_6 = \left\{ \begin{array}{c} 0 \le z_1 \le h, \\ 0 \le z_2 \le z_1, \\ 0 \le \hat{x}_1 \le h - z_1, \\ 0 \le \hat{x}_2 \le \hat{x}_1. \end{array} \right\}.$$

Hence, the integral I_i takes the form

$$I_i = \lim_{\varepsilon \to 0} \sum_{j=1}^{6} \int_{\substack{D_j \\ \|z\| \ge \varepsilon}} H_{loc}(\hat{x}, z + \hat{x})\, k_{loc}(\hat{x}, z + \hat{x})\, d\hat{x} dz.$$

Let T_d^h denote the scaled d–dimensional unit simplex

$$T_d^h := \left\{ \hat{x} \in \mathbb{R}_+^d \mid |\hat{x}|_{l^1} < h \right\}$$

and $R_d := (0,1)^d$. The following coordinate transforms map $(\eta, w_1, w_2, w_3) \in R_1 \times T_3^h$ onto the domain D_j.

$$\begin{pmatrix} z_1^{(1)} \\ z_2^{(1)} \\ \hat{x}_1^{(1)} \\ \hat{x}_2^{(1)} \end{pmatrix} = \begin{pmatrix} -w_1\eta \\ -w_1 \\ w_1 + w_2 + w_3 \\ w_1 + w_2 \end{pmatrix}, \quad \begin{pmatrix} z_1^{(2)} \\ z_2^{(2)} \\ \hat{x}_1^{(2)} \\ \hat{x}_2^{(2)} \end{pmatrix} = \begin{pmatrix} w_1\eta \\ w_1(\eta - 1) \\ w_1(1-\eta) + w_2 + w_3 \\ w_1(1-\eta) + w_2 \end{pmatrix},$$

$$\begin{pmatrix} z_1^{(3)} \\ z_2^{(3)} \\ \hat{x}_1^{(3)} \\ \hat{x}_2^{(3)} \end{pmatrix} = \begin{pmatrix} -w_1 \\ -w_1\eta \\ w_1 + w_2 + w_3 \\ w_1\eta + w_2 \end{pmatrix}, \quad \begin{pmatrix} z_1^{(4)} \\ z_2^{(4)} \\ \hat{x}_1^{(4)} \\ \hat{x}_2^{(4)} \end{pmatrix} = \begin{pmatrix} w_1\eta \\ w_1 \\ w_1(1-\eta) + w_2 + w_3 \\ w_2 \end{pmatrix},$$

$$\begin{pmatrix} z_1^{(5)} \\ z_2^{(5)} \\ \hat{x}_1^{(5)} \\ \hat{x}_2^{(5)} \end{pmatrix} = \begin{pmatrix} -w_1\eta \\ w_1(1-\eta) \\ w_1 + w_2 + w_3 \\ w_2 \end{pmatrix}, \quad \begin{pmatrix} z_1^{(6)} \\ z_2^{(6)} \\ \hat{x}_1^{(6)} \\ \hat{x}_2^{(6)} \end{pmatrix} = \begin{pmatrix} w_1 \\ w_1\eta \\ w_2 + w_3 \\ w_2 \end{pmatrix}.$$

The determinant of the Jacobi matrix equals w_1 and thus, the integral I_i can be written in the form

$$I_i = \int_{R_1} \int_{T_3^h} w_1 \sum_{j=1}^{6} H_{loc}\left(\hat{x}^{(j)}, z^{(j)} + \hat{x}^{(j)}\right) k_{loc}\left(\hat{x}^{(j)}, z^{(j)} + \hat{x}^{(j)}\right) dw d\eta. \tag{13}$$

It was shown in [12], [11], [15] that the integrand above is analytic. This implies that standard cubature techniques with properly chosen orders approximates (13) to any required accuracy.

THE CASE OF A COMMON EDGE

In this section we consider the computation of

$$I_e := \int_{K_1 \times K_2} H(x,y)\, k(x,y)\, dy dx.$$

The transformations in the case of a common edge are similar as for identical panels. However, the character of the singularity slightly changes. Let K_1, K_2 be two surface triangles which share exactly one edge. Let the parameter domains are given by $\Delta_1^h := \Delta^h$ and $\Delta_2^h := \left\{\hat{y} \in \mathbb{R}^2 \mid (\hat{y}_1, -\hat{y}_2)^T \in \Delta^h\right\}$. Let the mapping $\kappa : \overline{\Delta_1^h \cup \Delta_2^h} \to \overline{K_1 \cup K_2}$ be bi-Lipschitz continuous. For $\hat{x} \in \Delta_1^h$ and $\hat{y} \in \Delta_2^h$, the integrand in local coordinates is given by

$$\begin{aligned} k_{loc}(\hat{x},\hat{y}) &: = k(\kappa(\hat{x}), \kappa(\hat{y})) \\ H_{loc}(\hat{x},\hat{y}) &: = H(\kappa(\hat{x}), \kappa(\hat{y}))\, g_{K_1}(\hat{x})\, g_{K_2}(\hat{y}) \end{aligned}$$

and I_e takes the form

$$I_e := \int_0^h \int_0^{\hat{x}_1} \int_0^h \int_{-\hat{y}_1}^0 H_{loc}(\hat{x},\hat{y})\, k_{loc}(\hat{x},\hat{y})\, d\hat{y} d\hat{x}.$$

Due to the fact that κ is bi-Lipschitz continuous, the kernel function behaves singular if and only if $\hat{x} = \hat{y}$, i.e., $\hat{x}_1 - \hat{y}_1 = \hat{x}_2 = \hat{y}_2 = 0$. The location of the singularity is fixed by introducing one-dimensional relative coordinates $\hat{y}_1 = z_1 + \hat{x}_1$, $\hat{y}_2 = z_2$, $\hat{x}_2 = z_3$. The integration domain can be described as the union of the following six four-dimensional domains

$$D_1 = \left\{ \begin{array}{c} 0 \le z_3 \le h \\ -h \le z_1 \le -z_3 \\ -h - z_1 \le z_2 \le 0 \\ -z_1 - z_2 \le \hat{x}_1 \le h \end{array} \right\}, \quad D_2 = \left\{ \begin{array}{c} 0 \le z_3 \le h \\ -z_3 \le z_1 \le 0 \\ -h - z_1 \le z_2 \le -z_3 - z_1 \\ -z_1 - z_2 \le \hat{x}_1 \le h \end{array} \right\},$$

$$D_3 = \left\{ \begin{array}{c} 0 \le z_3 \le h \\ -z_3 \le z_1 \le 0 \\ -z_3 - z_1 \le z_2 \le 0 \\ z_3 \le \hat{x}_1 \le h \end{array} \right\}, \quad D_4 = \left\{ \begin{array}{c} 0 \le z_3 \le h \\ 0 \le z_1 \le h - z_3 \\ -h \le z_2 \le -z_3 - z_1 \\ -z_2 - z_1 \le \hat{x}_1 \le h - z_1 \end{array} \right\},$$

$$D_5 = \left\{ \begin{array}{c} 0 \le z_3 \le h \\ 0 \le z_1 \le h - z_3 \\ -z_3 - z_1 \le z_2 \le -z_3 \\ z_3 \le \hat{x}_1 \le h - z_1 \end{array} \right\}, \quad D_6 = \left\{ \begin{array}{c} 0 \le z_3 \le h \\ 0 \le z_1 \le h - z_3 \\ -z_3 \le z_2 \le 0 \\ z_3 \le \hat{x}_1 \le h - z_1 \end{array} \right\}.$$

The integral in these coordinates has the representation

$$I_e := \sum_{j=1}^{6} \int_{D_j} H_{loc}\left(\binom{z_1+\hat{x}_1}{z_1}, \binom{\hat{x}_1}{z_3}\right) k_{loc}\left(\binom{z_1+\hat{x}_1}{z_1}, \binom{\hat{x}_1}{z_3}\right) d\hat{x}_1 dz.$$

The following coordinate transforms map the coordinates $(\eta, w) \in R_2 \times T_2^h$ onto the domain of integration.

$$\begin{pmatrix} \hat{z}_1^{(1)} \\ z_2^{(1)} \\ z_3^{(1)} \\ \hat{x}_1^{(1)} \end{pmatrix} = \begin{pmatrix} -w_1\eta_1 \\ w_1(\eta_1 - 1) \\ w_1\eta_2\eta_1 \\ w_1 + w_2 \end{pmatrix}, \quad \begin{pmatrix} z_1^{(2)} \\ z_2^{(2)} \\ z_3^{(2)} \\ \hat{x}_1^{(2)} \end{pmatrix} = \begin{pmatrix} -w_1\eta_2\eta_1 \\ w_1(\eta_2\eta_1 - 1) \\ w_1\eta_1 \\ w_1 + w_2 \end{pmatrix},$$

$$\begin{pmatrix} \hat{z}_1^{(3)} \\ z_2^{(3)} \\ z_3^{(3)} \\ \hat{x}_1^{(3)} \end{pmatrix} = \begin{pmatrix} w_1(\eta_1 - 1) \\ -w_1\eta_2\eta_1 \\ w_1 \\ w_1 + w_2 \end{pmatrix}, \quad \begin{pmatrix} \hat{z}_1^{(4)} \\ z_2^{(4)} \\ z_3^{(4)} \\ \hat{x}_1^{(4)} \end{pmatrix} = \begin{pmatrix} w_1(\eta_1 - \eta_2\eta_1) \\ -w_1 \\ w_1\eta_2\eta_1 \\ w_1(1 - \eta_1 + \eta_2\eta_1) + w_2 \end{pmatrix},$$

$$\begin{pmatrix} \hat{z}_1^{(5)} \\ z_2^{(5)} \\ z_3^{(5)} \\ \hat{x}_1^{(5)} \end{pmatrix} = \begin{pmatrix} w_1(1 - \eta_2\eta_1) \\ -w_1\eta_1 \\ w_1\eta_2\eta_1 \\ w_1\eta_2\eta_1 + w_2 \end{pmatrix}, \quad \begin{pmatrix} \hat{z}_1^{(6)} \\ z_2^{(6)} \\ z_3^{(6)} \\ \hat{x}_1^{(6)} \end{pmatrix} = \begin{pmatrix} w_1(1 - \eta_1) \\ -w_1\eta_2\eta_1 \\ w_1\eta_1 \\ w_1\eta_1 + w_2 \end{pmatrix}.$$

The determinant of the Jacobi matrix is given for any of the transformations above by w_1^2. Analogously as in [12], [11], [15] one shows that the integrand of the following representation of I_e is analytic.

$$I_e := \int_{R_2}\int_{T_2^h} \eta_1 w_1^2 \sum_{j=1}^{6} H_{loc}\left(\begin{pmatrix} z_1^{(j)} + \hat{x}_1^{(j)} \\ z_1^{(j)} \end{pmatrix}, \begin{pmatrix} \hat{x}_1^{(j)} \\ z_3^{(j)} \end{pmatrix}\right) k_{loc}\left(\begin{pmatrix} z_1^{(j)} + \hat{x}_1^{(j)} \\ z_1^{(j)} \end{pmatrix}, \begin{pmatrix} \hat{x}_1^{(j)} \\ z_3^{(j)} \end{pmatrix}\right) dw d\eta. \tag{14}$$

Again, this integral can be approximated to any desired accuracy by standard cubature methods with properly chosen order.

Finally, we have to discuss the case that the panels have a common vertex. This is done in the following section.

THE CASE OF A COMMON VERTEX

Let K_1, K_2 be two surface triangles which have exactly one common point. The reference domains are given by $\Delta_1^h := \Delta^h$ and $\Delta_2^h := \left\{\hat{y} \in \mathbb{R}^2 \mid -\hat{y} \in \Delta^h\right\}$. By our assumptions on the surface and the triangulation we may assume that there is a bi-Lipschitz-continuous mapping $\kappa : \overline{\Delta_1^h \cup \Delta_2^h} \to \overline{K_1 \cup K_2}$ which is analytic on each reference element. The integral in local coordinates is given by

$$I_p = \int_0^h \int_0^{\hat{x}_1} \int_{-h}^0 \int_{\hat{y}_1}^0 H_{loc}(\hat{x}, \hat{y})\, k_{loc}(\hat{x}, \hat{y})\, d\hat{y} d\hat{x},$$

where we adopt the notations of the previous section. We split the domain of integration according to

$$I_p = \int_0^h \int_0^{\hat{x}_1} \int_{-\hat{x}_1}^0 \int_{\hat{y}_1}^0 H_{loc}(\hat{x}, \hat{y})\, k_{loc}(\hat{x}, \hat{y})\, d\hat{y} d\hat{x} + \int_{-h}^0 \int_{\hat{y}_1}^0 \int_0^{-\hat{y}_1} \int_0^{\hat{x}_1} H_{loc}(\hat{x}, \hat{y})\, k_{loc}(\hat{x}, \hat{y})\, d\hat{y} d\hat{x}.$$

The following transformations maps $(\eta, w_1, w_2, w_3) \in T_1^h \times R_3$ onto the domains of integration.

$$\begin{pmatrix} \hat{x}_1^{(1)} \\ \hat{x}_2^{(1)} \\ \hat{y}_1^{(1)} \\ \hat{y}_2^{(1)} \end{pmatrix} = \begin{pmatrix} \eta \\ w_2\eta \\ -w_1\eta \\ -w_3w_1\eta \end{pmatrix}, \qquad \begin{pmatrix} \hat{x}_1^{(2)} \\ \hat{x}_2^{(2)} \\ \hat{y}_1^{(2)} \\ \hat{y}_2^{(2)} \end{pmatrix} = \begin{pmatrix} w_1\eta \\ w_3w_1\eta \\ -\eta \\ -w_2\eta \end{pmatrix}.$$

In the new coordinates, the integral takes the form

$$I_p = \int_{R_3} \int_{T_1^h} w_1\eta^3 \sum_{i=1}^{2} H_{loc}\left(\hat{x}^{(i)}, \hat{y}^{(i)}\right) k_{loc}\left(\hat{x}^{(i)}, \hat{y}^{(i)}\right) d\eta dw. \tag{15}$$

In [12], [11], [15] it was shown that the integrand above is analytic and thus, can be approximated to any required accuracy with standard methods of properly chosen order.

It remains to consider the farfield case where $\overline{K_1} \cap \overline{K_2} = \emptyset$ holds. However, in this case the integrand is analytic and can be approximated by standard cubature methods with properly chosen order to any desired accuracy. In the following section we briefly sketch the cubature error analysis in appropriate norms.

APPROXIMATION OF THE INTEGRALS AND CUBATURE ERROR ANALYSIS

As mentioned above the integrands of the final integral representations are analytic and can be approximated by standard cubature formulae. The domains of integration are given by tensor products of the form $R_{d_1} \times T_{d_2}^h$ with $d_1 + d_2 = 4$. We will use tensor product Gauß formulae for the integration over R_{d_1} while for the variables contained in $T_{d_2}^h$ we employ abstract cubature formulae which are exact for polynomials of a predescribed degree. The reason for working with abstract formulae of a certain degree of exactness is that, especially for the h-version of the BEM, the use of Gauß-Legendre tensor formulae is sub-optimal on $T_{d_2}^h$ for practical problem sizes. However, the question which rule is optimal or convenient depends on the specific implementation and might be different for different situations. We refer to [2] and [17] for formulae which are exact for polynomials of certain orders on the domains $T_{d_2}^h$.

The estimates for the (local) cubature error are used to adapt the degrees of exactness to the consistency and stability requirements of the Galerkin discretization. Therefore, we will present the error estimates in terms of Sobolev norms since the consistency analysis for Galerkin methods is conveniently expressed in such norms.

In order to avoid too many technicalities, we restrict to triangulations which are quasi-uniform and non-degenerate

$$\begin{aligned} \max\{\operatorname{diam} K : K \in \tau\} &:= h, \\ \min\{\operatorname{diam} B_K : K \in \tau\} &\geq \rho h, \end{aligned}$$

where B_K denotes the largest ball contained in K and the constant ρ does not depend on K.

Furthermore, we assume that the inverse estimate holds which we will use in the form that, for integers l, m with $-\infty < l \le m$, there exists a constant $C = C(l, m)$ such that, for all $v \in V_{k,p}(K)$, the estimate

$$\|v\|_{W^{l,p}} \le C h^{m-l+2/p-1} \|v\|_m$$

holds.

We will now come to the cubature error estimates. Let $D \subset \mathbb{R}^d$ be a domain. A cubature rule M_D^γ has exactness degree γ if, for all $f \in \prod_\gamma (D)$,

$$E_D^\gamma [f] := \left| \int_D f(x)\, dx - M_D^\gamma [f] \right| = 0$$

holds. The domains under consideration are of the form $R_{d_1} \times T_{d_2}^h$ with $d_1 + d_2 = 4$. The notation $Q^\gamma [f]$ stands for the tensor cubature method $G_{R_{d_1}}^{\gamma_1} M_{T_{d_2}^h}^{\gamma_2}$ where $G_{R_{d_1}}^{\gamma_1}$ denotes the tensor Gauß formula composed of properly scaled one-dimensional Gauß-Legendre formulae $G_{(0,1)}^{\gamma_1}$ of order γ_1.

In [13] we have already considered the farfield case. Hence, we will concentrate here on the singular cases. We begin with the case of identical panels $K_1 = K_2$. The representation (13) is approximated by $G_{R_1}^{\gamma_1} M_{T_3^h}^{\gamma_2} [g_i]$ where $g_i(\eta, w)$ is an abbreviation of the integrand of (13). The following theorem concerns the cubature error.

Theorem 4 *The cubature error is given by* $E_{R_1 \times T_3^h}^\gamma := \left| I_i - G_{R_1}^{\gamma_1} M_{T_3^h}^{\gamma_2} [g_i] \right|$. *Let* $0 \le s, t$ *and* p *denote the polynomial degree of the trial space. In the case of* $H(x, y) = A(x, y)$ *we have*

$$E_{R_1 \times T_3^h}^\gamma \le Ch \left((1+\lambda)^{-\gamma_1 - 1} \|u\|_{m_1} \|v\|_{m_2} + h^{\gamma_2 + 1 - s - t} \|u\|_{p-s} \|v\|_{p-t} \right)$$

with a constant λ *independent of the discretization parameters* h *and* p *and the orders* m_j *of the operators* D_j *defined at the beginning of the previous section.*

For $H(x, y) = B(x, y)$ *we obtain*

$$E_{R_1 \times T_3^h}^\gamma \le Ch \left((1+\lambda)^{-\gamma_1 - 1} \|f\|_{m_3} \|v\|_{m_4} + h^{\gamma_2 + 1 - t} \|f\|_{\gamma_2 + 1} \|v\|_{p-t} \right).$$

Proof. The proof is essentially the same as the proof of [15]. The only difference is the estimate of the trial functions u and v in suitable ellipses $\mathcal{E}$ of analyticity. The arguments of u and v range in intervals I of length h. Hence, we obtain

$$u(x) = \sum_{j=0}^{p} \sum_{|\alpha| = j} \frac{1}{\alpha!} u^{(\alpha)}(y) (x-y)^\alpha$$

for any y of the integration interval and $x \in \mathcal{E}$. In combination with the inverse estimate we obtain for $\operatorname{dist}(\partial\mathcal{E}, I) = O(1)$:

$$\max_{x \in \mathcal{E}} |u(x)| \le \sum_{j=0}^{p} \frac{1}{j!} |u|_{W^{j,\infty}} (\operatorname{dist}(\partial\mathcal{E}, I))^j \le Ch^{-1} \|u\|_p$$

and for $\operatorname{dist}(\partial\mathcal{E}, I) = O(h)$:

$$\max_{x \in \mathcal{E}} |u(x)| \le Ch^{-1} \sum_{j=0}^{p} |u|_j h^j \le Ch^{-1} \sum_{j=0}^{p} h^{-j} \|u\|_0 h^j = Ch^{-1} \|u\|_0 .$$

If D_j is not the identity but a differentiating operator the estimate can be obtained in the same fashion but the inverse inequality has to be applied in the form of $|u|_j \leq Ch^{j-2} \|u\|_1$. ■

Next we consider the case that the panels K_1 and K_2 share exactly one edge. We approximate the integral in (14) by $G_{R_2}^{\gamma_1} M_{T_2^h}^{\gamma_2} [g_e]$ where $g_e(\eta, w)$ denotes the integrand of (14). The following theorem concerns the cubature error.

Theorem 5 *Let $s, t \geq 0$. For $H = A$, the cubature error $E_{R_2 \times T_2^h}^{\gamma} := \left| I_e - G_{R_2}^{\gamma_1} M_{T_2^h}^{\gamma_2} [g_e] \right|$ can be estimated by*

$$E_{R_2 \times T_2^h}^{\gamma} \leq C \left((1+\lambda)^{-\gamma_1 - 1} \|u\|_{m_1} \|v\|_{m_2} + h^{\gamma_2 + 1 - s - t} \|u\|_{p-s} \|v\|_{p-t} \right)$$

while for $H = B$ we obtain

$$E_{R_2 \times T_2^h}^{\gamma} \leq C \left((1+\lambda)^{-\gamma_1 - 1} \|f\|_{m_3} \|v\|_{m_4} + h^{\gamma_2 + 1 - t} \|f\|_{\gamma_2 + 1} \|v\|_{p-t} \right).$$

Proof. The proof is essentially the same as worked out in [15] with the same modifications as pointed out in the previous proof. ■

It remains to discuss the case that K_1 and K_2 share exactly one vertex. We approximate the integral representation (15) by $G_{R_3}^{\gamma_1} M_{T_1^h}^{\gamma_2} [g_p]$ where g_p denotes the integrand of (15). The cubature error is estimated in the following

Theorem 6 *Let $s, t \geq 0$. For $H = A$, the cubature error $E_{R_3 \times T_1^h}^{\gamma} := \left| I_p - G_{R_3}^{\gamma_1} M_{T_1^h}^{\gamma_2} [g_p] \right|$ can be estimated by*

$$E_{R_3 \times T_1^h}^{\gamma} \leq C \left((1+\lambda)^{-\gamma_1 - 1} \|u\|_{m_1} \|v\|_{m_2} + h^{\gamma_2 + 1 - s - t} \|u\|_{p-s} \|v\|_{p-t} \right)$$

while, for $H = B$, we get

$$E_{R_3 \times T_1^h}^{\gamma} \leq C \left((1+\lambda)^{-\gamma_1 - 1} \|f\|_{m_3} \|v\|_{m_4} + h^{\gamma_2 + 1 - t} \|f\|_{\gamma_2 + 1} \|v\|_{p-t} \right).$$

Proof. Again, the proof of this theorem is analogous as worked out in [15]. ■

The regular integrals a_K and F_K are transformed to the scaled master element T_2^h and approximated by a formula $Q_{T_2^h}^{\gamma}$ with exactness degree γ. The following theorem concerns the arising cubature error.

Theorem 7 *Let $s, t \geq 0$. For the approximation of $a_K(u, v)$, the error $E_{T_2^h}^{\gamma} := \left| a_K - Q_{T_2^h} \right|$ can be estimated by*

$$E_{T_2^h}^{\gamma} \leq C h^{1+\gamma-s-t} \|u\|_{p-s} \|v\|_{p-t}$$

while for the approximation of $F_K(v)$ we get

$$E_{T_2^h}^{\gamma} \leq C h^{1+\gamma-t} \|u\|_{p-t} \|f\|_{\gamma+1}.$$

Proof. The proof of the first assertion is straightforward, while the second one was proved in [13, Theorem 6].

Remark 8 *The regularity assumptions on f in the estimates above can be relaxed such that the required order of differentiability is independent of the degree of exactness (see [13, Theorem 5]).*

The effect of numerical cubature is treated as an approximation of the discrete (bi-) linear forms. The cubature orders have to be chosen such that the asymptotic convergence of the Galerkin method is preserved. In [13], the effect of the local cubature error to the global discretization process was investigated thoroughly in view of stability and consistency. We can use these results directly to determine the required cubature orders. They will depend on the polynomial degree of approximation p, the step size h, the norm in which the error is measured, and are possibly different for different variables.

For simplicity, we assume here full regularity of the variational problem, i.e., the right-hand side of (1) is such that

$$\|u - u_G\|_{s_1 - t} \leq C h^{p+1+t-s_1} \|f\|_{p+1-2(s_1+s_2)}$$

is satisfied for $t \in [0, p+1-s_1]$. In the following, the orders γ of the cubature formulae defined above are given such that the error $u - \tilde{u}_G$ converges with the same asymptotic rate as the Galerkin solution u_G. The notation $\lceil x \rceil$ stands for the smallest integer larger than or equal to x.

- The bilinear form a_K is approximated by Q_K^γ with
$$\gamma = \lceil 2(p - s_1) \rceil .$$
- "Case of identical panels": the bilinear form $a_{K \times K}$ is approximated by $Q^\gamma_{R_1 \times T_3^h}$ with
$$\gamma_1 = \lceil (p - 2s_1 + t + m_2) |\log h| - 1 \rceil , \qquad \gamma_2 = \lceil 2(p - s_1) \rceil . \tag{16}$$
- "Case of a common edge": the bilinear form $a_{K_x \times K_y}$ is approximated by $Q^\gamma_{R_2 \times T_2^h}$ with
$$\gamma_1 = \lceil (p + 1 - 2s_1 + t + m_2) |\log h| - 1 \rceil , \qquad \gamma_2 = \lceil 2(p - s_1) \rceil . \tag{17}$$
- "Case of a common point": the bilinear form $a_{K_x \times K_y}$ is approximated by $Q_{R_3 \times T_1^h}$ with
$$\gamma_1 = \lceil (p + 1 - 2s_1 + t + m_2) |\log h| - 1 \rceil , \qquad \gamma_2 = \lceil 2(p - s_1) \rceil . \tag{18}$$

The degrees of exactness for the approximation of the linear form F_K, $F_{K_1 \times K_2}$ are the same as the analog orders of the bilinear form. The cubature orders for the farfield cases can be found in [13].

REFERENCES

[1] ADAMS, R.: Sobolev Spaces. Academic Press, N.Y., 1975.

[2] COOLS, R. and RABINOWITZ, P.: Monomial Cubature Rules since "Stroud": A Compilation. J. Comput. Appl. Math. 48:309-326, 1993.

[3] GENG, P., ODEN, J., and DEMKOWICZ L.: Solution of Exterior Acoustic Problems by the Boundary Element Method at High Wave Numbers, Error Estimation and Parallel Computation. Technical Report 95-09, TICAM, The University of Texas at Austin, 1995.

[4] HACKBUSCH, W. and SAUTER, S.A.: On the Efficient Use of the Galerkin Method to Solve Fredholm Integral Equations. Applications of Mathematics 38(4-5):301-322, 1993.

[5] HAN, H.: A Boundary Element Method for Signorini Problems in Three Dimensions. Numer. Math., 60:63-76, 1991.

[6] HAN, H.: The Boundary Element Method for Solving Variational Inequalities. Contemp. Math. 163, 1994.

[7] HAN, H.: The Boundary Integro-Differential Equations of Three-Dimensional Neuman Problem in Linear Elasticity. Numer. Math. 68(2):269-281, 1994.

[8] LAGE, C.: Software Development for Boundary Element Mehtods: Analysis and Design of Efficient Techniques (in German). PhD thesis, Lehrstuhl Prakt. Math., Universität Kiel, 1995.

[9] LAGE, C. and SAUTER, S.: Analysis and Software Design Aspects of Advanced Algorithms in Boundary Element Methods. Wiley, planned for 1996.

[10] NEDELEC, J.C.: Integral Equations with Non Integrable Kernels. Integral Equations Oper. Theory 5:562-572, 1982.

[11] V. PETERSDORFF, T. and SCHWAB, C.: Fully Discrete Multiscale Galerkin BEM. Technical Report 95-08, Seminar for Applied Mathematics, ETH Zürich, 1995.

[12] SAUTER, S.A.: Über die effiziente Verwendung des Galerkinverfahrens zur Lösung Fredholmscher Integralgleichungen. PhD thesis, Lehrstuhl. f. Prakt. Math., Universität Kiel, 1992.

[13] SAUTER, S.A. and KRAPP, A.: On the Effect of Numerical Integration in the Galerkin Boundary Element Method.Technical Report 95-4, Lehrstuhl Praktische Mathematik, Universität Kiel, Germany, 1995, to appear in Numer. Math.

[14] SAUTER, S.A. and SCHWAB, C.: Realization of hp-Galerkin BEM in 3-d. In HACKBUSCH, W. and WITTUM, G., editors, BEM : Implementation and Analysis of Advanced Algorithms, Proceedings of the 12th GAMM-Seminar, Kiel. Verlag Vieweg, 1996.

[15] SAUTER, S.A. and SCHWAB, C.: Quadrature for hp-Galerkin BEM in 3-d. Technical report 96-02, Seminar for Appl. Math., ETH Zürich, 1996.

[16] SCHWAB, C. and WENDLAND, W.: Kernel Properties and Representations of Boundary Integral Operators. Math. Nachr. 156:187–218, 1992.

[17] STROUD A.H.: Approximate Calculations of Multiple Integrals. Prentice Hall, Englewood Cliffs, 1973.

A GALERKIN-TYPE BOUNDARY ELEMENT IMPLEMENTATION FOR 3D ELASTICITY PROBLEMS BY USING A COMPUTER ALGEBRA SYSTEM

Heiko Andrä
Institute of Solid Mechanics
University of Karlsruhe (TH)
Kaiserstraße 12
76128 Karlsruhe, Germany

SUMMARY

A Galerkin approximation of both strongly and hypersingular boundary integral equation is considered for the solution of a mixed boundary value problem (b.v.p.) in 3D elasticity. The evaluation of singular double integrals is one of the most difficult parts within the implementation of such boundary element methods. A new integration method for the cases of coincident elements as well as edge-adjacent and vertex-adjacent elements, leads to explicitly given regular integrand functions which can be integrated by the standard quadrature rules. Because the effort for the analytical evaluation of the integrands of the element stiffness matrices increases with the complexity of the kernel functions and with the polynomial degree of the shape functions, the computer algebra system (c.a.s.) "Maple V" [1] was used for this task.

INTRODUCTION: BOUNDARY VALUE PROBLEM

We consider a self-conjugate and strongly elliptic b.v.p. of second order with constant coefficients in a bounded (simply connected) domain $\Omega \subset \mathbb{R}^3$ with a LIPSCHITZ boundary $\Gamma = \Gamma_1 \cup \Gamma_2$, $\Gamma_1 \cap \Gamma_2 = \emptyset$, mes $\Gamma_1 > 0$, with given DIRICHLET data on Γ_1, and NEUMANN data on Γ_2. For the sake of simplicity we restrict ourselves to the case of a homogeneous right-hand side. This b.v.p. can be written in abstract form:

$$\boldsymbol{P}\boldsymbol{u} = 0 \qquad \text{in } \Omega, \tag{1}$$

$$\boldsymbol{R}_j(\gamma \boldsymbol{u}) = \boldsymbol{g}_j \qquad \text{on } \Gamma_j \quad (j = 1, 2), \tag{2}$$

[1]MAPLE is a registered trademark of Waterloo Maple Software

where $\gamma\phi = \left(\phi|_{\partial\Omega}, \frac{\partial\phi}{\partial n}\Big|_{\partial\Omega}\right)$ is the trace operator. The operator $P = P(D) = \sum_{\alpha=2} a_\alpha D^\alpha$ is a $N \times N$ matrix of linear elliptic differential operators and R_j are linear tangential differential operators of order $j-1$. Furthermore we assume that everything is real in order to keep the notation as simple as possible. In linear elasticity the operator P has the form

$$P_{ki} = c_{kjil}\frac{\partial}{\partial x_j}\frac{\partial}{\partial x_l} = c_{klij}\frac{\partial}{\partial x_l}\frac{\partial}{\partial x_j} = c_{ijkl}\frac{\partial}{\partial x_j}\frac{\partial}{\partial x_l} = P_{ik}$$

where c_{ijkl} is the tensor of the moduli of elasticity, which is symmetric with respect to the first and second pairs of subscripts. The CAUCHY stress tensor is defined by $\sigma_{ij} = c_{ijkl}u_{k,l}$. In an isotropic medium the number of different elastic constants is reduced to two:

$$c_{ijkl} = \lambda\delta_{ij}\delta_{lk} + \mu(\delta_{il}\delta_{jk} + \delta_{ik}\delta_{jl}).$$

The constants λ and μ are called LAMÉ constants. The usual mechanical boundary conditions are given by

$$(R_1(\gamma u))_i \equiv u_i|_{\partial\Omega}, \qquad i = 1,2,3; \tag{3}$$
$$(R_2(\gamma u))_i \equiv \sigma_{ij}(u)n_j u_{l,k}|_{\partial\Omega} = c_{ijkl}n_j u_{l,k}, \qquad i,j = 1,2,3, \tag{4}$$

where $n = (n_1, n_2, n_3)$ is the external unit normal vector.

FUNDAMENTAL SOLUTIONS

For the boundary element method we need the knowledge of a fundamental solution U of P. A vector of distributions $U \in (\mathscr{D}'(\mathbb{R}^3))^N$ is called a fundamental solution of the differential operator P with constant coefficients if

$$PU = I\delta_0, \tag{5}$$

where δ_a is the Dirac measure at $a \in \mathbb{R}$ which is defined by $\langle\delta_a, \phi\rangle = \phi(a) \quad \forall\phi \in C^\infty(\mathbb{R})$. Next, we consider the scheme of the construction of fundamental solutions proposed by HÖRMANDER [8, 9]:

1. We compute the symbolic determinant of $P(D)$ and find the scalar fundamental solution G satisfying

$$\det P(D)G = \delta_0. \tag{6}$$

In linear elasticity, we obtain

$$\det P = \epsilon_{mnl}P_{m1}P_{n2}P_{l3}, \tag{7}$$

where ϵ_{mnl} is the permutation tensor or the LEVI-CIVITA symbol. Obviously, by using $\det P(D) = \sum_{\alpha=6} b_\alpha D^\alpha =: P^{(6)}(D)$ there exists such a fundamental solution

G for every $P^{(6)}(D)$ with the following property (see [8, 17]). The function G can be represented in the form

$$G(\boldsymbol{x}) = g(\boldsymbol{x})\|\boldsymbol{x}\|^3 \in \mathscr{D}'(\mathbb{R}^3), \tag{8}$$

where $g(\boldsymbol{x})$ is a positive homogeneous and analytic function of degree 0 ($g(c\boldsymbol{x}) = g(\boldsymbol{x}) \quad \forall c > 0$).

2. We construct the cofactor matrix $\boldsymbol{P}^{co}$, such that $\boldsymbol{P}^{co}\boldsymbol{P} = \boldsymbol{P}\boldsymbol{P}^{co} = \det \boldsymbol{P} \cdot \boldsymbol{I}$. In linear elasticity, the cofactor matrix is defined by

$$P^{co}_{ij} = \frac{1}{2}\epsilon_{ikl}\epsilon_{jnm}P_{km}P_{ln}. \tag{9}$$

Hence the matrix $\boldsymbol{U} = \boldsymbol{P}^{co}\boldsymbol{I}G$ is the fundamental solution, because of

$$\boldsymbol{P}\boldsymbol{U} = \boldsymbol{P}\boldsymbol{P}^{co}\boldsymbol{I}G = \delta_0\boldsymbol{I}. \tag{10}$$

By using the properties of G (see (8)) and by introducing polar coordinates we obtain $G = \tilde{g}(\sin\phi, \cos\phi, \sin\theta, \cos\theta)r^3$. Now it can be easily shown that the fundamental solution $\boldsymbol{U}$ can be represented in the form

$$\boldsymbol{U}(\boldsymbol{x}) = \tilde{\boldsymbol{e}}(\sin\phi, \cos\phi, \sin\theta, \cos\theta)r^{-1} = \boldsymbol{e}(\boldsymbol{x})\|\boldsymbol{x}\|^{-1},$$

where $\boldsymbol{e}$ is also a positive homogeneous and analytic function of degree 0.

The presented method is well suitable for the computation of fundamental solutions by using a c.a.s. The implementation of MapleV source code is on the way and will be reported elsewhere. Otherwise, by applying the FOURIER transformation to (5),

$$\boldsymbol{P}(\boldsymbol{\xi})\hat{\boldsymbol{U}}(\boldsymbol{\xi}) = \boldsymbol{I}.$$

The fundamental solution is formally defined by inverse FOURIER transformation

$$\boldsymbol{U}(\boldsymbol{x}) = \frac{1}{2\pi^3}\int_{\mathbb{R}^3} \boldsymbol{P}^{-1}(\boldsymbol{\xi})\exp(i\boldsymbol{\xi}^T\boldsymbol{x})d\boldsymbol{\xi} = \frac{1}{2\pi^3}\int_{\mathbb{R}^3} \boldsymbol{P}^{co}(\boldsymbol{\xi})(\det \boldsymbol{P})^{-1}\exp(i\boldsymbol{\xi}^T\boldsymbol{x})d\boldsymbol{\xi}. \tag{11}$$

It can easily be shown that U_{ij} satisfy

$$c_{ijkl}U_{km,lj}(\boldsymbol{x}) + \delta_{im}\delta_{\boldsymbol{x}} = 0. \tag{12}$$

The proof of (12) is as follows

$$\begin{aligned} c_{ijkl}U_{km,lj}(\boldsymbol{x}) &= -\frac{1}{2\pi^3}\int_{\mathbb{R}^3} c_{ijkl}P^{co}_{km}(\det \boldsymbol{P})^{-1}\xi_i\xi_j\exp(i\boldsymbol{\xi}^T\boldsymbol{x})d\boldsymbol{\xi} \\ &= -\frac{1}{2\pi^3}\int_{\mathbb{R}^3} P_{ik}P^{co}_{km}(\det \boldsymbol{P})^{-1}\exp(i\boldsymbol{\xi}^T\boldsymbol{x})d\boldsymbol{\xi}. \end{aligned}$$

The assertion follows from $P_{ik}P^{co}_{km}(\det \boldsymbol{P})^{-1} = \delta_{im}$.
The details of integration of (11) are explained in MURA [13]. The volume element $d\boldsymbol{\xi}$ is defined as $d\boldsymbol{\xi} = \xi^2 d\xi dS$, where $\xi = (\xi_i, \xi_i)^{\frac{1}{2}}$ and dS is the surface element on the unit sphere S^2 in the $\boldsymbol{\xi}$-space, centered at the origin of the coordinates ξ_i. As seen in (9) and

(7), $P_{ij}(\boldsymbol{\xi})$ and $\det \boldsymbol{P}(\boldsymbol{\xi})$ are homogeneous polynomials of degree four and six, respectively. Therefore, by denoting $\boldsymbol{\xi} = \xi\bar{\boldsymbol{\xi}}$, $\boldsymbol{x} = r\bar{\boldsymbol{x}}$, we can write (11) as

$$U_{ij}(\boldsymbol{x}) = \frac{1}{8\pi^2}\int_{S^2} \delta(r\bar{\boldsymbol{\xi}}^T\bar{\boldsymbol{x}})P_{ij}^{co}(\bar{\boldsymbol{\xi}})(\det \boldsymbol{P}(\bar{\boldsymbol{\xi}}))^{-1}dS(\bar{\boldsymbol{\xi}})$$
$$= \frac{1}{8\pi^2 r}\oint_{S^1} P_{ij}^{co}(\bar{\boldsymbol{\xi}})(\det \boldsymbol{P}(\bar{\boldsymbol{\xi}}))^{-1}d\phi,$$

where $S^1 \subset S^2$ is a circle, which lies on the plane perpendicular to $\boldsymbol{x}$. Also this second method by applying the FOURIER transformation seems to be appropriate for the c.a.s. MapleV, which have got a library "fourier" with functions for the FOURIER transformation "*fourier*" and the inverse FOURIER transformation "*invfourier*". However these functions are not sufficiently powerful.
Much effort has been devoted to deriving explicit expressions of fundamental solutions in elasticity [13]. The fundamental solution for isotropic materials was found by Lord Kelvin [5, 13]:

$$U_{ij}(\boldsymbol{x}, \boldsymbol{y}-\boldsymbol{x}) = \frac{1}{16\pi\mu(1-\nu)}\left(\frac{1}{r}\right)[(3-4\nu)\delta_{ij} + r_{,i}r_{,j}],$$

$r_i = y_i - x_i$, $r = \|\boldsymbol{r}\|$, $r_{,i} = \frac{\partial r}{\partial y_i} = \frac{r_i}{r}$. Explicit expressions for transversely isotropic materials are given in [13].

THE BOUNDARY INTEGRAL EQUATIONS

We derive the boundary integral equations by the so called direct approach for a general b.v.p. in linear elasticity. This method can be generalized to general elliptic b.v.p. of even order [4, 18]. The operator $\boldsymbol{P}$ in linear elasticity is formally self-conjugate with respect to the scalar produkt $(\boldsymbol{u}, \boldsymbol{v}) = \sum_{i=1}^{3} u_i v_i$. By using the relation $\sigma_{ij,i}u_i = (\sigma_{ij}u_i)_{,j} - \sigma_{ij}u_{i,j}$, the second GREEN Formula (BETTI's reciprocity formula)

$$\int_\Omega (\boldsymbol{Pu}\boldsymbol{v} - \boldsymbol{u}\boldsymbol{Pv})d\boldsymbol{x} = \int_{\partial\Omega} (\sigma_{ij}(\boldsymbol{u})v_i n_j - u_i\sigma_{ij}(\boldsymbol{v})n_j d\boldsymbol{s}(\boldsymbol{x}) \qquad (13)$$

may be derived in the standard manner. From (13) with $\boldsymbol{v}(\boldsymbol{y}) = \boldsymbol{U}(\boldsymbol{x}, \boldsymbol{y}-\boldsymbol{x})$, where $\boldsymbol{U}(\boldsymbol{x}, \boldsymbol{y}-\boldsymbol{x})$ is the fundamental solution, and $\boldsymbol{Pu} = \boldsymbol{0}$ for the solution $\boldsymbol{u}$ of the b.v.p., we obtain the GREEN representation formula (SOMIGLIANA formula)

$$\boldsymbol{u}(\boldsymbol{x}) = \int_\Gamma [\boldsymbol{R}_1(\gamma\boldsymbol{U}(\boldsymbol{x}, \boldsymbol{y}-\boldsymbol{x}))\boldsymbol{R}_2(\gamma\boldsymbol{u}(\boldsymbol{y})) - \boldsymbol{R}_2(\gamma\boldsymbol{U}(\boldsymbol{x}, \boldsymbol{y}-\boldsymbol{x}))\boldsymbol{R}_1(\gamma\boldsymbol{u}(\boldsymbol{y}))]\, d\boldsymbol{s}(\boldsymbol{y}) \qquad (14)$$

for all $\boldsymbol{x} \in \Omega$. We denote by $V_\Gamma = \operatorname{Im}\gamma$ the space of CAUCHY data. The space of CAUCHY data can be determined by using e.g. the following trace theorem [19]: The trace operator $\boldsymbol{\gamma}$ is a bounded linear mapping

$$\gamma : \mathbb{H}^s(\Omega) \longrightarrow \mathbb{H}^{s-\frac{1}{2}}(\partial\Omega) \times \mathbb{H}^{s-1-\frac{1}{2}}(\partial\Omega)$$

for all $s > \frac{3}{2}$ and $\Gamma \in C^{k,\kappa}$ with $k + \kappa \geq s + 1$. (We use the notation $\mathbb{H}^s := (H^s)^3$ for SOBOLEV spaces.) If we extend g_1 and g_2 with zero on Γ to $\binom{\tilde{v}}{\tilde{t}} \in V_\Gamma$, then the CAUCHY data can be written in the form $\binom{\hat{v}}{\hat{t}} := \binom{v}{t} + \binom{\tilde{v}}{\tilde{t}}$ with

$$\begin{pmatrix} v \\ t \end{pmatrix} \in \overset{\circ}{V}_\Gamma := \left\{ \begin{pmatrix} \hat{v} \\ \hat{t} \end{pmatrix} \in V_\Gamma : \hat{v}|_{\Gamma_1} = 0, \hat{t}|_{\Gamma_2} = 0 \right\}.$$

Then the CAUCHY data in (2) can be expressed by $\binom{\hat{v}}{\hat{t}} = R\gamma u$, where $R = \binom{R_1}{R_2}$. The system $\{R_1, R_2\}$, where R_j are defined by (3) and (4), is called *normal* [12], p. 344, i. e. $\operatorname{Im} R$ is isomorphic with the space of CAUCHY data $V_\Gamma = \operatorname{Im} \gamma$. Then the CAUCHY data can be expressed vice versa by $\gamma u = R^{-1}\binom{\hat{v}}{\hat{t}}$. Then application of R to both sides of (14) yields the overdetermined system of equations

$$\begin{pmatrix} \hat{v} \\ \hat{t} \end{pmatrix}(x) = R\gamma \left\{ \int_\Gamma \left[R_1(\gamma U(x, y - x))\hat{t}(y) - R_2(\gamma U(x, y - x))\hat{v}(y) \right] ds(y) \right\} =: C_1 \begin{pmatrix} \hat{v} \\ \hat{t} \end{pmatrix}(x), \tag{15}$$

where $C_1 := \frac{1}{2}(I + S)$ is the CALDERÓN projector with $S = \begin{pmatrix} -K & V \\ D & K' \end{pmatrix}$. The boundary integral operators are defined by

$$\begin{aligned}
V\psi &= 2 \int_\Gamma R_1(\gamma U(x, y - x))\psi(y)ds(y), \\
K\varphi &= 2 \text{ v.p.} \int_\Gamma R_2(\gamma U(x, y - x))\varphi(y)ds(y), \\
K'\psi &= 2 \text{ v.p.} \int_\Gamma R_{2,x}(\gamma_x U(y, x - y))\psi(y)ds(y), \\
D\varphi &= -2 \text{ p.f. } R_{2,x}\gamma_x \int_{\Gamma \setminus B_\varepsilon(x)} R_2(\gamma U(x, y - x))\varphi(y)ds(y).
\end{aligned}$$

In the last equation, we denoted by $B_\varepsilon(x) = \{y : \ |y - x| < \varepsilon(\varepsilon > 0)\}$ the ε-ball about $x \in \mathbb{R}^3$. Explicit formulae of the kernels of these boundary integral operators for isotropic materials are given e. g. in [2]. The application of the tangential differential operator R_2 is a formal operation, which can be performed by the c.a.s. to avoid mistakes.

Theorem 1 (Theorem on involution) *Let* $\binom{\hat{v}}{\hat{t}} \in \mathbb{H}^{s-\frac{1}{2}}(\partial\Omega) \times \mathbb{H}^{s-\frac{3}{2}}(\partial\Omega)$, $s > \frac{5}{2}$ *and* $s \neq \frac{1}{2} +$ *integer , or* $s \in (\frac{1}{2}, \frac{3}{2})$. *Then* $S^2 = I$ *and* $S\binom{\hat{v}}{\hat{t}} = \binom{\hat{v}}{\hat{t}}$, *if and only if* $\binom{\hat{v}}{\hat{t}}$ *are* CAUCHY *data of first order of a function* $u \in \mathbb{H}^s(\Omega)$ *which satisfies the equation* $Pu = 0$ *in* Ω.

Proof. For the proof in the case $s > \frac{5}{2}$, and $s \neq \frac{1}{2} +$ integer , we refer to [12]. By using the mapping properties of S on LIPSCHITZ domains [3] the assertion in the case $s \in (\frac{1}{2}, \frac{3}{2})$ can be obtained in the same manner. □

Now we formulate GALERKIN's method for $S\binom{\hat{v}}{\hat{t}} = \binom{\hat{v}}{\hat{t}}$. We introduce the bilinear form

$$a\left((I - S)\begin{pmatrix} \hat{v} \\ \hat{t} \end{pmatrix}, \begin{pmatrix} \phi \\ \psi \end{pmatrix}\right) := \left\langle (I - S)\begin{pmatrix} \hat{v} \\ \hat{t} \end{pmatrix}, \begin{pmatrix} \psi \\ \phi \end{pmatrix} \right\rangle_{V_\Gamma \times V'_\Gamma}$$

The triangulations on Γ are obtained as images of regular quasiuniform triangulations of the parameter region V_l. All boundary elements of one surface piece S_l can be obtained by the composition of $\boldsymbol{F}^{(l)}$ and one linear mapping $\boldsymbol{A}^{(l)}$ together with translations $\boldsymbol{a}_i$ applied to a reference element

$$\hat{\pi} = \{z \in \mathbb{R}^2 : 0 < z_1 < 1, 0 < z_2 < z_1\}.$$

Then

$$\pi_i^h = \chi^{(i)}(\hat{\pi}) = \boldsymbol{F}^{(l)} \circ \boldsymbol{A}^{(l)}(\hat{\pi} + \boldsymbol{a}_i)$$

defines the in general curved elements of S_l. Each element π_i^h belongs to one and only one surface piece S_l. Thus we omit the index l at $\pi_i^h = \chi^{(i)}(\hat{\pi})$. The approximating boundary element trial spaces are now defined by lifting finite element functions from the parameter partitions of $\bar{V}_l$ onto Γ via $(\chi^{(i)})^{-1}$. As approximating subspace V_h of V_Γ we take-piecewise polynomials of different degree for the approximation of $\boldsymbol{v}$ or $\boldsymbol{t}$.

Theorem 1 (Galerkin Scheme (exact integration))
Find $\binom{v_h}{t_h} \in V_h^* \subset V_\Gamma^* = \{\binom{v}{t} \in V_\Gamma : \binom{v-\bar{v}}{t-\bar{t}} \in \overset{\circ}{V}_\Gamma\}$ *such that*

$$a\left((\boldsymbol{I} - \boldsymbol{S})\begin{pmatrix} \boldsymbol{v}_h \\ \boldsymbol{t}_h \end{pmatrix}, \begin{pmatrix} \boldsymbol{\phi} \\ \boldsymbol{\psi} \end{pmatrix}\right) = 0 \qquad \forall \begin{pmatrix} \boldsymbol{\phi} \\ \boldsymbol{\psi} \end{pmatrix} \in \overset{(i)}{V}_h,$$

where some appropriate choices of $\overset{(i)}{V}_h$ *are given by*

$$\overset{(1)}{V}_h = \left\{\begin{pmatrix} \boldsymbol{v}_h \\ \boldsymbol{t}_h \end{pmatrix} \in V_h : \boldsymbol{v}_h \equiv \boldsymbol{0}\right\} \qquad \text{(strongly singular equation)},$$

$$\overset{(2)}{V}_h = \left\{\begin{pmatrix} \boldsymbol{v}_h \\ \boldsymbol{t}_h \end{pmatrix} \in V_h : \boldsymbol{t}_h \equiv \boldsymbol{0}\right\} \qquad \text{(hypersingular equation)},$$

$$\overset{(3)}{V}_h = \overset{\circ}{V}_h \qquad \text{(symmetric Galerkin method)}.$$

In the last case a definite and nonsymmetric (or a symmetric and nondefinite) linear system is obtained. Fast solvers are available for this system [11]. Asymptotic error estimates of optimal order can be proved in each case of $\overset{(i)}{V}_h$; see [18], Theorem 9, p. 534, for the general case. For generating the GALERKIN system we propose a direct computation of CAUCHY singular integrals, however a regularization of finite part integrals if the hypersingular operater $\boldsymbol{D}$ is included in the scheme. (In [10] a direct computational method of hypersingular kernels is used.)

HYPERSINGULAR OPERATORS

For generating the GALERKIN system, the following finite part integrals have to be evaluated

$$\langle \mathbf{D}_{22}\varphi_h^{(l)}, \varphi_h^{(k)} \rangle_{\boldsymbol{H}^{-\frac{1}{2}}(\Gamma_2)\times \boldsymbol{H}^{\frac{1}{2}}(\Gamma_2)} = \int_{\Gamma_2} \text{p.f.} \int_{\Gamma_2} [\varphi_h^{(l)}(\boldsymbol{x})]^T \mathbf{S}(\boldsymbol{x}, \boldsymbol{y} - \boldsymbol{x}) \varphi_h^{(k)}(\boldsymbol{y})\, ds(\boldsymbol{y})\, ds(\boldsymbol{x}).$$

The shape functions $\varphi_h^{(.)}(.)$ have to be linear functions or polynomials of higher degree. This integral can be regularized by using the identity

$$\text{p.f.} \int_\Gamma \mathbf{S}(x, y - x)\, ds(y) = 0 \qquad \forall x \in \Gamma.$$

The result is a sum of integrals with a reduced order of singularity

$$\begin{aligned}
&\int_{\Gamma_2} \text{p.f.} \int_{\Gamma_2} [\varphi_h^{(l)}(x)]^T \mathbf{S}(y, x) \varphi_h^{(k)}(y)\, ds(y)\, ds(x) \\
&= \int_{\Gamma_2} \text{p.f.} \int_{\Gamma} [\varphi_h^{(l)}(x)]^T \mathbf{S}(y, x) [\varphi_h^{(k)}(y) - \varphi_h^{(k)}(x)]\, ds(y)\, ds(x) \\
&= \int_{\text{supp } \varphi_h^{(l)}} \text{p.f.} \int_{\text{supp } \varphi_h^{(k)}} [\varphi_h^{(l)}(x)]^T \mathbf{S}(y, x) [\varphi_h^{(k)}(y) - \varphi^{(k)}(x)]\, ds(y)\, ds(x) \qquad (16) \\
&- \int_{\text{supp } \varphi_h^{(l)} \cap \text{supp } \varphi_h^{(k)}} \text{p.f.} \int_{\Gamma \setminus \text{supp } \varphi_h^{(k)}} [\varphi_h^{(l)}(x)]^T \mathbf{S}(y, x) \varphi_h^{(k)}(x)\, ds(y)\, ds(x).
\end{aligned}$$

This regularization is easy to implement, but leads to an additional computational effort. Otherwise the following basic property of all boundary integral operators on Γ, provided $\Gamma \in C^\infty$, can be exploited [16], p. 239: *After appropriate integrations by parts, boundary integral operators associated with regular elliptic boundary value problems can be represented as compositions of tangential derivatives and* CAUCHY *singular integral operators.* For the sake of simplicity we consider only the hypersingular operator of potential theory: Using STOKES' theorem and simple calculations we obtain [5]

$$\begin{aligned}
-\frac{1}{2}\boldsymbol{D}\phi &= \text{p.f.} \int_\Gamma \frac{\partial^2}{\partial n(x) \partial n(y)} \frac{1}{|y - x|} \phi(y) ds(y) \\
&= -\text{p.f.} \int_\Gamma \frac{\partial^2}{\partial s_k(x) \partial s_k(y)} \frac{1}{|y - x|} \phi(y) ds(y) = \text{p.f.} \int_\Gamma \frac{\partial}{\partial s_k(x)} \frac{1}{|y - x|} \frac{\partial \phi}{\partial s_k(y)} ds(y),
\end{aligned} \qquad (17)$$

where $\dfrac{\partial}{\partial s_k(z)} = \epsilon_{ijk} n_i \dfrac{\partial}{\partial z_j}$ and $n_k \dfrac{\partial}{\partial s_k(z)} = 0$; moreover,

$$\int_\Gamma \text{p.f.} \int_\Gamma \psi(x) \frac{\partial}{\partial s_k(x)} \frac{1}{|y - x|} \frac{\partial \phi}{\partial s_k(y)} ds(y) ds(x) = -\int_\Gamma \int_\Gamma \frac{1}{|y - x|} \frac{\partial \psi}{\partial s_k(x)} \frac{\partial \phi}{\partial s_k(y)} ds(y) ds(x).$$

This result can be proved in a different manner [14]. The representation correspondend to (17) is given in [5], pp. 321–322, for the elasticity kernel.

INTEGRATION OF SINGULAR AND REGULAR INTEGRALS

The components k_n of the kernel functions (and also other kernel functions, which arise from any integral equation formulation of a strongly elliptic boundary value problem) can be written in the form

$$k_n(x, y - x) = \sum_{\kappa=0}^{\kappa_{max}} \varrho^{-2-n-\kappa} \sum_{|\alpha| \ge \kappa} c_\alpha(x) (y - x)^\alpha, \qquad n = -1, 0, \qquad (18)$$

where $2+n+\kappa$ is an odd number and $\varrho^2 = (\boldsymbol{y}-\boldsymbol{x})\cdot(\boldsymbol{y}-\boldsymbol{x})$. The coefficients $c_\alpha(\boldsymbol{x})$ are supposed to be piecewise analytic. Therefore, we have to compute integrals of the type

$$I = \int_{\pi_i^h} \text{p.f.} \int_{\pi_j^h \setminus B_\epsilon(\boldsymbol{x})} k_n(\boldsymbol{x}, \boldsymbol{y}-\boldsymbol{x}) \Phi((\chi^{(i)})^{-1}(\boldsymbol{x}), (\chi^{(j)})^{-1}(\boldsymbol{y})) ds(\boldsymbol{y}) ds(\boldsymbol{x}), \tag{19}$$

where

$$\Phi(\boldsymbol{\xi}, \boldsymbol{\eta}) = \sum_{|\alpha|>0} \frac{1}{\alpha!} (D_2^\alpha \Phi)(\boldsymbol{\xi}, \boldsymbol{\xi})(\boldsymbol{\eta}-\boldsymbol{\xi})^\alpha. \tag{20}$$

The cases of (1) coincident elements, (2) edge-adjacent elements, and (3) vertex-adjacent elements are potentially singular. In all of the singular cases the integration method [2] leads to sums of regular integrals. Similar integration methods are described in [15, 6, 10]. We introduce the notations

$$\tilde{k}_n(\boldsymbol{\xi}, \boldsymbol{\eta}-\boldsymbol{\xi}) = k_n(\chi(\boldsymbol{\xi}), \chi(\boldsymbol{\eta}) - \chi(\boldsymbol{\xi})), \tag{21}$$

$$\tilde{\Phi}(\boldsymbol{\xi}, \boldsymbol{\eta}) := |J(\chi)(\boldsymbol{\xi})||J(\chi)(\boldsymbol{\eta})| \begin{cases} \Phi(\boldsymbol{\xi}, \boldsymbol{\eta}), & n \le 0 \\ \Phi(\boldsymbol{\xi}, \boldsymbol{\eta}) - \Phi(\boldsymbol{\xi}, \boldsymbol{\xi}), & n = 1 \end{cases}, \tag{22}$$

where J denotes the JACOBIAN determinant. Here the results are given without proof, because the proofs can be found in [2].

Theorem 2 *Let the following suppositions be fulfilled:*

1. $\pi_i^h = \pi_j^h = \chi(\hat{\pi})$, *where* χ *is analytic.*
2. *The kernel function* k_n *is of type* (18), *and* $\tilde{k}_n$ *is defined by* (21).
3. *The shape functions* Φ $(\tilde{\Phi})$ *are piecewise polynomials (analytic).*

Then the singular integral (19) *is equal to the following* regular *integral (which has bounded integrands)*

$$\begin{aligned} I = \int_0^1 \int_0^1 \int_0^1 \int_0^1 \Big\{ & (\tilde{K}_n \tilde{\Phi}) \left((1-w_1)\tilde{\xi}_1 \tbinom{1}{\tilde{\xi}_2}, w_1 \tbinom{1}{w_2} \right) \\ & + (\tilde{K}_n \tilde{\Phi}) \left((1-w_1)\tilde{\xi}_1 \tbinom{1}{\tilde{\xi}_2} + w_1(1-w_2)\tbinom{1}{1}, w_1 \tbinom{w_2}{w_2 - 1} \right) \\ & + (\tilde{K}_n \tilde{\Phi}) \left((1-w_1)\tilde{\xi}_1 \tbinom{1}{\tilde{\xi}_2} + w_1(1-w_2)\tbinom{1}{0}, w_1 \tbinom{w_2}{1} \right) \Big\} \\ & \times (1-w_1)^2 w_1 \tilde{\xi}_1 d\tilde{\xi}_2 d\tilde{\xi}_1 dw_1 dw_2, \end{aligned} \tag{23}$$

where $(\tilde{K}_n \tilde{\Phi})(\boldsymbol{\xi}, \boldsymbol{\eta}) := \tilde{k}_n(\boldsymbol{\xi}, \boldsymbol{\eta}) \tilde{\Phi}(\boldsymbol{\xi}, \boldsymbol{\xi}+\boldsymbol{\eta}) + \tilde{k}_n(\boldsymbol{\xi}+\boldsymbol{\eta}, -\boldsymbol{\eta}) \tilde{\Phi}(\boldsymbol{\xi}+\boldsymbol{\eta}, \boldsymbol{\xi})$.

Theorem 3 *Let the mapping* χ *be linear, and let the suppositions of Theorem 2 be fulfilled. Then the integrands of* (23) *are polynomials in* ξ_1, ξ_2, w_1 *and the integrations of* (23) *can be performed fully analytically.*

Theorem 4 *Let $x \in \pi_i^h$ and $y \in \pi_j^h$ be two elements with adjacent edges, such that*

$$\chi^{(i)}(0) = \chi^{(j)}(0) \text{ and } \chi^{(i)}(\tbinom{1}{1}) = \chi^{(j)}(\tbinom{1}{0}),$$

and let all other suppositions of Theorem 2 be fulfilled. Then the singular integral (19) *is equal to the following* regular *integral (which has bounded integrands)*

$$\begin{aligned} I = \int_0^1 \int_0^1 \int_0^1 \int_{v_2}^1 v_2^2 \Big\{ u_1 \Big[(\tilde{k}_n \tilde{\Phi})(\xi^{(1)}, \eta^{(1)}) + (\tilde{k}_n \tilde{\Phi})(\eta^{(1)}, \xi^{(1)}) \Big] \\ + (\tilde{k}_n \tilde{\Phi})(\xi^{(2)}, \eta^{(2)}) + (\tilde{k}_n \tilde{\Phi})(\eta^{(2)}, \xi^{(2)}) \Big\} d\xi_1 dv_2 du_1 du_2 \end{aligned} \tag{24}$$

where $\xi^{(1)} = \binom{\xi_1}{v_2}$, $\eta^{(1)} = \binom{\xi_1 + v_2 u_1 (u_2 - 1)}{v_2 u_1 u_2}$, $\xi^{(2)} = \binom{\xi_1}{v_2(1-u_2)}$, $\eta^{(2)} = \binom{\xi_1 + v_2(u_1 - 1)}{v_2 u_1}$, *and* $(\tilde{k}_n \tilde{\Phi})(\xi, \eta) := \tilde{k}_n(\xi, \eta - \xi) \tilde{\Phi}(\xi, \eta)$ *with* $\chi(\xi) = \chi^{(i)}(\xi)$, $\chi(\eta) = \chi^{(j)}(\eta)$ *(see* (21), (22)*).*

Theorem 5 *Let π_i^h and π_j^h be two elements with one adjacent vertex, such that*

$$\chi^{(i)}(0) = \chi^{(j)}(0),$$

and let all other suppositions of Theorem 2 be fulfilled. Then the singular integral (19) *is equal to the following* regular *integral (which has bounded integrands)*

$$I = \int_0^1 \int_0^1 \int_0^1 \int_0^1 u_1^3 u_3 \left\{ (\tilde{k}_n \tilde{\Phi})(\xi, \eta) + (\tilde{k}_n \tilde{\Phi})(\eta, \xi) \right\} du_1 du_2 du_3 du_4, \tag{25}$$

where $\xi = u_1 \binom{1}{u_2}$, $\eta = u_1 u_3 \binom{1}{u_4}$, *and* $(\tilde{k}_n \tilde{\Phi})(\xi, \eta) := \tilde{k}_n(\xi, \eta - \xi) \tilde{\Phi}(\xi, \eta)$ *with* $\chi(\xi) = \chi^{(i)}(\xi)$, $\chi(\eta) = \chi^{(j)}(\eta)$ *(see* (21),(22)*).*

Let us briefly consider the regular and *nearly* singular case. By using the triangular coordinates $\xi_1 = u_1$, $\xi_2 = u_1 u_2$, $\eta_1 = v_1$, $\eta_2 = v_1 v_2$ we obtain integrals of the type

$$I = \int_0^1 \int_0^1 \int_0^1 \int_0^1 u_1 v_1 \left(\tilde{k}_n \tilde{\Phi} \right) \left(\binom{u_1}{u_1 u_2}, \binom{v_1}{v_1 v_2} \right) dv_2 dv_1 du_2 du_1.$$

All of the regular integrand functions in Theorems 3–5 can be integrated numerically by the standard GAUSS-LEGENDRE quadrature rule (where we used 4 upto 6 knots per dimension in our test problems). Estimations of the quadrature error are proved in [2].

IMPLEMENTATION BY USING MAPLE V

Because the effort of the analytical evaluation of the integrands of the element stiffness matrices increases with the complexity of the kernel functions and with the order of the shape functions, the c.a.s. Maple V [1] was employed for this task. In this section a Maple source for assistance in implementing the element stiffness matrices in "C" or "Fortran77" is illustrated.

```
# Type of shape function Psi: const, linear, or quadratic
PSI=linear
# Type of shape function Phi: const, linear, or quadratic
PHI=quadratic
# Type of kernel function: potential or elasticity
KERNEL=elasticity
Compute_U=yes
Compute_T=yes
Compute_S=yes
# Generation optimized source code (optimize=yes or no):
optimize=yes
```

Figure 1: Input file

The main program calls the procedure **input**, which scans the input file, and the procedures **i_ident** (integrand in the case of coincident elements), **i_edge** (integrand in the edge-adjacent case), **i_corner** (integrand in the vertex-adjacent case), and **i_reg** (integrand in the regular case). The latter four procedures have a similar structure and generate "C" and "Fortran77" source code for the evaluation of the integrands of each component $ke[1..4, 1..9, 1..9]$, where the first index specifies the kernel (see Fig. 2), of the element stiffness matrices. These procedures call the following routines:
(1) **Mapping**. Evaluation of the affine-linear and linear variable transformations.
(2) **Shape / density_function**. Evaluation of the standard nodal basis functions.
(3) **Kernel_function**. Evaluation of the kernel function.
Now we restrict ourselves to problems on polygonal bounded domains. In the case of coincident elements all of the four integrations may be performed analytically, if constant or linear shape functions are used. The three inner integrations over polynomials are performed by the standard function *"int"* of MapleV. The resulting outer integral was decomposed into integrals I_k of the type

$$\int_0^1 \frac{q_n(w)}{\sqrt{a^2w^2+bw+c^2}^m} dw, \qquad (26)$$

where q_n is a polynomial of degee n and a, b, c are constants. First the general integral (26) was evaluated. Then the integrals I_k were obtained by simple insertion of the constants. Otherwise, when shape functions of higher degree included, there may be stability problems, see [7]. In the other singular cases (edge-adjacent, vertex-adjacent), only the inner integrations can evaluated analytically. The remaining numerical integration(s), which can be performed by using standard GAUSSIAN quadrature have to be implemented manually.

CONCLUDING REMARKS

Differing from the well-known collocation approach, the GALERKIN BEM for 3D elasticity was implemented. The most difficult part, the integration of singular double integrals, within the implementation was performed by using the MapleV program *GBEMINT*. The source code of *GBEMINT* is appropriate for a rather wide range of integral kernels.

REFERENCES

[1] M. L. Abell and J. P. Braselton. *The Maple V handbook.* AP Professional, Boston, 1994.

[2] H. Andrä and E. Schnack. *Integration of Singular Galerkin-Type Boundary Element Integrals For 3D Elasticity Problems.* Preprint, Institute of Solid Mechanics, Karlsruhe University, 1995. (to appear in "Numerische Mathematik").

[3] M. Costabel. Boundary integral operators on lipschitz domains: Elementary results. *SIAM J. Math. Anal.*, 19(3):613–626, 1988.

[4] M. Costabel and W. L. Wendland. Strong ellipticity of boundary integral operators. *J. Reine Angew. Math.*, 372:34–63, 1986.

[5] V. D. Kupradze (Editor), T. G. Gegelia, M. O. Basheleishvili, and T. V. Burchuladze. *Three-dimensional Problems of the Mathematical Theory of Elasticity and Thermoelasticity.* North-Holland Publ. Comp., Amsterdam, New York, Oxford, 1979.

[6] W. Hackbusch and S. Sauter. On the efficient use of the Galerkin method to solve Fredholm integral equations. *Applications of Mathematics*, 38:301–322, 1993.

[7] W. Hackbusch and S. Sauter. On numerical cubature of nearly singular surface integrals arising in BEM collocation. *Computing*, 52:139–159, 1994.

[8] L. Hörmander. *The Analysis of Linear Partial Differential Operators*, volume I. Springer, Berlin, Heidelberg, New York, Tokyo, 1983.

[9] T. Westphal Jr., J. Tomás Pereira, and C. S. de Barcellos. On general fundamental solutions of some linear elliptic differential operators. *Eng. Anal. with Bound. Elements (to appear)*, 1996.

[10] C. Lage. *A Parallel Implementation of 3-d BEM with Focus on Panel Clustering.* PhD thesis, Inst. F. Prakt. Math., Universität Kiel, 1995. (in German).

[11] U. Langer. Parallel iterative solution of symmetric coupled FE/BE-equations via domain decomposition. *Contemp. Math.*, 157(217):335–344, 1992.

[12] S. G. Mikhlin, N. F. Morozov, and M. V. Paukshto. *The Integral Equations of the Theory of Elasticity*, volume 135 of *Teubner-Texte zur Mathematik.* Teubner, Stuttgart, Leipzig, 1995.

[13] T. Mura. *Micromechanics of defects in solids.* Nijhoff Publ., Hague, Boston, London, 1982.

[14] J. C. Nedelec. Integral equations with non integrable kernels. *Integral Equations*, 5:563–572, 1982.

[15] S. A. Sauter. *Über die effiziente Verwendung von Galerkinverfahren zur Lösung Fredholmscher Integralgleichungen.* Dissertation, Universität zu Kiel, 1992.

[16] A. H. Schatz, V. Thomée, and W. L. Wendland. *Mathematical Theory of Finite and Boundary Element Methods.* Birkhäuser, 1990.

[17] B.-W. Schulze and G. Wildenhain. *Methoden der Potentialtheorie für elliptische Differentialgleichungen beliebiger Ordnung.* Birkhäuser, Basel, Stuttgart, 1977.

[18] W. L. Wendland. Strongly elliptic boundary integral equations. In M. Powl A. Iserles, editor, *The State of the Art in Numerical Analysis.* Oxford University Press, Oxford, 1987.

[19] J. Wloka. *Partielle Differentialgleichungen (Partial differential equations).* B. G. Teubner, Stuttgart, 1982.

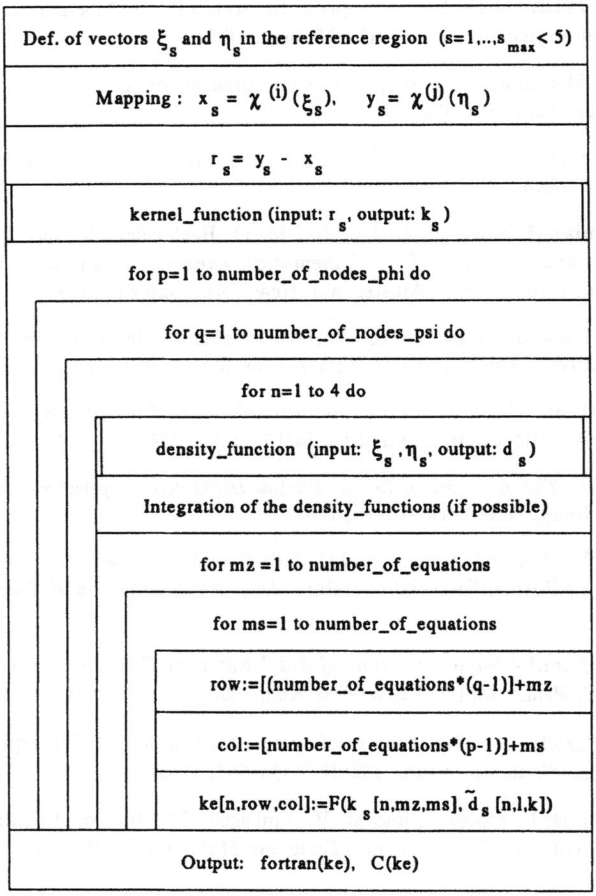

Figure 2: Structure of the procedures **i_ident, i_corner, i_edge, i_reg**

PRECONDITIONING BOUNDARY ELEMENT EQUATIONS

Ke CHEN
Computational Mathematics Group,
Department of Mathematical Sciences,
The University of Liverpool, Liverpool L69 3BX, England.
(Email: K.Chen@liverpool.ac.uk)

SUMMARY

In this paper we discuss several sparse matrix based preconditioners suitable for preconditioning boundary element equations. All preconditioners involve only $O(n)$ nonzeros. We provide a framework for constructing operator splitting based preconditioners and use it to analyze a class of sparse preconditioners. For singular integral equations, a more efficient preconditioner is proposed that has a band-2 structure.

BOUNDARY ELEMENTS AND DENSE LINEAR SYSTEMS

Let $\Omega \in R^2$ denote a closed domain that may be interior and bounded, or exterior and unbounded, and $\Gamma = \partial\Omega$ be its (finite part) boundary that can be parameterised by $p = (x, y) = (x(s), y(s)), a \le s \le b$. The R^3 case can be studied similarly. Then a boundary integral equation that usually arises from reformulating a PDE in Ω can be written as

$$U(p) - \int_\Gamma \bar{k}(p,q)U(q)dS_q = f(p), \qquad p \in \Gamma, \tag{1}$$

or
$$U(s) - \int_a^b k(s,t)U(t)dt = f(s), \qquad s \in [a,b], \tag{2}$$

or simply
$$(I - \mathcal{K})u = f. \tag{3}$$

To solve the above equation numerically, we divide the boundary Γ (interval $[a, b]$) into m boundary elements (non-intersecting subintervals $E_i = [s_{i-1}, s_i]$). On each interval E_i, we may either approximate the unknown u by an interpolating polynomial of order η that leads to a collocation method, or apply a quadrature method of η nodes that gives rise to the Nyström method. Both discretization methods approximate equation (3) by

$$(I - \mathcal{K}_n)u_n = f, \tag{4}$$

where we can write

$$\mathcal{K}_n u = \mathcal{K}_n u_n = \sum_{j=1}^{m} \left[\sum_{i=1}^{\eta} w_i k(s, t_{ji}) u_{ji} \right], \quad u_n(t_{ji}) = U(t_{ji}) = u_{ji}, \quad \text{and} \quad n = m\eta.$$

We use vector u to denote u_{ji}'s at all nodes. By a collocation step in equation (4), we obtain a linear system of equations

$$(I - K)u = f, \qquad \text{or} \qquad Au = f, \tag{5}$$

where matrices K and A are dense and unsymmetric (in general). The conditioning of A depends on the smoothness of kernel function $k(s,t)$. A strong singularity (as $t \to s$) leads to non-compactness of operator $\mathcal{K}$ and renders equation (5) difficult to solve by iterative methods without preconditioning.

PRECONDITIONING TECHNIQUES

For equation (5) i.e. $Au = f$, denote a preconditioner by M. Then a (left) preconditioned system can be written as $MAu = Mf$. In general, two requirements are imposed. Firstly a system $Mx = y$ for any vectors x, y should be efficiently solved in less than $O(n^2)$ operations or $O(n)$ operations if a more sophisticated method like the panel clustering is also used). This normally means that M or its inverse must be sparse and so it natural to seek sparse preconditioners. Secondly the preconditioned matrix MA should possess some better properties than A e.g. eigenvalues are more clustered. Often we hope the condition number of MA to be smaller than that of A. In this sense, the best preconditioner should approach the inverse of A.

Indeed, both requirements are reflected (at least in part) in the preconditioning techniques discussed below. Each case will be illustrated using a small n.

TWO GRID BASED SPARSE COLUMN PRECONDITIONERS

The success of this preconditioner, due to Yan [14], follows from the facts that a $n \times n$ sparse column matrix M with m nonzero long columns has its inverse of the identical sparsity and a system such as $Mx = y$ for $x, y \in R^n$ can be solved in only $(n-m)m$ operations. For example, with $n = 9$,

$$M = \begin{pmatrix} B_{11} & B_{12} & B_{13} \\ B_{21} & B_{22} & B_{23} \\ B_{31} & B_{32} & B_{33} \end{pmatrix} = \begin{pmatrix} \times & & \times & & & \times & & & \times \\ & \times & \times & & & \times & & & \times \\ & & \boxed{\times} & & & \boxed{\times} & & & \boxed{\times} \\ & & \times & \times & & \times & & & \times \\ & & \times & & \times & \times & & & \times \\ & & \boxed{\times} & & & \boxed{\times} & & & \boxed{\times} \\ & & \times & & & \times & \times & & \times \\ & & \times & & & \times & & \times & \times \\ & & \boxed{\times} & & & \boxed{\times} & & & \boxed{\times} \end{pmatrix},$$

where each B_{ij} denotes a 3×3 block matrix. Formally, let $n = \eta m$ with η, m integers. Then from matrix $K = (k_1, k_2, \cdots, k_n)$ of equation (5), construct column vectors by

$$k_i' = \begin{cases} k_i, & \text{if } i = \ell\eta, \quad 1 \le \ell \le m, \\ 0, & \text{otherwise,} \end{cases}$$

and define a new matrix by $K' = (k_1', k_2', \cdots, k_n')$. Then we use $M = A' = (I + \eta K')$ as a preconditioner. It may be expected that A' is 'close' to $A = A_n$ because $A' \approx A_m$ which is the corresponding discrete matrix with m nodes.

It can be shown that both the inverse of M and its LU decomposition are of the same sparsity structure. Therefore the solution of $Mx = y$ reduces to that of $M^* x^* = y^*$ where M^* is a $m \times m$ principal submatrix of M (i.e. of those boxed elements in the above case of $n = 9$ with $m = 3$). Formally define a restriction operator R_m from n-dimensional (complex) space $\mathbf{C}^n$ to $\mathbf{C}^m$ by $R_m u = (u_\eta, u_{\eta 2}, \cdots, u_{\eta m})^T$ for $u = (u_1, \cdots, u_n)^T \in \mathbf{C}^n$ and a prolongation operator P_m from $\mathbf{C}^m$ to $\mathbf{C}^n$ by $P_m v = (v_1', \cdots, v_n')^T$ for $v = (v_1, \cdots, v_m)^T \in \mathbf{C}^m$ where $v_k' = v_k$ for $k = \eta\ell$ otherwise $v_k' = 0$.

Algorithm 1 (Two-grid for $Mx = y$)

1. *Solve for v_m from an $m \times m$ system $(I + \eta K_m) v_m = R_m y$, where K_m is the submatrix extracted from K', corresponding to M^*.*

2. *Evaluate $x = y - (I + \eta K' P_m) v_m$.*

MESH NEIGHBOUR BASED APPROXIMATE INVERSE

The starting point in the mesh neighbour preconditioner of Vavasis [13] is that we hope to find a matrix such that $AM \approx I$ and this M should have its diagonal elements and immediate neighbours possess more importance as A usually comes from singular integral equations. In particular, assume that M is a quasi-tridiagonal matrix; for $n = 9$ we have

$$M = \begin{pmatrix} \times & \times & & & & & & & \times \\ \times & \times & \times & & & & & & \\ & \times & \times & \times & & & & & \\ & & \times & \times & \times & & & & \\ & & & \times & \times & \times & & & \\ & & & & \times & \times & \times & & \\ & & & & & \times & \times & \times & \\ & & & & & & \times & \times & \times \\ \times & & & & & & & \times & \times \end{pmatrix}. \tag{6}$$

Then to find $M = (p_1, p_2, \cdots, p_n)$ in terms of $A = (a_1, a_2, \cdots, a_n)$, a direct approach is used. Let $AM = I \iff Ap_i = e_i$. Then, as p_i only has three nonzero positions i.e. $p_{i_1}, p_{i_2}, p_{i_3}$, an approximate solution is obtained by solving a 3×3 system

$$\begin{pmatrix} A_{i_1 i_1} & A_{i_1 i_2} & A_{i_1 i_3} \\ A_{i_2 i_1} & A_{i_2 i_2} & A_{i_2 i_3} \\ A_{i_3 i_1} & A_{i_3 i_2} & A_{i_3 i_3} \end{pmatrix} \begin{pmatrix} p_{i_1} \\ p_{i_2} \\ p_{i_3} \end{pmatrix} = \begin{pmatrix} 0 \\ 1 \\ 0 \end{pmatrix},$$

where $i_1 = i-1, i_2 = i, i_3 = i+1$ for column $i = 2, \cdots, n-1$, $i_1 = n, i_2 = 1, i_3 = 2$ for $i = 1$ and $i_1 = n-1, i_2 = n, i_3 = 1$ for $i = n$ due to the wrap around nature of M.

Algorithm 2 (Mesh neighbours for $x = My$)

1. *(once and only once) For each column $i = 1, \cdots, n$,*

   set i_1, i_2, i_3, and form the 3×3 submatrix W as above. Set $e = (0,1,0)^T$. Then solve a 3×3 system $Wc = e$. Set $p_i = c$ (the i-th column of M).

2. *Evaluate $x = My$.*

APPROXIMATE INVERSION TECHNIQUES

Sparse approximate inverse preconditioners have been proposed for preconditioning sparse linear systems [7] and [10]. They are here adapted for applications to dense linear systems $Au = f$.

The problem is to construct approximations, M, to the inverse of matrix A for which $\|AM - I\|$ is small in some norm. Once such an M has been constructed, we solve the preconditioned linear system $AMw = f$, $u = Mw$.

To specify matrix M, denote $\mathcal{N} = \{1, 2, \cdots, n\}$, let $\mathcal{S}$ be a given set of (i,j)'s with $(i,j) \in \mathcal{N}$, and $\mathcal{G}_\mathcal{S}$ be the space of all $n \times n$ matrices that have entries in positions indexed by $\mathcal{S}$. For each column index $j = 1, \cdots, n$, define its row indexes $\mathcal{S}_j = \{i : (i,j) \in \mathcal{S}\}$ and let its vector space $\mathcal{G}_{\mathcal{S}_j}$ contain all vectors that have entries in positions indexed by $\mathcal{S}_j$. Then the approximate inverse M is calculated by solving the least squares problem (LS)

$$\min_{M \in \mathcal{G}_\mathcal{S}} \|AM - I\|_F^2 = \sum_{j=1}^{n} \min_{m_j \in \mathcal{G}_{\mathcal{S}_j}} \|Am_j - e_j\|_2^2, \tag{7}$$

where $M = [m_1, \cdots, m_n]$ and $I = [e_1, \cdots, e_n]$. In theory, such a least squares problem can be posed in any norm but the Frobenius norm leads to an easier solution. The full problem of finding M is reduced to n standard least squares problems.

For singular integral equations, it seems reasonable to select $\mathcal{S}$ from the same sparsity pattern as equation (6).

Algorithm 3 (Approximate inversion for $x = My$)

1. *(once and only once) For each column $i = 1, \cdots, n$,*

   set i_1, i_2, i_3, and form the $n \times 3$ submatrix W as above. Set $e = e_i$ the i-th unit vector. Then solve a $n \times 3$ LS system $Wc = e$ by the QR method. Set $p_i = c$ (the i-th column of M).

2. *Evaluate $x = My$.*

SPARSE LU DECOMPOSITIONS

Similar to the idea of mesh neighbour based approximate inverses, the work of [2] assumes that preconditioner M should have its diagonal elements and immediate neighbours possess more importance for the same reason that A usually comes from singular integral equations; see also [5]. When $n = 9$, such a matrix M is of the same sparsity as in equation (6).

Here we form M explicitly from A with $M_{ij} = A_{ij}$ for $|i-j| \le 1$ and $i = 1, \cdots, n$, $M_{1n} = A_{1n}$ and $M_{n1} = A_{n1}$. As M^{-1} cannot maintain any sparsity property of M, it is proposed, in order to solve $Mx = y$ for $x, y \in R^n$, to decompose M as $M = LU$, where L and U are sparse triangular matrices. To illustrate, for $n = 9$, we have

$$M = LU = \begin{pmatrix} P_{1,1} & & & & & & & & \\ P_{2,1} & P_{2,2} & & & & & & & \\ & P_{3,2} & P_{3,3} & & & & & & \\ & & P_{4,3} & P_{4,4} & & & & & \\ & & & P_{5,4} & P_{5,5} & & & & \\ & & & & P_{6,5} & P_{6,6} & & & \\ & & & & & P_{7,6} & P_{7,7} & & \\ & & & & & & P_{8,7} & P_{8,8} & \\ P_{9,1} & P_{9,2} & P_{9,3} & P_{9,4} & P_{9,5} & P_{9,6} & P_{9,7} & P_{9,8} & P_{9,9} \end{pmatrix}$$
$$\times \begin{pmatrix} 1 & P_{1,2} & & & & & & & P_{1,9} \\ & 1 & P_{2,3} & & & & & & P_{2,9} \\ & & 1 & P_{3,4} & & & & & P_{3,9} \\ & & & 1 & P_{4,5} & & & & P_{4,9} \\ & & & & 1 & P_{5,6} & & & P_{5,9} \\ & & & & & 1 & P_{6,7} & & P_{6,9} \\ & & & & & & 1 & P_{7,8} & P_{7,9} \\ & & & & & & & 1 & P_{8,9} \\ & & & & & & & & 1 \end{pmatrix}.$$

A general algorithm is given as follows

Algorithm 4 (Sparse LU for $Mx = y$)

1. (once and only once) Calculation of $M = LU$ factors i.e. $[L + U - I] = (P_{ij})$:

$P_{1,1} = M_{1,1}, \quad P_{2,1} = M_{2,1}, \quad P_{n,1} = M_{n,1},$
$P_{1,2} = M_{1,2}/M_{1,1}, \quad P_{1,n} = M_{1,n}/M_{1,1}, \quad P_{2,2} = M_{2,2} - M_{2,1}P_{1,2},$
$P_{n,2} = -M_{n,1}P_{1,2}, \quad s = M_{n,1}P_{1,n}$

FOR $i = 2 : n-2$

> $P_{i,i+1} = M_{i,i+1}/P_{i,i}, \qquad P_{i,n} = -M_{i,i-1}P_{i-1,n}/P_{i,i}$
> $P_{i+1,i+1} = M_{i+1,i+1} - M_{i+1,i}P_{i,i+1}, \qquad P_{n,i+1} = -P_{n,i}P_{i,i+1}$
> $s = s + P_{n,i}P_{i,n}$

END

$P_{n,n-1} = P_{n,n-1} + M_{n,n-1}, \quad P_{n-1,n} = (M_{n-1,n} - M_{n-1,n-2}P_{n-2,n})/P_{n-1,n-1},$
$P_{n,n} = M_{n,n} - P_{n,n-1}P_{n-1,n} - s.$

2. Forward substitution $Lz = y$:

	$z_1 = y_1/P_{1,1},$	$s = P_{n,1},$
FOR $i = 2 : n-1$		
	$z_i = (y_i - P_{i,i-1}z_{i-1})/P_{i,i},$	$s = s + P_{n,i}z_i$
END		
	$z_n = (y_n - s)/P_{n,n}.$	

3. Backward substitution $Ux = z$*:*

$x_n = z_n,$	$x_{n-1} = z_{n-1} - P_{n-1,n}x_n,$
FOR $i = n-2 : -1 : 1$	
	$x_i = z_i - P_{i,i+1}x_{i+1} - P_{i,n}x_n,$
END	

To understand and analyze the above otherwise heuristic sparse preconditioners appeared in the literature, we now consider operator splittings or preconditioning of singular integral operators.

OPERATOR SPLITTING STRATEGIES

Here we introduce the idea of operator splitting that was originated in [5]. Use the partition previously defined, $[a,b] = \bigcup_{i=1}^{m} E_i$. Accordingly we can partition variable U and vector u as follows $U = (U_1, U_2, \cdots, U_m)^T$ and $u = (u_1, u_2, \cdots, u_m)^T$.

SPLITTING STRATEGY 1

Operator $\mathcal{K}$ can be partitioned into a matrix form as for vectors and further we can observe that all singularities of $\mathcal{K}$ are contained in the following operator

$$\bar{\mathcal{K}} = \begin{pmatrix} \mathcal{K}_{1,1} & \mathcal{K}_{1,2} & & & \mathcal{K}_{1,m} \\ \mathcal{K}_{2,1} & \mathcal{K}_{2,2} & \mathcal{K}_{2,3} & & \\ & \mathcal{K}_{3,2} & \ddots & \ddots & \\ & & \ddots & \ddots & \mathcal{K}_{m-1,m} \\ \mathcal{K}_{m,1} & & & \mathcal{K}_{m,m-1} & \mathcal{K}_{m,m} \end{pmatrix}.$$

The corresponding matrix out of K is

$$\bar{K} = \begin{pmatrix} K_{1,1} & K_{1,2} & & & K_{1,m} \\ K_{2,1} & K_{2,2} & K_{2,3} & & \\ & K_{3,2} & \ddots & \ddots & \\ & & \ddots & \ddots & K_{m-1,m} \\ K_{m,1} & & & K_{m,m-1} & K_{m,m} \end{pmatrix}.$$

Define operators $\mathcal{D} = I - \bar{\mathcal{K}}$ and $\mathcal{C} = \mathcal{K} - \bar{\mathcal{K}}$, and correspondingly matrices $D = I - \bar{K}$ and $C = K - \bar{K}$. Then it can be shown that operator $\mathcal{D}$ is bounded and $\mathcal{C}$ is compact under certain conditions. So operator $\mathcal{D}^{-1}\mathcal{C}$ is also compact. Thus the solution of $Au = f$ is reduced to that of $[I - D^{-1}C]u = D^{-1}f$, which can be solved by an iterative method efficiently. Here $M = D$, a block quasi-tridiagonal matrix, is then used as a preconditioner and the solution of $Mx = y$ should similarly follow Algorithm 4.

SPLITTING STRATEGY 2

Recall that for many numerical methods collocation occurs at nodal points. Therefore we may surround all collocation points by suitable boundary domains, defining an intermediate operator, and split above partitions further in order to have preconditioning matrices with an even simpler structure.

Define an ε-cover for all nodal points by $I_i = [s_i^-, s_i^+]$ for node $i = 1, \cdots, n$, where $s_i^\pm = s_i \pm \varepsilon$. Note that s_0 and s_n refer to the same point due to the closed boundary Γ. Therefore a new partition for the boundary (or interval $[a,b]$) is $[a,b] = \bigcup_{i=1}^n (I_i \bigcup I_i^0)$, where $I_i^0 = (s_{i-1}^+, s_i^-)$. Let $\Omega_\varepsilon = \bigcup_{i=1}^n I_i$ and $\Omega_o = \bigcup_{i=1}^n I_i^0$. Then $[a,b] = \Omega_\varepsilon \bigcup \Omega_0$ and $\Omega_\varepsilon \bigcap \Omega_0 = 0$.

Then we may regard Ω_ε as the collocation space. So accordingly all singularities of $\mathcal{K}$, depending on how nodes are numbered, are contained in either

$$\bar{\mathcal{K}} = \begin{pmatrix} \mathcal{K}_{1,1} & & & & \mathcal{K}_{1,m} \\ \mathcal{K}_{2,1} & \mathcal{K}_{2,2} & & & \\ & \mathcal{K}_{3,2} & \ddots & & \\ & & \ddots & \ddots & \\ & & & \mathcal{K}_{m,m-1} & \mathcal{K}_{m,m} \end{pmatrix} \quad \text{or } \bar{\mathcal{K}} = \begin{pmatrix} \mathcal{K}_{1,1} & \mathcal{K}_{1,2} & & & \\ & \mathcal{K}_{2,2} & \mathcal{K}_{2,3} & & \\ & & \ddots & \ddots & \\ & & & \ddots & \mathcal{K}_{m-1,m} \\ \mathcal{K}_{m,1} & & & & \mathcal{K}_{m,m} \end{pmatrix}.$$

The corresponding matrix out of K is either

$$\bar{K} = \begin{pmatrix} K_{1,1} & & & & K_{1,m} \\ K_{2,1} & K_{2,2} & & & \\ & K_{3,2} & \ddots & & \\ & & \ddots & \ddots & \\ & & & K_{m,m-1} & K_{m,m} \end{pmatrix} \quad \text{or } \bar{K} = \begin{pmatrix} K_{1,1} & K_{1,2} & & & \\ & K_{2,2} & K_{2,3} & & \\ & & \ddots & \ddots & \\ & & & \ddots & K_{m-1,m} \\ K_{m,1} & & & & K_{m,m} \end{pmatrix}.$$

As in Strategy 1, we use $M = D = I - \bar{K}$ as the preconditioner, which is a quasi-bidiagonal matrix. To illustrate, for $n = 4$, we can decompose M as follows (case 1)

$$M = LU = \begin{pmatrix} M_{1,1} & & & \\ M_{2,1} & M_{2,2} & & \\ & M_{3,2} & M_{3,3} & \\ & & M_{4,3} & P_{4,4} \end{pmatrix} \begin{pmatrix} 1 & & & P_{1,4} \\ & 1 & & P_{2,4} \\ & & 1 & P_{3,4} \\ & & & 1 \end{pmatrix}.$$

Similarly, for $n = 4$ and depending on how nodes are numbered, we may also decompose

M as follows (case 2)

$$M = LU = \begin{pmatrix} 1 & & & \\ & 1 & & \\ & & 1 & \\ P_{4,1} & P_{4,2} & P_{4,3} & 1 \end{pmatrix} \begin{pmatrix} M_{1,1} & M_{1,2} & & \\ & M_{2,2} & M_{2,3} & \\ & & M_{3,3} & M_{3,4} \\ & & & P_{4,4} \end{pmatrix}.$$

Algorithm 5 (New sparse LU for $Mx = y$)

1. *(once and only once) Calculation of $M = LU$ factors i.e. $[L + U - I] = (P_{ij})$:*

Splitting case 1	*Splitting case 2*
$P_{1,1} = M_{1,1}, \quad P_{2,1} = M_{2,1},$	$P_{1,1} = M_{1,1}, \quad P_{1,2} = M_{1,2},$
$P_{1,n} = M_{1,n}/P_{1,1}$	$P_{n,1} = M_{n,1}/P_{1,1}$
FOR $i = 2 : n-1$	*FOR* $i = 2 : n-1$
$P_{i,i} = M_{i,i}, \quad P_{i,i-1} = M_{i,i-1},$	$P_{i,i} = M_{i,i}, \quad P_{i,i+1} = M_{i,i+1},$
$P_{i,n} = -P_{i,i-1}P_{i-1,n}/P_{i,i}$	$P_{n,i} = -P_{i-1,i}P_{n,i-1}/P_{i,i}$
END	*END*
$P_{n,n-1} = M_{n,n-1},$	$P_{n-1,n} = M_{n-1,n},$
$P_{n,n} = M_{n,n} - P_{n,n-1}P_{n-1,n}.$	$P_{n,n} = M_{n,n} - P_{n,n-1}P_{n-1,n}.$

2. *Forward substitution $Lz = y$:*

Splitting case 1	*Splitting case 2*
$z_1 = y_1/P_{1,1},$	$s = 0,$
FOR $i = 2 : n$	*FOR* $i = 1 : n-1$
$z_i = (y_i - P_{i,i-1}z_{i-1})/P_{i,i},$	$z_i = y_i, \quad s = s + P_{n,i}z_i,$
END.	*END*
	$z_n = y_n - s.$

3. *Backward substitution $Ux = z$:*

Splitting case 1	*Splitting case 2*
$x_n = z_n,$	$x_n = z_n/P_{n,n},$
FOR $i = n-1 : -1 : 1$	*FOR* $i = n-1 : -1 : 1$
$x_i = z_i - P_{i,n}x_n,$	$x_i = (z_i - P_{i,i+1}x_{i+1})/P_{i,i},$
END	*END*

SPARSE PRECONDITIONERS FOR INTEGRAL OPERATORS

Corresponding to every preconditioning technique for matrices, there must exist an underlying preconditioning strategy for the integral operator. But it is not always obvious to work out such a strategy. We now use the theory of the above section to identify the operator splittings implied in the algorithms presented.

Firstly we see that the preconditioner of Algorithm 1 uses the following splitting $\mathcal{K} = \mathcal{D} + \mathcal{C}$, where $\mathcal{D} = I - \tilde{\mathcal{K}}$ with $\tilde{\mathcal{K}} = (\tilde{\mathcal{K}}_{ij})$ and $\tilde{\mathcal{K}}_{ij} = \eta \int_{E_j/\eta} k(s,t)u_j(t)dt \approx \int_{E_j} k(s,t)u_j(t)dt = \mathcal{K}_{ij}$. Theoretically $\mathcal{C}$ is not compact but $\mathcal{C}$ should approach 0 if $m, n \to \infty$. Practically η should not be too large (then the method becomes more expensive if $m \approx n$).

Secondly, for the preconditioners of both Algorithms 2 and 4, the underlying splittings would be identical to that in the last section for piecewise constant approximations or the panel method (mid-point rule). The reason is that both were proposed based on assumptions on singular integrals rather than operators.

Thirdly, for Algorithms 2 and 3, it appears impossible to identify any operator splittings. The truth of Algorithm 3 still eludes us. But for Algorithm 2, it turns out that this is possible and the method splits a diagonal operator; refer to [6].

ALGORITHM 1 FOR FIRSTKIND EQUATIONS

Algorithms 2-5 all apply to singular integral equations of the first kind but Algorithm 1 is not immediately applicable. In [14], for the case where part of the integral operator may be reducible to a circulant matrix, it is suggested to apply the fast Fourier transform for this part to obtain a diagonal matrix. Then the situation resembles the second kind equation case.

Instead of carrying out more case studies, in view of our discussions on singularity separation, we propose a more general method here to enable Algorithm 1 to be applicable to first kind equations. First decompose matrix $K = K_1 + K_2$ and let $K_1 = (k_1, k_2, \cdots, k_n)$, where K_1 is the diagonal matrix of K. Then construct new column vectors by

$$k_i' = \begin{cases} k_i, & \text{if } i = \ell\eta, \quad 1 \le \ell \le m, \\ 0, & \text{otherwise,} \end{cases}$$

and define a new matrix by $K' = (k_1', k_2', \cdots, k_n')$. Then we can use $M = A' = K_1 + (\eta - 1)K'$ as a preconditioner. As before it may be expected that A' is 'close' to $A = A_n$ because $A' \approx A_m$ which is the corresponding discrete matrix with m nodes.

NUMERICAL RESULTS

The first problem to be tested has a weak singularity (see [14])

$$\text{Problem 1}: \qquad u(s) + \gamma \int_{-\pi}^{\pi} u(t)b(s,t)dt = f(s), \qquad s \in [-\pi, \pi]$$

which arises from the solution of the exterior Neumann's problem of the Laplace equation (when $\gamma = 1$) over an elliptic boundary $p(s) = (\cos(s), \sin(s)/4)$. Here the kernel function

Table 1: Convergence results of Problem 1 ($\gamma = 10$)

Method	A1 (2-grid)	A2 (Mesh)	A3 (Inverse)	A4 (LU 3)	A5 (LU 2)
N = 16	4	5	4	8	9
N = 32	6	6	7	10	7
N = 64	37	7	7	11	8
N =128	52	8	9	13	8
N =256	60	8	9	13	10
N =512	62	10	9	13	10
N =1024	62	10	10	14	11

is

$$b(s,t) = \frac{(p(t)-p(s)) \cdot n(p(t))|p'(t)|}{\pi |p(t)-p(s)|^2} = \frac{4}{\pi(17 - 15\cos(t+s))}.$$

The parameter γ is included to vary the difficulty of the problem. We specifically choose

$$f(s) = |\sin(s)| + \frac{2\gamma}{15\pi}\left\{4\cos(s)\log\frac{17+15\cos(s)}{17-15\cos(s)} + 17\sin(s)\tan^{-1}\frac{15\sin(s)}{8}\right\}$$

so that $u(s) = |\sin(s)|$.

The second problem possesses a Cauchy singularity (see [5])

$$\text{Problem 2 :} \qquad \begin{cases} \dfrac{1}{\pi}\displaystyle\int_{-1}^{1}\frac{w(t)\phi(t)}{t-x}dt + \int_{-1}^{1}\frac{(t^2-x^2)^2}{t^2+x^2}w(t)\phi(t)dt = f(x), \\ \dfrac{1}{\pi}\displaystyle\int_{-1}^{1} w(t)\phi(t)dt = 0, \qquad\qquad x \in (-1,1), \end{cases}$$

which has the exact solution $\phi(x) = x|x|$.

Both problems are discretized by the Nyström method using uniform nodes for Problem 1 and Chebyshev nodes for Problem 2. The conjugate gradient iterative method to the normal equation (CGN) is adopted; see [11]. The tolerance for residual errors (RMS norm) is set to be TOL $\in [10^{-8},\ 10^{-2}]$ so that it is of the same magnitude as the discretization error (measured using a direct method). Results are shown in Tables 1–2 of comparisons of six preconditioners, where A1 means Algorithm 1 and so on.

Although further and more extensive tests are needed to compare these preconditioners, our preliminary conclusion is that for singular boundary element equations, the most robust preconditioners (Algorithms 2 and 5) should be based on singularity separations.

REFERENCES

[1] Amini S. and Chen K. (1989) Conjugate gradient method for second kind integral equations – applications to the exterior acoustic problem. *Engineering Analysis with Boundary Elements*, **6**(2), 72–77.

Table 2: Convergence results of Problem 2

Method	A1 (2-grid)	A2 (Mesh)	A3 (Inverse)	A4 (LU 3)	A5 (LU 2)
N = 16	13	9	10	8	9
N = 32	17	11	18	11	10
N = 64	24	12	30	17	11
N =128	32	13	39	23	13
N =256	45	14	43	27	14
N =512	60	14	46	30	14
N =1024	69	16	47	32	15

[2] Amini S. and Maines N. (1994) Iterative solutions of boundary integral-equations. In : *Proceedings of the 14th conference on Boundary Element Methods*, ed. C. A. Brebbia, Ch.65, 193–200.

[3] Brebbia C. A. et al (1984) *Boundary element techniques – theory and applications*, Springer-Verlag.

[4] Canning F. (1992) Sparse approximation for solving integral equations with oscillatory kernels. *SIAM J. Sci. Stat. Comp.*, **13**, 71–87.

[5] Chen K. (1994) Efficient iterative solution of linear systems from discretizing singular integral equations. *Elec. Tran. Numer. Anal.*, **2**, 76–91.

[6] Chen K. (1996) On preconditioning dense linear systems from singular integral equations. Submitted for publication.

[7] Cosgrave J. D. F., Diaz J. C. and Griewank A. (1992) Approximate inverse preconditioners for sparse linear systems, *Int. J. Compt. Math.*, **44**, 91-110.

[8] Hackbusch W. (1985) *Multigrid methods and applications.* Springer-Verlag.

[9] Hemker P. W. and Schippers H. (1981) Multigrid methods for the solution of Fredholm integral equations of the second kind. *Math. Comp.*, **36**, 215–232.

[10] Huckle T. and Grote M. (1994) A new approach to parallel preconditioning with sparse approximate inverses, *Preprint.*

[11] Nachtigal N. M., Reddy S. C. and L. N. Trefethen (1992) How fast are nonsymmetric matrix iterations?. *SIAM J. Matr. Anal. Appl.*, **13**(3), 778–795.

[12] Partridge P.W., Brebbia C. A. and Wrobel L. C. (1992) *The dual reciprocity boundary element method.* CMP Southampton, UK.

[13] Vavasis S. (1992) Preconditioning for boundary integral equations. *SIAM J. Matr. Anal. Appl.*, **13**(3), 905–925.

[14] Yan Y. (1994) Sparse preconditioned iterative methods for dense linear systems. *SIAM J. Sci. Comp.*, **15**(5), 1190–1200.

LOCAL A-POSTERIORI ERROR ESTIMATORS FOR THE DISCRETIZATION OF BOUNDARY INTEGRAL EQUATIONS

Birgit Faermann
Lehrstuhl für Praktische Mathematik
Mathematisches Seminar Bereich II
Christian-Albrechts-Universität zu Kiel
Hermann-Rodewald-Str. 3/1
24098 Kiel, Germany

SUMMARY

In this paper we present local a-posteriori error estimators for the Galerkin and for the collocation discretization of boundary integral equations. These error estimators are introduced and investigated by Babuška-Rheinboldt [2] for finite element methods. We transfer them from finite element methods onto boundary element methods and show that they are reliable and efficient for a wide class of integral operators under relatively weak assumptions. These local error estimators base on the computable residual and can be used for controlling of adaptive mesh refinement.

1 INTRODUCTION

Adaptive methods are already well established in the finite element field. In boundary element methods (BEM), on the other hand, only a few theoretical and numerical results concerning adaptive procedures exist.

In this paper we consider an abstract boundary integral equation defined on the Lipschitz-continuous boundary $\Gamma = \partial\Omega$ of a domain $\Omega \subseteq \mathbb{R}^d$. In the h-version of BEM, the unknown solution u of such an integral equation is approximated by a discrete solution u_Δ associated to a mesh Δ on the boundary Γ. The discretization error $e = u_\Delta - u$ will be reduced by mesh refinement. Adaptive mesh refinement is needed, if the error e is not uniformly distributed over the mesh Δ. In this case, we want to refine the mesh only there, where the local error is large. Therefore, we need information about the error distribution. Our aim is to find local quantities $\{\delta_q\}_{q\in\Xi}$ associated to the elements of the mesh Δ, such that

$$C_1 \sum_{q\in\Xi} \delta_q^2 \;\leq\; |\!|\!| u_\Delta - u |\!|\!|^2 \;\leq\; C_2 \sum_{q\in\Xi} \delta_q^2\,, \tag{1}$$

where $|\!|\!| \cdot |\!|\!|$ is an appropriate global norm of the error, and C_1, C_2 should be constants independent of the solution u and the mesh Δ. We call the $\{\delta_q\}_{q\in\Xi}$ *local a-posteriori error estimators*, or briefly (local) error estimators, if they satisfy (1) and if their definition is based on the discrete solution u_Δ. Local error estimators are called *reliable*, if they satisfy the upper estimate in (1) with a constant C_2 independent u and Δ, and they are called *efficient*, if they satisfy the lower estimate in (1) with a constant C_1 independent u and Δ.

For boundary element methods, only few authors have investigated local a-posteriori error estimates of the form (1), see for example [3,4,11,12,14,15,16]. In particular, the efficiency is problematic in the cited papers.

In this paper we present local error estimators for the Galerkin and for the collocation discretization of boundary integral equations. These error estimators are introduced and investigated by Babuška-Rheinboldt [2] for finite element methods (FEM), and we call them BR error estimators. We transfer them from FEM onto BEM and show that they are reliable and efficient for a wide class of integral operators under relatively weak assumptions. In particular, no inverse estimates are needed. The BR error estimators base on the computable residual and can be used for controlling of adaptive mesh refinement.

An outline of this paper is as follows. In section 2 we introduce the function spaces used later on and present an inequality, which is important for the lower estimate in (1). In section 3 we introduce the BR error estimators for Galerkin boundary element methods and show that they satisfy (1). In section 4 we transfer the BR error estimators from the Galerkin method onto the collocation. In section 5 we present numerical results, which confirm our theoretical results.

2 THE NORM OF $H^t(\Gamma)$ AND FUNCTIONS WITH DISJOINT SUPPORTS

Let Γ be the boundary of a bounded and simply connected Lipschitz domain $\Omega \subseteq \mathbb{R}^d$. For brevity, we restrict ourselves to the 2-dimensional case ($d = 2$). The curve Γ is identified with the interval $[0,1]$ via a fixed one-periodic parametrization $\gamma : \mathbb{R} \to \Gamma$, and functions $v : \Gamma \to \mathbb{K}$ are identified with one-periodic functions $v : \mathbb{R} \to \mathbb{K}$.
For $t \geq 0$, we define the Sobolev space $H^t(\Gamma)$ by

$$H^t(\Gamma) \;:=\; \{v \in L^2(\Gamma) : \|v\|_t < \infty\}, \quad \text{where}$$

$$\|v\|_t^2 \;=\; \|v\|_{H^t(\Gamma)}^2 \;:=\; |\hat{v}_0|^2 + \sum_{\nu\in\mathbb{Z}\setminus\{0\}} |\hat{v}_\nu|^2\, |2\pi\nu|^{2t} \tag{2}$$

with the Fourier coefficients

$$\hat{v}_\nu \;=\; \mathcal{F}_\nu(v) \;:=\; \int_0^1 v(t)\, e^{-2\pi i\nu t}\, dt \;=\; \overline{\langle e^{2\pi i\nu\cdot}, v\rangle_{L^2(\Gamma)}} \qquad \text{for } \nu \in \mathbb{Z}\,.$$

For negative $t < 0$, we define $H^t(\Gamma)$ to be the dual space of $H^{-t}(\Gamma)$ and extend the definition of the Fourier coefficients by

$$\hat{v}_\nu \;:=\; \overline{v(e^{2\pi i\nu\cdot})} \qquad \text{for } \nu \in \mathbb{Z}\,,\; v \in H^t(\Gamma)\,,\; t<0\,.$$

Then, the dual norm $\|\cdot\|_{H^t(\Gamma)}$ coincides with the norm given by (2), i.e.,

$$\|v\|^2_{H^t(\Gamma)} \;:=\; \sup_{\substack{\varphi\in H^{-t}(\Gamma)\\ \varphi\neq 0}} \frac{|v(\varphi)|}{\|\varphi\|_{-t}} \;=\; |\hat{v}_0|^2 + \sum_{\nu\in\mathbb{Z}\setminus\{0\}} |\hat{v}_\nu|^2\,|2\pi\nu|^{2t} \qquad \text{for } t<0\,.$$

For $t \in \mathbb{R}$, $H^t(\Gamma)$ forms a Hilbert space with the scalar product

$$\langle v,w\rangle_t \;=\; \hat{v}_0\,\overline{\hat{w}_0} + \sum_{\nu\in\mathbb{Z}\setminus\{0\}} \hat{v}_\nu\,\overline{\hat{w}_\nu}\,|2\pi\nu|^{2t} \qquad \text{for } v,w\in H^t(\Gamma)\,.$$

This scalar product is also well-defined for $v \in H^{t+s}(\Gamma)$ and $w \in H^{t-s}(\Gamma)$ $(t,s\in\mathbb{R})$.

The following lemma, which relies on interpolation theory, is discussed in [7] and proved in [6].

Lemma 1 *Let* $t > -\frac{1}{2}$. *For any arbitrary finite index set* I *and for all functions* $\{v_q\}_{q\in I} \subseteq H^{\max\{t,0\}}(\Gamma)$ *with pairwise disjoint supports, i.e.,*

$$\operatorname{supp}(v_q)\cap\operatorname{supp}(v_p) \;=\; \emptyset \qquad \text{for all } q,p\in I \text{ with } q\neq p\,,$$

there holds

$$\Big\|\sum_{q\in I} v_q\Big\|^2_{H^t(\Gamma)} \;\leq\; C_t^{supp} \sum_{q\in I} \|v_q\|^2_{H^t(\Gamma)}$$

with a constant C_t^{supp} *independent of the index set* I *and the functions* $\{v_q\}_{q\in I}$ *with pairwise disjoint supports.*

In Lemma 1, we need for $t\in(-\frac{1}{2},0)$ some additional technical assumptions (see [7]). We do not mention them here, since Lemma 1 is only needed for the proof of Theorem 3 and the technical assumptions are satisfied in the situation of Theorem 3.

3 ERROR ESTIMATORS FOR THE GALERKIN METHOD

In this chapter, we consider a bijective and continuous operator

$$A : H^\alpha(\Gamma) \to H^{-\alpha}(\Gamma), \qquad \alpha\in\mathbb{R}\,,$$

of order 2α. For a given right-hand side $g \in H^{-\alpha}(\Gamma)$, we search for the solution $u \in H^\alpha(\Gamma)$ of the abstract equation

$$Au \;=\; g \qquad \text{on } \Gamma\,. \tag{3}$$

Equation (3) might be an integral equation of the first kind, i.e.,

$$A\,u(x) \;=\; \int_\Gamma k(x,\xi)\,u(\xi)\,d\xi \qquad \text{for } x\in\Gamma\,,$$

or an integral equation of the second kind, i.e.,

$$Au(x) \;=\; u(x) - \int_\Gamma k(x,\xi)\, u(\xi)\, d\xi \qquad \text{for } x \in \Gamma\,.$$

For example, the following integral operators arise, if one transforms the 2-dimensional Laplace problem into a boundary integral equation (see, e.g., [8]):

$$Au(x) \;=\; -\frac{1}{2\pi}\int_\Gamma \log|x-\xi|\, u(\xi)\, d\xi\,, \qquad \alpha = -\frac{1}{2}\,, \tag{4}$$

$$Au(x) \;=\; u(x) \mp \frac{1}{\pi}\int_\Gamma \frac{\langle x-\xi, n(\xi)\rangle}{|x-\xi|^2}\, u(\xi)\, d\xi\,, \qquad \alpha = 0 \tag{5}$$

$$Au(x) \;=\; -\frac{1}{2\pi}\int_\Gamma \frac{\partial^2 \log|x-\xi|}{\partial n_x \partial n_\xi}\, u(\xi)\, d\xi\,, \qquad \alpha = \frac{1}{2}\,,$$

with the outer normal direction $n:\Gamma \to \mathbb{R}^3$ and with the normal derivative $\frac{\partial}{\partial n}$. The integral operator in (4) corresponds to the single-layer potential, and the integral operator in (5) corresponds to the double-layer potential of the Laplace problem.
Further examples of boundary integral operators are, e.g., given in [1,5,10] and [8,§8.5].

The abstract Problem (3) is equivalent to the following variational problem:
Find $u \in H^\alpha(\Gamma)$ such that

$$B(u,v) \;:=\; \langle Au, v\rangle_0 \;=\; \langle g, v\rangle_0 \qquad \text{for all } v \in H^\alpha(\Gamma)\,, \tag{6}$$

where the bilinear form $B: H^\alpha(\Gamma)\times H^\alpha(\Gamma) \to \mathbf{K}$ is associated to the operator A. We assume that A is $H^\alpha(\Gamma)$-elliptic, i.e., there exists a constant $C_{ell} > 0$ such that

$$|B(v,v)| \;=\; |\langle Av, v\rangle_0| \;\ge\; C_{ell}\,\|v\|_\alpha^2 \qquad \text{for all } v \in H^\alpha(\Gamma)\,. \tag{7}$$

For the following, it would be sufficient if A satisfies a Gårding inequality, which is considerably weaker then (7) (see [6,7]). However, we restrict here to condition (7) in order to simplify the notation.

For the discretization of problem (6), we introduce a mesh Δ on $\Gamma = \gamma([0,1])$ by

$$\Delta \;=\; \left\{ \gamma([q_{\nu-1}, q_\nu]) \,:\, \nu \in \{1,\ldots,n\} \right\}.$$

and extend the set of mesh points

$$\Xi \;=\; \Xi_\Delta \;=\; \{q_1,\ldots,q_n\} \quad \text{with} \quad 0 < q_1 < q_2 < \ldots < q_n \le 1\,,\; q_0 = q_n - 1\,,$$

one-periodic to a sequence of mesh points $\{q_\nu\}_{\nu\in\mathbf{Z}}$. S_Δ^m, the space of one-periodic smoothest spline functions of degree $m \in \mathbb{N}_0$ corresponding to the mesh Δ, is used as trial space for the Galerkin method. We choose m with $m + \frac{1}{2} > \alpha$, which implies that $S_\Delta^m \subseteq H^\alpha(\Gamma)$ (see, e.g., [13]). By [9, Kap. 6, §3.3], S_Δ^m has a basis $\{B_q\}_{q\in\Xi} = \{B_{q_\nu}\}_{\nu=1}^n$ of B-splines with the following properties: The basis function B_{q_ν} has a small support

$$\operatorname{supp}(B_{q_\nu}) \;=\; [q_\nu, q_{\nu+m+1}]\,, \tag{8}$$

and the basis $\{B_q\}_{q\in\Xi}$ is a partition of unity, i.e.,

$$\sum_{q\in\Xi_\Delta} B_q(x) \;=\; 1 \qquad \text{for almost all } x \in \Gamma\,. \tag{9}$$

We only want to consider meshes Δ satisfying a technical stability property. Therefor, we introduce a family of meshes $\mathcal{M}$ containing all these meshes Δ. We assume that there exists a constant $C_\alpha^{stab} > 0$, which may depend only on $\mathcal{M}$, such that the following stability property holds

$$\forall v \in H^{\max\{\alpha,0\}}(\Gamma)\ \forall \Delta \in \mathcal{M}\ \exists v_\Delta \in S_\Delta^m \quad \sum_{q \in \Xi_\Delta} \|B_q(v - v_\Delta)\|_\alpha^2 \leq C_\alpha^{stab}\ \|v\|_\alpha^2. \tag{10}$$

In (10), it is important that the constant $C_\alpha^{stab} > 0$ is independent of $\Delta \in \mathcal{M}$. This technical assumption is discussed in [6,7,15]. For example, the family of K-meshes

$$\mathcal{M} = \mathcal{M}_K = \{\Delta :\ \Delta \text{ is a K-mesh on } \Gamma\}\,.$$

satisfies (10) for $\alpha = \frac{1}{2}$ and piecewise linear and continuous trial functions (i.e., $m = 1$). We call Δ a K-mesh, $K \geq 1$, if the sizes of neighbouring elements are in a reasonable ratio, i.e.,

$$K^{-1} \leq \frac{|\gamma([q_{\nu-1}, q_\nu])|}{|\gamma([q_\nu, q_{\nu+1}])|} \leq K \qquad \text{for all } \nu \in \{1, \ldots, n\}\,, \tag{11}$$

where $|\gamma([q_{\nu-1}, q_\nu])|$ denotes the length of the element $\gamma([q_{\nu-1}, q_\nu]) \subseteq \Gamma$. The restriction on K-meshes is not very restrictive. K-meshes allow adaptive refinement, since they may contain small elements as well as large elements.

In the following, let $u \in H^\alpha(\Gamma)$ be the solution of problem (3) respectively (6). Then the discrete Galerkin formulation of problem (6) reads as: Find $u_\Delta \in S_\Delta^m$ such that

$$\begin{aligned} B(u_\Delta, v_\Delta) &= \langle g, v_\Delta \rangle_0 \\ &\overset{(3,6)}{=} B(u, v_\Delta) \qquad \text{for all } v_\Delta \in S_\Delta^m\,. \end{aligned} \tag{12}$$

Our aim is to estimate the unknown Galerkin error

$$e := u_\Delta - u \in H^\alpha(\Gamma)$$

by computable local quantities, and, thereby, to obtain information on the error distribution over the mesh. Since the residual

$$r := Ae = Au_\Delta - g \in H^{-\alpha}(\Gamma)\,,$$

is computable after the calculation of $u_\Delta \in S_\Delta^m$, local quantities of r are suitable for controlling the mesh refinement.
Babuška-Rheinboldt [2] introduced and investigated the following quantities:

Definition 2 *For the B-splines $\{B_q\}_{q \in \Xi_\Delta}$ and for $q \in \Xi_\Delta$, we define*

$$\delta_q = \delta_{q,\Delta} := \sup_{\substack{v \in H^{\max\{\alpha,0\}}(\Gamma) \\ B_q v \neq 0}} \frac{|B(e, B_q v)|}{\|B_q v\|_\alpha} \overset{(6)}{=} \sup_{\substack{v \in H^{\max\{\alpha,0\}}(\Gamma) \\ B_q v \neq 0}} \frac{|\langle r, B_q v \rangle_0|}{\|B_q v\|_\alpha}\,. \tag{13}$$

We call $\{\delta_q\}_{q \in \Xi_\Delta}$ the Babuška-Rheinboldt error estimators (BR error estimators).

δ_q is a local quantity of the residual r, since δ_q depends only on the local behaviour of r on the small support of the basis function B_q.
Babuška-Rheinboldt [2] showed that these $\{\delta_q\}_{q\in\Xi_\Delta}$ are error estimators for finite element methods. Now we show that they are also local error estimators for boundary element methods:

Theorem 3 *Assume that the operator $A : H^\alpha(\Gamma) \to H^{-\alpha}(\Gamma)$ and the family of meshes $\mathcal{M}$ satisfy the properties mentioned above with the constants C_{ell} and C_α^{stab} given by (7) and (10). Then, we obtain for any solution $u \in H^{\max\{\alpha,0\}}(\Gamma)$ and for all meshes $\Delta \in \mathcal{M}$ the following estimate for the Galerkin error $e \in H^{\max\{\alpha,0\}}(\Gamma)$:*

$$C_1 \sum_{q\in\Xi} \delta_q^2 \;\leq\; \|e\|_\alpha^2 \;\leq\; C_2 \sum_{q\in\Xi} \delta_q^2 \,, \tag{14}$$

where the constant $C_2 := C_\alpha^{stab} C_{ell}^{-2}$ is independent of u and $\Delta \in \mathcal{M}$. For $\alpha > -\frac{1}{2}$, the lower estimate in (14) holds with the constant $C_1^{-1} := (m+2)\|A\|^2_{H^{-\alpha}\leftarrow H^\alpha} C_\alpha^{supp}$ independent of u and Δ, where C_α^{supp} is the constant from Lemma 1.
For $\alpha \leq -\frac{1}{2}$, we get (14) with the quantity $C_1^{-1} = C_{1,\Delta}^{-1} := \|A\|^2_{H^{-\alpha}\leftarrow H^\alpha}\, n$, $n := dim(S_\Delta^m)$, depending on Δ.

Remark 4 As mentioned above the results of Theorem 3 are also valid if A satisfies a weaker Gårding inequality. Moreover, we obtain the same results for the 3-dimensional case, when Γ is the surface of a domain $\Omega \subseteq \mathbb{R}^3$ (see [6,7]).

Proof of Theorem 3.
(i) First, we show the upper estimate in (14). Due to the Galerkin problem (12), any $v_\Delta \in S_\Delta^m$ satisfies

$$\begin{aligned} |B(e,e)| &\overset{(12)}{=} |B(e,e-v_\Delta)| \overset{(9)}{=} \Big|\sum_{q\in\Xi_\Delta} B(e, B_q(e-v_\Delta))\Big| \\ &\leq \sum_{q\in\Xi_\Delta} \delta_q\, \|B_q(e-v_\Delta)\|_\alpha \qquad \text{due to the definition of } \delta_q \\ &\leq \sqrt{\sum_{q\in\Xi_\Delta} \|B_q(e-v_\Delta)\|_\alpha^2}\, \sqrt{\sum_{q\in\Xi_\Delta} \delta_q^2}\,. \end{aligned} \tag{15}$$

Since the family of meshes $\mathcal{M}$ satisfies the stability property (10), there exists $v_\Delta \in S_\Delta^m$ with

$$\sum_{q\in\Xi_\Delta} \|B_q(e-v_\Delta)\|_\alpha^2 \;\leq\; C_\alpha^{stab}\, \|e\|_\alpha^2.$$

This inserted in (15) yields together with the ellipticity property (7) the estimate

$$C_{ell}\, \|e\|_\alpha^2 \overset{(7)}{\leq} |B(e,e)| \;\leq\; \sqrt{C_\alpha^{stab}}\, \|e\|_\alpha \sqrt{\sum_{q\in\Xi_\Delta} \delta_q^2}\,,$$

which yields the upper estimate in (14).
(ii) Consider the partition $I_1, I_2, \ldots, I_{m+2}$ of the set of mesh points $\Xi = \{q_1, \ldots, q_n\}$ defined by $I_\nu := \{q_\nu, q_{\nu+(m+2)}, q_{\nu+2(m+2)}, q_{\nu+3(m+2)}, \ldots\} \cap \Xi$. Because of the support condition (8), there holds for any $\nu \in \{1, \ldots, m+2\}$ that

$$\mathrm{supp}(B_p) \cap \mathrm{supp}(B_q) \;=\; \emptyset \qquad \text{for all } p, q \in I_\nu \text{ with } p \neq q\,, \tag{16}$$

i.e., the supports of the basis functions $\{B_q\}_{q\in I_\nu}$ are pairwise disjoint. Let Γ_ν be the union of these pairwise disjoint supports, i.e.,

$$\Gamma_\nu \;=\; \bigcup_{q\in I_\nu} \operatorname{supp}(B_q)\,.$$

Using (16), one can show that

$$\sum_{q\in I_\nu} \delta_q^2 \;:=\; \sum_{q\in I_\nu} \sup_{\substack{v\in H^{\max\{\alpha,0\}}(\Gamma)\\ B_q v\neq 0}} \frac{|\langle r, B_q v\rangle_0|^2}{\|B_q v\|_\alpha^2} \;\stackrel{!}{\leq}\; \sup_{\substack{w\in H^{\max\{\alpha,0\}}(\Gamma)\\ \operatorname{supp}(w)\cap\Gamma_\nu\neq\emptyset}} \frac{\left|\sum_{q\in I_\nu}\langle r, B_q w\rangle_0\right|^2}{\sum_{q\in I_\nu}\|B_q w\|_\alpha^2}\,. \qquad (17)$$

The exact proof of (17) is given in [6,7]. By means of (17) we get

$$\begin{aligned}
\sum_{q\in I_\nu} \delta_q^2 \;&\stackrel{(17)}{\leq}\; \sup_{\substack{w\in H^{\max\{\alpha,0\}}(\Gamma)\\ \operatorname{supp}(w)\cap\Gamma_\nu\neq\emptyset}} \frac{\left|\langle Ae, \sum_{q\in I_\nu} B_q w\rangle_0\right|^2}{\sum_{q\in I_\nu}\|B_q w\|_\alpha^2} \\
&\leq\; \|A\|_{H^{-\alpha}\leftarrow H^\alpha}^2\, \|e\|_\alpha^2 \sup_{\substack{w\in H^{\max\{\alpha,0\}}(\Gamma)\\ \operatorname{supp}(w)\cap\Gamma_\nu\neq\emptyset}} \frac{\|\sum_{q\in I_\nu} B_q w\|_\alpha^2}{\sum_{q\in I_\nu}\|B_q w\|_\alpha^2}\,. \qquad (18)
\end{aligned}$$

For $\alpha > -\frac{1}{2}$, the supremum in (18) is bounded by C_α^{supp} (see Lemma 1), since the supports of the functions $\{B_q w\}_{q\in I_\nu}$ are pairwise disjoint. Thus, we obtain

$$\sum_{q\in\Xi_\Delta} \delta_q^2 \;=\; \sum_{\nu=1}^{m+2}\sum_{q\in I_\nu} \delta_q^2 \;\stackrel{(18)}{\leq}\; (m+2)\, \|A\|_{H^{-\alpha}\leftarrow H^\alpha}^2\, \|e\|_\alpha^2\, C_\alpha^{supp}\,,$$

which shows the lower estimate in (14) for $\alpha > -\frac{1}{2}$.
Now let $\alpha \leq -\frac{1}{2}$. For $w \in H^{\max\{\alpha,0\}}(\Gamma) = L^2(\Gamma)$ we get

$$\begin{aligned}
\Big\|\sum_{q\in I_\nu} B_q w\Big\|_\alpha^2 \;&=\; \sum_{q,p\in I_\nu} \langle B_q w, B_p w\rangle_\alpha \;\leq\; \sum_{q,p\in I_\nu} \|B_q w\|_\alpha \|B_p w\|_\alpha \\
&=\; \Big(\sum_{q\in I_\nu} \|B_q w\|_\alpha\Big)^2 \;\leq\; |I_\nu| \sum_{q\in I_\nu} \|B_q w\|_\alpha^2\,. \qquad (19)
\end{aligned}$$

Hence, the supremum in (18) is bounded by $|I_\nu|$, and we obtain using (18) and (19)

$$\sum_{q\in\Xi_\Delta} \delta_q^2 \;=\; \sum_{\nu=1}^{m+2}\sum_{q\in I_\nu} \delta_q^2 \;\leq\; \sum_{\nu=1}^{m+2} \|A\|_{H^{-\alpha}\leftarrow H^\alpha}^2\, \|e\|_\alpha^2\, |I_\nu| \;=\; \|A\|_{H^{-\alpha}\leftarrow H^\alpha}^2\, \|e\|_\alpha^2\, n\,,$$

which proves the lower estimate in (14) for $\alpha \leq -\frac{1}{2}$. □

Theorem 3 shows that the BR error estimators $\{\delta_q\}_{q\in\Xi_\Delta}$ estimate the Galerkin error e. In general, the $\{\delta_q\}_{q\in\Xi_\Delta}$ are not computable, since they are defined by a supremum, see (13). Thus, we need computable bounds for the BR error estimators, which we can use for the controlling of an adaptive mesh refinement process.
Due to the definition of δ_q, it is straightforward to find computable lower bounds.

Theorem 5 *Let Δ be an arbitrary mesh on Γ, and let $q \in \Xi_\Delta$. Then, any $v \in H^{\max\{\alpha,0\}}(\Gamma)$ with $B_q v \neq 0$ satisfies*

$$\delta_q \;\geq\; \frac{|\int_\Gamma r\, \overline{B_q v}\, dx|}{\|B_q v\|_\alpha}\,.$$

In particular, the choice $v = 1$ and $v = B_q$ leads to

$$\delta_q \geq \frac{|\int_\Gamma r\, \overline{B_q}\, dx|}{\|B_q\|_\alpha} \quad \text{and} \quad \delta_q \;\geq\; \frac{|\int_\Gamma r\, \overline{B_q^2}\, dx|}{\|B_q^2\|_\alpha}\,.$$

The determination of a computable upper bound is more involved. For that, we need the local Sobolev-Slobodeckij norm $\|\cdot\|_{H^s(I)}$, for $s \geq 0$ and $I \subseteq \mathbb{R}$ with $|I| < 1$, given by

$$\|v\|^2_{H^s(I)} \;=\; \sum_{\nu \in \{0,\cdots,k\} \cup \{s\}} |v|^2_{H^\nu(I)}\,,$$

where $s \geq 0$ is decomposed as $s = k + \theta$, with $k = [s] \in \mathbb{N}_0$, $\theta \in [0,1)$, and with the semi-norms

$$|v|^2_{H^\nu(I)} \;:=\; \|v^{(\nu)}\|^2_{L^2(I)} \;=\; \int_I |v^{(\nu)}(x)|^2\, dx\,, \qquad \text{if } \nu \in \mathbb{N}_0 \quad \text{and}$$

$$|v|^2_{H^s(I)} \;:=\; \int_I \int_I \frac{|v^{(k)}(x) - v^{(k)}(y)|^2}{|x-y|^{1+2\theta}}\, dx\, dy\,, \qquad \text{if } s \notin \mathbb{N}_0\,.$$

By [7], we get the following upper bounds for δ_q:

Theorem 6 *Let $\{B_q\}_{q\in\Xi} = \{B_{q_\nu}\}_{\nu=1}^n$ be the B-splines basis of S^m_Δ.*
(a) *Let $\alpha < 0$, $s := -\alpha$, and let Δ be an arbitrary mesh on Γ. For $q \in \Xi_\Delta$, we assume that $h_q := |\mathrm{supp}(B_q)|$ is small enough, more precisely that $2(2+[s])h_q \leq 1$. Then, there holds for the BR error estimators $\{\delta_q\}_{q\in\Xi_\Delta}$ that*

$$\delta_q^2 \leq C \sum_{\nu \in \{0,\cdots,[s]\} \cup \{s\}} h_q^{2(\nu - s)}\, |r|^2_{H^\nu(\mathrm{supp}(B_q))} \;=:\; C\, \|r\|^2_{H^s_w(\mathrm{supp}(B_q))} \tag{20}$$

with a constant C independent of the solution $u \in H^\alpha(\Gamma)$, of the mesh Δ and of $q \in \Xi_\Delta$. In particular, we have for $s \in (0,1)$

$$\delta_q^2 \;\leq\; C\left(h_q^{-2s}\, \|r\|^2_{L^2(\mathrm{supp}(B_q))} \;+\; |r|^2_{H^s(\mathrm{supp}(B_q))}\right).$$

(b) *Let $\alpha \geq 0$. Let the neighbourhood $\mathrm{supp}(B_{q_\nu})^*$ of $\mathrm{supp}(B_{q_\nu}) \overset{(8)}{=} [q_\nu, q_{\nu+1+m}]$ given by*

$$\mathrm{supp}(B_{q_\nu})^* \;:=\; [q_{\nu-M}, q_{\nu+1+m+M}] \quad \text{with} \quad M := \begin{cases} \frac{m}{2} & \text{for even } m \\ \frac{m+1}{2} & \text{for odd } m \end{cases}.$$

Then, there exists a constant $C = C(K)$, for $K > 1$, such that the estimate

$$\delta_q^2 \;\leq\; C\, h_q^{2\alpha}\, \|r\|^2_{L^2(\mathrm{supp}(B_q)^*)}$$

holds for any solution $u \in H^\alpha(\Gamma)$, for all K-meshes Δ (see (11)), and all $q \in \Xi_\Delta$, provided that $r \in L^2(\Gamma)$.

We use in (20) the notation $\|\cdot\|_{H^s_w(\text{supp}(B_q))}$, since $\|\cdot\|_{H^s_w(\text{supp}(B_q))}$ is a "weighted" local norm with differently weighted semi norms in its definition.
Theorem 6 shows that

$$\tilde{\delta_q} \;:=\; \begin{cases} \|r\|_{H^{|\alpha|}_w(\text{supp}(B_q))} & \text{for } \alpha < 0 \\ h_q^\alpha \|r\|_{L^2(\text{supp}(B_q)^*)} & \text{for } \alpha \geq 0 \end{cases}$$

is an upper bound of the BR error estimator δ_q – except a constant independent of u, $\triangle$ and q. After the calculation of the Galerkin solution $u_\triangle \in S^m_\triangle$ and the residual $r = Au_\triangle - g \in H^{-\alpha}(\Gamma)$, the local quantity $\tilde{\delta_q}$ of the residual is approximately computable using a quadrature method. These computable upper bounds $\tilde{\delta_q}$ and the computable lower bounds $|\int_\Gamma r\overline{B_q v}dx| \;/\; \|B_q v\|_\alpha$ (see Theorem 5) of the BR error estimators can be used for the controlling of mesh refinement processes.

4 ERROR ESTIMATORS FOR THE COLLOCATION METHOD

The collocation method is a further important method for the discretization of boundary integral equations. In practical implementations this method is often preferred, since the generation of the linear system is much simpler compared to the Galerkin method. However, the theory is more sophisticated and there are still many open questions concerning the convergence of the collocation method (see, e.g., [5]). The best convergence results for nonuniform meshes go back to Arnold-Wendland [1]. In [1] the 2-dimensional nodal collocation method with odd degree splines is characterized as a modified Galerkin method. Using the Arnold-Wendland technique, we transfer the BR error estimators from the Galerkin method onto the collocation.

Before we may apply the collocation method onto the abstract equation (3), we have to insure that there are continuous functions on both sides of equation (3), i.e., that the image of the operator A is a subset of $C(\Gamma)$. Therefore, we assume in the following that

$$A : H^{k+\alpha}(\Gamma) \to H^{k-\alpha}(\Gamma) \quad \text{is a bijective and continuous operator} \tag{21}$$

of order $2\alpha \in \mathbb{R}$, where $k \geq 1$ is an integer with $k - \alpha > \frac{1}{2}$. Moreover, we assume that A satisfies the ellipticity condition

$$|\langle Av, v\rangle_k| \;\geq\; C_{ell}\, \|v\|^2_{k+\alpha} \qquad \text{for all } v \in H^{k+\alpha}(\Gamma) \tag{22}$$

with a constant C_{ell} independent of v. Analogously to section 3, it would be sufficient for the following if A satisfies a weaker Gårding inequality (see [6]). Again, we restrict here to condition (22) in order to simplify the notation.

In the following, let $u \in H^{k+\alpha}(\Gamma)$ be the solution of the equation

$$Au \;=\; g \qquad \text{on } \Gamma \tag{23}$$

for a given right-hand side $g \in H^{k-\alpha}(\Gamma)$. The collocation method for (23) reads as: Find $u_\triangle \in S^m_\triangle$ such that the equations

$$Au_\triangle(q_\nu) \;=\; g(q_\nu) \qquad \text{for all } \nu \in \{1, \ldots, n\} \tag{24}$$

are satisfied for the mesh points $\Xi_\Delta = \{q_1, \ldots, q_n\}$. If we fix the odd spline degree $m = 2k-1$, which implies that $S_\Delta^m \subseteq H^{k+\alpha}(\Gamma)$, then this collocation method is equivalent to a modified Galerkin method.

Theorem 7 *[1] A spline function $u_\Delta \in S_\Delta^m$ satisfies the collocation equations (24) if and only if it satisfies the Galerkin equations*

$$\langle A_\Delta u_\Delta, v_\Delta \rangle_k \;=\; \langle A_\Delta u, v_\Delta \rangle_k \qquad \text{for all } v_\Delta \in S_\Delta^m ,$$

where the Operator A_Δ is given by

$$A_\Delta := (Id - J + J_\Delta)\, A : H^{k+\alpha}(\Gamma) \to H^{k-\alpha}(\Gamma)$$

with the integral $Jv := \int_0^1 v(t)\,dt$ and its numerical counterpart, the trapezoidal rule $J_\Delta v := \sum_{\nu=1}^n d_\nu\, v(q_\nu)$, $d_\nu := \frac{1}{2}\,(q_{\nu+1} - q_{\nu-1})$.

Our aim is to estimate the unknown collocation error

$$e \;:=\; u_\Delta - u \;\in\; H^{k+\alpha}(\Gamma)$$

by local quantities of the computable residual

$$r \;:=\; Ae \;=\; Au_\Delta - g \;\in\; H^{k-\alpha}(\Gamma)\,.$$

Using Theorem 7, one can transfer the BR error estimators from the Galerkin method onto the collocation by the following definition:

Definition 8 *For the B-splines $\{B_q\}_{q\in\Xi_\Delta} \subseteq S_\Delta^m$ and for $q \in \Xi_\Delta$, we define*

$$\delta_q \;=\; \delta_{q,\Delta} \;:=\; \sup_{\substack{v\in H^{\max\{k+\alpha,0\}}(\Gamma)\\ B_q v\neq 0}} \frac{|\langle A_\Delta e, B_q v\rangle_k|}{\|B_q v\|_{k+\alpha}}\,. \tag{25}$$

By [6], we get the following a-posteriori estimate:

Theorem 9 *Let $A : H^{k+\alpha}(\Gamma) \to H^{k-\alpha}(\Gamma)$ be as above (with the constant C_{ell} given by (22)). Let $\mathcal{M}$ be a family of meshes satisfying the stability property (10) with the constant $C_{k+\alpha}^{stab}$, where the norm $\|\cdot\|_\alpha$ in (10) is replaced by the norm $\|\cdot\|_{k+\alpha}$. Then, we obtain for any solution $u \in H^{\max\{k+\alpha,0\}}(\Gamma)$ and for all meshes $\Delta \in \mathcal{M}$ the following estimate for the collocation error $e \in H^{\max\{k+\alpha,0\}}(\Gamma)$ and for the $\{\delta_q\}_{q\in\Xi_\Delta}$ given by (25):*

$$C_1 \sum_{q\in\Xi} \delta_q^2 \;\leq\; \|e\|_{k+\alpha}^2 \;\leq\; C_2 \sum_{q\in\Xi} \delta_q^2\,, \tag{26}$$

where the upper estimate holds if the mesh size $|\Delta|$ is small enough with the constant $C_2 := 4C_{k+\alpha}^{stab} C_{ell}^{-2}$ independent of u and $\Delta \in \mathcal{M}$. For $k+\alpha > -\frac{1}{2}$, the lower estimate in (26) holds with the constant $C_1^{-1} := (m+2) C_{k+\alpha}^{supp} C_{op}^2$ independent of u and Δ, where $C_{k+\alpha}^{supp}$ is the constant from Lemma 1 and $C_{op} := \sup_{\Delta \text{ mesh}} \|A_\Delta\|_{k-\alpha\leftarrow k+\alpha} < \infty$ (see [1]). For $k+\alpha \leq -\frac{1}{2}$, we get (26) with the quantity $C_1^{-1} = C_{1,\Delta}^{-1} := C_{op}^2 n$, $n := dim(S_\Delta^m)$, depending on Δ.

Remark 10 Theorem 9 shows that the $\{\delta_q\}_{q\in\Xi_\Delta}$ are reliable error estimators for the collocation method. Moreover, they are efficient if $k+\alpha > -\frac{1}{2}$. This efficiency result is better than in the Galerkin case. For example, if we use piecewise linear and continuous trial functions (i.e., $m=1$, $k=1$) for the collocation discretization of the single layer potential (4) (i.e., $\alpha = -\frac{1}{2}$), then we have efficiency, since $k+\alpha = \frac{1}{2}$. For $\alpha \leq -\frac{3}{2}$, we may increase k, and consequently the spline degree m, as long as the condition $k+\alpha > -\frac{1}{2}$ is satisfied. This increasing is possible as long as the conditions (21) and (22) do not fail to hold.

Analogously to the Galerkin case, we need computable bounds for the $\{\delta_q\}_{q\in\Xi_\Delta}$, which we can use for the controlling of an adaptive mesh refinement process.

Theorem 11 *[6,7] Let Δ be an arbitrary mesh on Γ. Then, there holds for the $\{\delta_q\}_{q\in\Xi_\Delta}$ given by (25) and for $q \in \Xi_\Delta$: Any $v \in H^{\max\{k+\alpha,0\}}(\Gamma)$ with $B_q v \neq 0$ satisfies*

$$\delta_q \;\geq\; \frac{|\int_\Gamma r^{(k)}\,\overline{(B_q v)^{(k)}}\,d\Gamma|}{\|B_q v\|_{k+\alpha}}\,.$$

In particular, the choice $v=1$ and $v=B_q$ leads to

$$\delta_q \;\geq\; \frac{|\int_\Gamma r^{(k)}\,\overline{B_q^{(k)}}\,d\Gamma|}{\|B_q\|_{k+\alpha}} \qquad \textit{and} \qquad \delta_q \;\geq\; \frac{|\int_\Gamma r^{(k)}\,\overline{(B_q^2)^{(k)}}\,d\Gamma|}{\|B_q^2\|_{k+\alpha}}\,.$$

Moreover, due to the assumption $k-\alpha > \frac{1}{2}$ we get

$$\delta_q \;\leq\; C\,\|r\|_{H_w^{k-\alpha}(\mathrm{supp}(B_q))} \tag{27}$$

with a constant C independent of u, Δ and $q\in\Xi_\Delta$, and where $\|\cdot\|_{H_w^{k-\alpha}(\mathrm{supp}(B_q))}$ is the weighted local Sobolev-Slobodeckij norm given by (20).

After the calculation of the collocation residual $r = Au_\Delta - g \in H^{k-\alpha}(\Gamma)$, the bounds $|\int_\Gamma r^{(k)}\overline{(B_q v)^{(k)}}\,d\Gamma|/\|B_q v\|_{k+\alpha}$ and $\|r\|_{H_w^{k-\alpha}(\mathrm{supp}(B_q))}$ of δ_q are approximately computable using a quadrature method and they can be used for the controlling of mesh refinement processes.

5 NUMERICAL EXAMPLES

As test problem, we consider Symm's equation, i.e., equation (23) with the integral operator A given by (4), on the circle $\Gamma = \{x \in \mathbb{R}^2 : \|x\|_2 = R\}$ with radius $R<1$. We compute piecewise linear and continuous collocation approximations $u_\Delta \in S_\Delta^1$ (see (24)) for different solutions $u \in H^{k+\alpha}(\Gamma) = H^{\frac{1}{2}}(\Gamma)$.

The a-posteriori error estimate (26) and the estimate (27) are motivations for the following adaptive mesh refinement algorithm:

Adaptive mesh refinement Algorithm (A)

- Compute the collocation solution $u_\Delta \in S^1_\Delta$ associated to a given mesh $\Delta = \{\gamma([q_{\nu-1}, q_\nu]) : \nu \in \{1, \dots, n\}\}$.
- Compute the residual $r \in H^{k-\alpha}(\Gamma)$ and compute the local quantities

$$\tilde{\delta}_\nu \; := \; \|r\|_{H_w^{k-\alpha}([q_{\nu-1},q_\nu])} \; = \; \|r\|_{H_w^{3/2}([q_{\nu-1},q_\nu])}$$

  using a quadrature method.
- Stop the refinement process, if $\sum_{\nu=1}^n \tilde{\delta}_\nu^2$ is small enough.
  Otherwise refine an element $[q_{\nu-1}, q_\nu]$ by halving if

$$\tilde{\delta}_\nu \; \geq \; \theta \max_{\mu=1}^{n} \tilde{\delta}_\mu \, ,$$

  where $\theta \in [0,1]$ is a fixed suitable refinement parameter. Start this process again with the refined mesh.

For details of implementation, we refer to [6].

The eigenfunctions of the single layer potential

$$u(x) \; := \; \cos(2\pi k x) \, , \qquad x \in [0,1] \approx \Gamma$$

for frequencies $k \in \mathbb{Z} \setminus \{0\}$, with the right-hand sides $g := Au = \frac{R}{2|k|} u$ are regular test examples. The adaptive mesh refinement Algorithm (A) applied to these regular test examples generates, as one should expect, in any refinement step a uniform mesh and the efficiency index

$$eff \; := \; \|e\|_{H^{1/2}(\Gamma)} \, / \sqrt{\sum_{\nu=1}^{n} \tilde{\delta}_\nu^2}$$

is reasonably bounded.

Singular test examples are needed to demonstrate the advantages of adaptive mesh refinement. Therefore, we consider the function

$$u_\varepsilon(x) \; := \; |x - x_\varepsilon|^{-2} \qquad \text{for } x \in \Gamma = \{x \in \mathbb{R}^2 \, : \, \|x\|_2 = R\}$$

with $x_\varepsilon := R(1+\varepsilon, 0)$ for $\varepsilon > 0$. u_ε is a smooth function on Γ, since $x_\varepsilon \notin \Gamma$. But u_ε has a strong peak in $x_0 := R(1,0) \in \Gamma$, see Figure 1, which shows the function values for $\varepsilon = 0.1$ and $R = 0.5$.

This nearly singular behaviour of u_ε increases for decreasing ε. We compute the right-hand side $g_\varepsilon := Au_\varepsilon$ approximately by means of a finite Fourier sum, since the Fourier representation of g_ε, which is explicitly given in [6], converges very fast.

The following numerical results are computed for $\varepsilon = 0.1$ and $R = 0.5$. Table 1 presents the norms $\|e\|_{L^2(\Gamma)}$ and $\|e\|_{H^{1/2}(\Gamma)}$ of the collocation error and the efficiency index eff for uniform meshes with $n = 16, 32, 64, 128, \cdots$ nodes. If we start the adaptive mesh refinement Algorithm (A) with a coarse uniform mesh with 16 nodes, then the Algorithm (with refinement parameter $\theta = \frac{1}{2}$) generates a sequence of adaptive meshes with $n = 16, 20, 24, 28, 32, \cdots$ nodes. The corresponding numerical results are given in Table 2. These adaptive meshes are very fine in the neighbourhood of the "singular" point x_0 and relatively coarse in the remaining part of Γ.

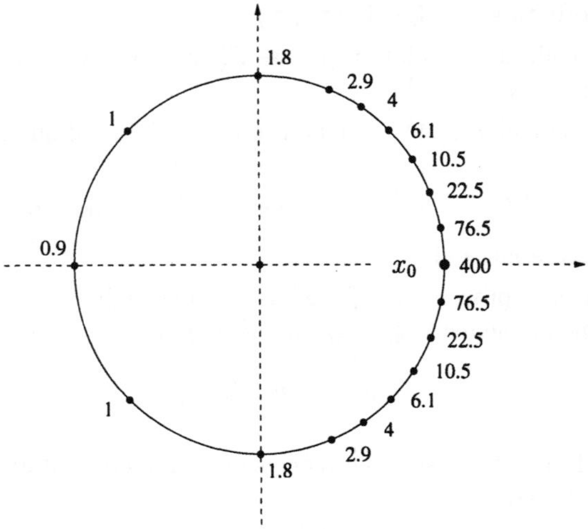

Figure 1: u_ε for $\varepsilon = 0.1$ and $R = 0.5$

A comparison between Table 1 and Table 2 shows the advantages of the adaptive mesh refinement. An adaptive mesh with 28 (respectively 90) nodes has an error magnitude, which would be reached only by a uniform mesh with 128 (respectively 512) nodes.

The efficiency index eff in Table 1 and Table 2 is reasonably bounded. This confirms our theoretical results of section 4.

Table 1:
Numerical results for uniform meshes

n	$\|e\|_{L^2(\Gamma)}$	$\|e\|_{H^{1/2}(\Gamma)}$	eff
16	25.42	95.98	1.04
32	5.86	28.44	1.49
64	3.39	33.33	0.82
128	1.43	18.37	0.86
256	0.39	5.30	1.65
512	0.10	1.26	1.69

Table 2: Numerical results for adapted meshes generated by Algorithm (A)

n	$\|e\|_{L^2(\Gamma)}$	$\|e\|_{H^{1/2}(\Gamma)}$	eff
16	25.42	95.98	1.04
20	5.86	28.50	1.49
24	3.40	33.36	0.82
28	1.55	18.55	0.85
32	0.867	7.75	1.50
44	0.529	6.26	3.03
56	0.430	8.20	4.13
62	0.225	2.72	1.57
80	0.142	1.30	1.23
90	0.103	1.06	1.36
116	0.053	0.652	0.89
134	0.039	0.592	0.98
168	0.023	0.296	0.77
199	0.018	0.259	0.83
241	0.014	0.218	0.89
292	0.010	0.195	0.91
308	0.009	0.158	0.89
377	0.007	0.129	0.89

Further numerical computations for Symm's equation on the curve $\Gamma = \partial\Omega$, where $\Omega = \{\, t(\cos\omega, \sin\omega) \;:\; t \in [0, \frac{1}{2}),\; \omega \in (0, \frac{3}{2}\pi) \,\}$ is three quarter of the disc of radius $\frac{1}{2}$ centered in the origin, are presented in [6].

REFERENCES

[1] Arnold, D.N., Wendland, W.L. (1983): On the asymptotic convergence of collocation methods. Math. Comput. 41, 349-381.

[2] Babuška, I., Rheinboldt, W.C. (1978): Error estimates for adaptive finite element computations. SIAM J. Num. Anal. Vol.15, 736-754.

[3] Carstensen C. (1996): Efficiency of a-posteriori BEM-error estimates for first kind integral equations on uniform meshes. Math. Comp., to appear.

[4] Carstensen C., Stephan, E.P. (1995): A-posteriori error estimates for boundary element methods. Math. Comp. 64, 483-500.

[5] Chen, G., Zhou, J. (1992): Boundary element methods. Academic Press London.

[6] Faermann, B. (1993): Lokale a-posteriori Fehlerschätzer bei der Diskretisierung von Randintegralgleichungen. PhD-thesis, University of Kiel.

[7] Faermann, B. (1996): Local a-posteriori error estimators for the Galerkin discretization of boundary integral equations (submitted)

[8] Hackbusch, W. (1995): Integral equations. Birkhäuser Basel.

[9] Hämmerlin, G., Hoffmann, K.H. (1989): Numerische Mathematik. Springer Berlin .

[10] Hsiao, G.C., Kopp, P., Wendland, W.L. (1984): A Galerkin collocation method for some integral equations of the first kind. Computing 25, 89-130.

[11] Rank, E. (1984): A-posteriori error estimates and adaptive refinement for some boundary integral element method. In: Proc. Int. Conf. on Accuracy Estimates and Adaptive Refinements in FE Computations ARFEC, Lisbon 1984, 55-64.

[12] Rank, E. (1986): Adaptivity and accuracy estimation for FEM and BEM. In: Accuracy Estimates and Adaptive Refinements in FE Computations (I. Babuška, O.C. Zienkiewicz, J. Gago and E.R. Oliveira, eds.), Wiley, New York.

[13] Schmidt, G. (1986): Splines und die näherungsweise Lösung von Pseudodifferentialgleichungen auf geschlossenen Kurven. Report R-MATH-09/86, Akademie der Wissenschaften der DDR, Berlin.

[14] Wendland, W.L., Yu, D. (1988): Adaptive BEM for strongly elliptic integral equations. Numer. Math. 53, 539-558.

[15] Yu, D. (1987): A-posteriori error estimates and adaptive approaches for some boundary element methods. In: Boundary Elements 9 (C.A. Brebbia, W.L. Wendland, G. Kuhn, eds.), Springer Berlin, 241-256.

[16] Yu, D. (1988): Self-adaptive boundary element methods. ZAMM 68, T435-T437.

ON SPLINE APPROXIMATION METHODS FOR SINGULAR INTEGRAL EQUATIONS OVER PLANE DOMAINS

Tilo Finck
Fakultät für Mathematik
Technische Universität Chemnitz-Zwickau
09107 Chemnitz, Germany

SUMMARY

A Banach algebra approach to the stability problem of spline approximation methods for a class of singular integral equations over plane domains is presented. Considering a concrete Galerkin method efficient necessary and sufficient criteria for the method to be stable can be derived.

INTRODUCTION

We provide stability considerations of spline approximation methods for a class of two-dimensional singular integral equations. On the one hand there is a close connection between the singular integral equations under consideration and certain boundary value problems (the problem of complex conjugation), as A. DZHURAEV pointed out in his book [3]. On the other hand the work should be regarded as a further step in establishing spline approximation methods for pseudodifferential operators on multi-dimensional manifolds.

The Cauchy-type singular operator is defined as follows. Let Ω be a bounded convex domain with a smooth boundary $\partial\Omega$. The restriction of a $L^2(\mathbb{R}^2)$-function onto Ω generates an operator p_Ω

$$p_\Omega : L^2(\mathbb{R}^2) \to L^2(\Omega).$$

The extension by zero of an $L^2(\Omega)$-function defines an operator L_Ω

$$L_\Omega : L^2(\Omega) \to L^2(\mathbb{R}^2).$$

The operator

$$(Sf)(x) = -\frac{1}{\pi}\int\limits_{\mathbb{R}^2} \frac{f(y)dy}{(x-y)^2}, \tag{1}$$

where $x = x_1+ix_2$ and $y = y_1+iy_2$ are points of $\mathbb{C}$ (identified with the plane $\mathbb{R}^2$) and the integral is understood as the Cauchy principal value, is bounded on the Hilbert space $L^2(\mathbb{R}^2)$ with $\|S\| = 1$ (see [3]). In the Hilbert space $L^2(\Omega)$ we consider the equation

$$Au = au + bS_\Omega u = f\,, \tag{2}$$

where the integral operator on the domain is defined by $S_\Omega = p_\Omega S L_\Omega$ and the coefficient functions a and b are supposed to be continuous on the closure $\bar{\Omega}$ of the domain. For the approximative solution of (2) one constructs a sequence of approximating equations

$$A_n u_n = f_n\,. \tag{3}$$

The stability investigation forms the basis of any numerical analysis. Remember that a sequence (A_n) of operators is called *stable* if there exists an n_0 such that all operators A_n with $n \geq n_0$ are invertible and if the norms of their inverses are uniformly bounded. We aim to present an algebraic approach to the stability problem which is based on local principles. For one–dimensional singular integral operators the power of a suited Banach algebra language (interpreting Fredholm, resp. stability, problems as invertibility problems) in both Fredholm and numerical analysis has been shown e.g. in [5], [9] and [13].

Let us mention that for the invertibility of the concrete operator in (2) it is necessary and sufficient that $|a(x)| > |b(x)|$ for all $x \in \bar{\Omega}$ (see [4]). This is just the condition for the strongly ellipticity of this operator. Actually we have investigated more complicate operators (from the algebra generated by the considered operators, by the adjoint operator S_Ω^* and the operator of complex conjugation). In this general case we don't assume the related operator A from (2) to be strongly elliptic, in contrast with the paper [11]. Using the local principle of Allen (see [1]) in the investigation of the invertibility problem related to the stability of (3) the crucial point is the identification of the local representatives at the boundary points of the domain Ω. In this sense the boundary $\partial\Omega$ forms a geometrical singularity for the approximation problem (3). Whereas in [10] there was presented a doubly indexed approximation method, combining a spline approximation and a finite section method for avoiding (numerical) singularities, we intend to modify the spline spaces in the plane used in [8]. Thereby the basic properties of the constructed spline projection are maintained, the local principle can be applied and the local representatives can be found.

It turns out that the operators studied in [8] represent (slightly modified) the local objects. At boundary points there arise Toeplitz opertors on half planes which are generated by functions (depending on the operator and the spline projection) with a discontinuity on the Torus $\mathbb{T}^2$. We could solve the invertibility problem for these operators up to now for a simple spline function only. In this case it turnes out that for the stability of the approximating operator sequence (3) the invertibility of the operator is sufficient and necessary (see Theorem 1). This is one of the reasons, why we consider this special case first. For lack of space we omit the proofs here, they will be published in [7]. For the general scheme see also [8] and [9].

Let F be the characteristic function $\chi_{[0,1)^2}$ of the square $[0,1) \times [0,1)$. For a fixed ϵ, $0 < \epsilon < 1$ and a constant $c > 0$ define the distance function $d : \mathbb{N} \to \mathbb{R}_+$ by $d(n) = c \cdot n^{-\epsilon}$ and put

$$\Omega_d := \{x \in \Omega \; : \; dist(x, \partial\Omega) > d(n)\}. \tag{4}$$

These domains depend via d on n. For $l = (l_1, l_2) \in \mathbb{Z}^2$ and a natural number $n \geq 1$ the function

$$\varphi_{nl}^F(x) = F(nx - l) \tag{5}$$

is the characteristic function of the square $[\frac{l_1}{n}, \frac{l_1+1}{n}) \times [\frac{l_2}{n}, \frac{l_2+1}{n})$. For fixed n there is only a finite number of functions (5) supported in Ω_d and we denote the linear span of these functions in $L^2(\Omega_d)$ by S_n^{F,Ω_d},i.e.

$$S_n^{F,\Omega_d} = \text{span}\{p_\Omega \varphi_{nl}^F \; : \; \text{supp}\, \varphi_{nl}^F \subset \Omega_d\}. \tag{6}$$

This space will be referred to as spline space. The related index set defines a projection $P_n^{\Omega_d}$ in the space $l^2_{\mathbb{Z}^2}$ of all sequences $(x_l)_{l\in\mathbb{Z}^2}$ of complex numbers (provided with the norm $\|(x_l)\| = (\sum_{l\in\mathbb{Z}^2} |x_l|^2)^{\frac{1}{2}}$) in a natural way,

$$\left(P_n^{\Omega_d}(x_l)\right)_k = \begin{cases} x_k & \text{if supp}\, \varphi_{nk}^F \subset \Omega_d \\ 0 & \text{else} \end{cases}. \tag{7}$$

As in [8] we let S_n^F refer to as the smallest closed subspace of $L^2(\mathbb{R}^2)$ containing all functions φ_{nk}^F with $l \in \mathbb{Z}^2$. From ([8], de Boor's estimate and Prop.5) we know that the operators

$$E_n^F \; : \; l^2_{\mathbb{Z}^2} \to S_n^F, \;\; (x_l)_{l\in\mathbb{Z}^2} \mapsto \sum_{l\in\mathbb{Z}^2} x_l \varphi_{nl}(x)\,, \tag{8}$$

assigning to each sequence a spline function on the plane, and the operator

$$P_n^F \; : \; L^2(\mathbb{R}^2) \to l^2_{\mathbb{Z}^2}, \;\; f \mapsto \Big(\int_{\mathbb{R}^2} f(\frac{\cdot + l}{n}) F(x) dx \Big)_{l\in\mathbb{Z}^2} \tag{9}$$

are bounded with $\|E_n^F\| \leq Cn^{-1}$ and $\|P_n^F\| \leq Cn$. Moreover a straightforward computation shows that $P_n^F E_n^F$ is the identy operator I on $l^2_{\mathbb{Z}^2}$. We define the operator

$$L_n^{F,F,\Omega_d} := p_\Omega E_n^F P_n^{\Omega_d} P_n^F L_\Omega\,. \tag{10}$$

The basic properties of this operator are summarized in the following lemma.

LEMMA 1 *The operator L_n^{F,F,Ω_d} is a bounded projection from $L^2(\Omega)$ onto the spline space S_n^{F,Ω_d} with the following properties:*

(P1) (strong convergence) $\lim_{n\to\infty} \|L_n^{F,F,\Omega_d} g - g\|_{L^2(\Omega)} = 0\,,$ *for all* $g \in L^2(\Omega)$.

(P2) *(commutator property)* $\lim_{n\to\infty} \|L_n^{F,F,\Omega_d} bI - bL_n^{F,F,\Omega_d}\|_{L^2(\Omega)} = 0$, *for all* $b \in C(\bar{\Omega})$.

(P3) *(extended commutator property) Let* f_n *be a sequence of continuous functions* $f_n \in \dot{C}(\mathbb{R}^2)$ *with* $f_n \equiv 1$ *on* $\Omega_{\frac{d}{2}}$ *and* $f_n \equiv 0$ *outside of* Ω. *Then*

$$\lim_{n\to\infty} \|f_n I E_n^F P_n^{\Omega_d} P_n^F - E_n^F P_n^{\Omega_d} P_n^F f_n I\| = 0 .$$

We specify the approximating equations (3) with

$$L_n^{F,F,\Omega_d}(aI + bS_\Omega)u_n = L_n^{F,F,\Omega_d} f \, , u_n \in S_n^{F,\Omega_d} . \tag{11}$$

If $(\cdot,\cdot)$ denotes the scalar product in $L^2(\Omega)$ equation (11) can be rewritten in the form

$$(au_n + bS_\Omega u_n, \varphi_{nl}^F) = (f, \varphi_{nl}^F) \text{ for all } \varphi_{nl}^F \in S_n^{F,\Omega_d} . \tag{12}$$

(12) describes the Galerkin method with S_n^{F,Ω_d} both as trial and test space. Solving the stability problem (following the general scheme from the next sections) we get

THEOREM 1 *The Galerkin method (12) is stable if and only if the operator* $A = aI + bS_\Omega$ *is invertible in* $L^2(\Omega)$.

REMARK 1
(1) Since the sequence (A_n) *with* $A_n = L_n^{F,F,\Omega_d}(aI + bS_\Omega)|_{S_n^{F,\Omega_d}}$ *converges to* A *in the strong operator topology the applicability of the approximation method (3) to the equation (2) is equivalent to the stability of the sequence* (A_n).

(2) For more complicate operators there appear further conditions on the coefficients.

GENERAL SPLINE SPACES AND SPLINE PROJECTIONS

We start with a prespline F, a premeasure G and a distance function $d : \mathbb{N} \to \mathbb{R}_+$ satisfying the following conditions

(H1) The function F is a compactly supported, Riemann on $\mathbb{R}^2$ with $\sum_{l\in\mathbb{Z}^2} F(x-l) = 0$ for all $x \in \mathbb{R}^2$.

(H2) The functional G on $L^2(\mathbb{R}^2)$ (resp. $R^2(\mathbb{R}^2)$) has a compact support and variation 1.

(H3) $|\hat{F}(x)||\hat{G}(x)| \leq Cx^{-2-\epsilon}$ $\quad 0 < \epsilon < 1$ ($\hat{F}$ is the Fourier transform $\hat{F}(z_1, z_2) = \int_{\mathbb{R}^2} e^{-2\pi(x_1 z_1 + x_2 z_2)} F(x_1, x_2) dx_1 dx_2$ of F)

(H4) $\lim_{n\to\infty} d(n) = 0$ and $\lim_{n\to\infty} nd(n) = \infty$.

Analogous to (6) we define the spline space as the linear span of all functions φ_{nl}^F, defined by (5), which are supported in Ω_d. The operator E_n^F is defined as in (8) and P_n^G we get by replacing the functional generated by F with the functional G,i.e.

$$P_n^G : L^2(\mathbb{R}^2) \to l^2_{\mathbb{Z}^2}, \; f \mapsto G\big(f(\frac{\cdot + l}{n})\big)_{l \in \mathbb{Z}^2} .$$

REMARK 2 *With the functionals G on the L^2-space we cover the Galerkin–Bubnov methods only. Using the Riemann space $R^2(\mathbb{R}^2)$ (see [8]) one can examine collocation methods without difficulties, too.*

It turns out that

- the operators $P_n^G E_n^F$ are independent of n and
- there is a trigonometric polynomial $\lambda^{F,G}$ such that $P_n^G E_n^F = T^o(\lambda^{F,G})$.

Here, $T^o(a)$ abbreviates the discrete convolution operator with the generating function a. For a bounded measurable function a on the torus $\mathbb{T}^2 = \mathbb{R}^2/\mathbb{Z}^2$ define its $k = (k_1, k_2)$th Fourier coefficients a_k by

$$a_{(k_1,k_2)} = \int\limits_{[0,1]} \int\limits_{[0,1]} a(e^{2\pi i s}, e^{2\pi i t}) e^{-2\pi i k_1 s} e^{-2\pi i k_2 t} ds\, dt .$$

The discrete convolution operator $T^o(a)$ acts on finitely supported sequences $(x_l) \in l^2_{\mathbb{Z}^2}$ as

$$T^o(a)(x_l) = (y_l) \text{ with } y_l = \sum_{k \in \mathbb{Z}^2} a_{l-k} x_k ,$$

and this operator extends by continuity to a bounded operator on all of $l^2_{\mathbb{Z}^2}$ which is denoted by $T^o(a)$ again (cf. [2], section 8).

Let the following additional hypothesis be fulfilled:

(H5) $\lambda^{F,G}$ is invertible on the torus $\mathbb{T}^2$.

Under this assumption $T^o(\lambda^{F,G})$ is invertible with $(T^o(\lambda^{F,G}))^{-1} = T^o((\lambda^{F,G})^{-1})$. Then the operator

$$L_n^{F,G,\Omega_d} := p_\Omega E_n^F P_n^{\Omega_d} (T^o(\lambda^{F,G}))^{-1} P_n^G L_\Omega \tag{13}$$

is a bounded projection from $L^2(\Omega)$ onto the spline space S_n^{F,Ω_d} and the properties (P1–3) read as follows

(P1') (strong convergence) $\lim\limits_{n \to \infty} \|L_n^{F,G,\Omega_d} g - g\|_{L^2(\Omega)} = 0$, for all $g \in L^2(\Omega)$.

(P2') (commutator property) $\lim\limits_{n \to \infty} \|L_n^{F,G,\Omega_d} bI - bL_n^{F,G,\Omega_d}\|_{L^2(\Omega)} = 0$, for all $b \in C(\bar{\Omega})$.

(P3') (extended commutator property) Let f_n be a sequence of continuous functions $f_n \in C(\dot{\mathbb{R}}^2)$ with $f_n \equiv 1$ on $\Omega_{\frac{d}{2}}$ and $f_n \equiv 0$ outside of Ω. Then

$$\lim_{n\to\infty} \| f_n I E_n^F P_n^{\Omega_d} (T^o(\lambda^{F,G}))^{-1} P_n^G - E_n^F P_n^{\Omega_d} (T^o(\lambda^{F,G}))^{-1} P_n^G f_n I \| = 0 .$$

Let us mention that the properties (P1') and (P2') can be easely derived from the representation

$$L_n^{F,G,\Omega_d} := p_\Omega E_n^F P_n^{\Omega_d} (E_n^F)^{-1} L_n^{F,G} L_\Omega ,$$

and the properties of the spline projection $L_n^{F,G} := E_n^F (T^o(\lambda^{F,G}))^{-1} P_n^G$ onto S_n^F (see [8], Theorem 3).

If we introduce the function $\sigma^{F,G}$ on the torus by

$$\sigma^{F,G}(y_1, y_2) = \sum_{(m_1,m_2)\in\mathbb{Z}^2} \frac{y_1 + m_1 - i(y_2 + m_2)}{y_1 + m_1 + i(y_2 + m_2)} \hat{F}(y+m)\hat{G}(y+m) , \tag{14}$$

where we wrote $y + m$ in place of $(y_1 + m_1, y_2 + m_2)$, the approximating sequence for the operator (2) is

$$L_n^{F,G,\Omega_d}(aI + bS_\Omega)|_{S_n^{F,\Omega}} = \left(L_n^{F,G,\Omega_d} aI + L_n^{F,G,\Omega_d} b p_\Omega E_n^F P_n^{\Omega_d} T^o(\sigma^{F,G}/\lambda^{F,G})(E_n^F)^{-1} \right)|_{S_n^{F,\Omega_d}} + G_n \tag{15}$$

with $\|G_n\| \to 0$ as n tends to infinity.

A BANACH ALGEBRA APPROACH TO STABILITY ANALYSIS

In [9] there was provided a well–described general scheme for studying the stability problem. We explain the main steps in our case.

Algebraization and Essentialization

Let $\mathcal{F}^{\Omega_d}$ be the set of all sequences (A_n) of bounded operators $A_n : S_n^{F,\Omega_d} \to S_n^{F,\Omega_d}$ for which the strong limits

$$W^1(A_n) := s - \lim A_n L_n^{F,G,\Omega_d} \text{ and } (W^1(A_n))^* := s - \lim (A_n L_n^{F,G,\Omega_d})^*$$

exist. Equipped with pointwise operations $(A_n)+(B_n) := (A_n+B_n)$, $(A_n)(B_n) := (A_n B_n)$, $(A_n)^* := (A_n^*)$ and the norm

$$\|(A_n)\| = \sup_n \|A_n L_n^{F,G,\Omega_d}\|$$

(that the supremum is finite follows from the Banach–Steinhaus theorem) $\mathcal{F}^{\Omega_d}$ becomes a Banach algebra, even a C*–algebra. Further we write $\mathcal{J}^{\Omega_d}$ for the set of all sequences $(L_n^{F,G,\Omega_d} K|_{S_n^{F,\Omega_d}} + G_n)$ with K running through the set $\mathcal{K}(L^2(\Omega))$ of compact operators in $L^2(\Omega)$ and $\|G_n\| \to 0$ as n tends to infinity. The set $\mathcal{J}^{\Omega_d}$ forms a closed two–sided ideal in $\mathcal{F}^{\Omega_d}$ and plays a similar role as the ideal of compact operators in the Fredholm theory.

PROPOSITION 1 *Let $(A_n) \in \mathcal{F}^{\Omega_d}$ and $W^1(A_n) =$s–lim $A_n L_n^{F,G,\Omega_d}$. Then the sequence (A_n) is stable if and only if the operator $W^1(A_n)$ is invertible and the coset $(A_n) + \mathcal{J}^{\Omega_d}$ is invertible in the quotient algebra $\mathcal{F}^{\Omega_d}/\mathcal{J}^{\Omega_d}$.*

Let $\phi^{\mathcal{J}^{\Omega_d}}$ denote the canonical homomorphism from $\mathcal{F}^{\Omega_d}$ onto $\mathcal{F}^{\Omega_d}/\mathcal{J}^{\Omega_d}$.

Localization

The algebra $\mathcal{F}^{\Omega_d}/\mathcal{J}^{\Omega_d}$ is too large to tackle the invertibility problem succesfully. We therefore construct a subalgebra $\mathcal{A}^{\Omega_d}$ of $\mathcal{F}^{\Omega_d}$ which contains all "interesting" sequences. Namely, let $\mathcal{A}^{\Omega_d}$ stand for the smallest closed *–subalgebra of $\mathcal{F}^{\Omega_d}$ which contains

(A1) the sequence $(L_n^{F,G,\Omega_d} S_\Omega|_{S_n^{F,\Omega_d}})$,

(A2) all sequences $(L_n^{F,F,\Omega_d} bI|_{S_n^{F,\Omega_d}})$ for $b \in C(\bar{\Omega})$ and

(A3) the ideal $\mathcal{J}^{\Omega_d}$.

The operator sequence (15) belongs to $\mathcal{A}^{\Omega_d}$ and we have $W^1(A_n) = A = aI + bS_\Omega$. The related quotient algebra $\mathcal{A}^{\Omega_d}/\mathcal{J}^{\Omega_d}$ possesses a rich center. Really, the discrete commutator property (P2) and the commutator relation $bIS_\Omega - S_\Omega bI \in \mathcal{K}(L^2(\Omega))$ imply that the coset $\phi^{\mathcal{J}^{\Omega_d}}(L_n^{F,F,\Omega_d} bI)$, $b \in C(\bar{\Omega})$ commutes with all elements from $\mathcal{A}^{\Omega_d}/\mathcal{J}^{\Omega_d}$. Thus, we can apply Allen's local principle (see [1]) to this quotient algebra with its central subalgebra $\mathcal{B} = \{\phi^{\mathcal{J}^{\Omega_d}}(L_n^{F,F,\Omega_d} bI) : b \in C(\bar{\Omega})\}$. The maximal ideal space of the commutative algebra $\mathcal{B}$ is homeomorph to $\bar{\Omega}$, the maximal ideal associated with $x \in \bar{\Omega}$ is

$$\{\phi^{\mathcal{J}^{\Omega_d}}(L_n^{F,F,\Omega_d} bI) : b \in C(\bar{\Omega}),\ b(x) = 0\}. \tag{16}$$

Let $\mathcal{J}_x^{\Omega_d}$ denote the smallest closed two–sided ideal of $\mathcal{A}^{\Omega_d}/\mathcal{J}^{\Omega_d}$ containing the ideal (16) and $\phi_x^{\mathcal{J}^{\Omega_d}}$ stand for the canonical homomorphism from $\mathcal{A}^{\Omega_d}$ onto the quotient algebra $\mathcal{A}_x^{\Omega_d} = (\mathcal{A}^{\Omega_d}/\mathcal{J}^{\Omega_d})/\mathcal{J}_x^{\Omega_d}$. Then the local principle states that

an element $\phi^{\mathcal{J}^{\Omega_d}}(A_n)$ is invertible in $\mathcal{A}^{\Omega_d}/\mathcal{J}^{\Omega_d}$ if and only if the cosets $\phi_x^{\mathcal{J}^{\Omega_d}}(A_n)$ are invertible for all $x \in \bar{\Omega}$. (17)

REMARK 3 *We actually look for the invertibility in the larger algebra $\mathcal{F}^{\Omega_d}/\mathcal{J}^{\Omega_d}$ (see Proposition 1). But $\mathcal{A}^{\Omega_d}/\mathcal{J}^{\Omega_d}$ is a C^*–subalgebra of this C^*–algebra and consequently inverse closed. If we consider the operator equation (2) in L^p–spaces some further effort is necessary to prove the inverse closedness*

Identification of the local algebras

We are required to study the invertibility in the so–called local algebras $\mathcal{A}_x^{\Omega_d}$. This will be done by means of locally equivalent representations (see [9]) which we are going to construct now. For $k = (k_1, k_2) \in \mathbb{Z}^2$ let V^k denote the discrete shift $T^o(t^k)$ in $l^2_{\mathbb{Z}^2}$, i.e.

$$V_k(x_l) = x_{l-k}, \quad l - k = (l_1 - k_1, l_2 - k_2).$$

For $r = (r_1, r_2) \in \mathbb{R}^2$ let $\{r\}$ refer to the point $(\{r_1\}, \{r_2\}) \in \mathbb{Z}^2$, where $\{r_1\}$ stands for the smallest integer which is greater or equal r_1.

PROPOSITION 2 *Let $x \in \Omega$ be an inner point of the convex domain Ω. Then there exist for all $(A_n) \in \mathcal{A}^{\Omega_d}$ the strong limits*

$$s\text{-}\lim V_{-\{nx\}}(E_n^F)^{-1} L_\Omega A_n p_\Omega E_n^F P_n^{\Omega_d} V_{\{nx\}} =: W_x^{\Omega_d}(A_n)\,. \tag{18}$$

*The mapping $W_x^{\Omega_d} : \mathcal{A}^{\Omega_d} \to \mathcal{L}(l^2_{\mathbb{Z}^2})$ is a continuous *-homomorphism which acts on the generating sequences (A1-3) as follows*

$$\begin{aligned} W_x^{\Omega_d}(L_n^{F,G,\Omega_d} S_\Omega|_{S_n^{F,\Omega_d}}) &= T^o(\sigma^{F,G}/\lambda^{F,G}) \\ W_x^{\Omega_d}(L_n^{F,F,\Omega_d} bI|_{S_n^{F,\Omega_d}}) &= b(x)I \text{ for all } b \in C(\bar{\Omega}) \text{ and} \\ W_x^{\Omega_d}(J_n) &= 0 \text{ for all } (J_n) \in \mathcal{J}^{\Omega_d}\,. \end{aligned}$$

$W_x^{\Omega_d}(A_n)$ depends on the coset $\phi_x^{\mathcal{J}^{\Omega_d}}(A_n)$ only and $\phi_x^{\mathcal{J}^{\Omega_d}}(A_n)$ is invertible in $\mathcal{A}_x^{\Omega_d}$ if and only if $W_x^{\Omega_d}(A_n)$ is invertible.

The construction of the homomorphism for the boundary points needs more technical details. Let x_d denote the nearest point to x of the set $\Omega_{d+\frac{d}{n}}$ (which is defined analogous to Ω_d). Let $\gamma_1 x_1 + \gamma_2 x_2 = c$ be the equation of the tangent to the boundary $\partial\Omega$ at x such that the domain Ω lies in the half plane $\gamma_1 x_1 + \gamma_2 x_2 \geq c$. We define the (rotated) half plane grid related to the localization point x

$$\mathbb{Z}_x^2 := \{(k_1, k_2) \in \mathbb{Z}^2 \,:\, \gamma_1 k_1 + \gamma_2 k_2 \geq 0\}\,.$$

Let P_x stand for the associated projection in $l^2_{\mathbb{Z}^2}$ and $im\, P_x$ for its image.

PROPOSITION 3 *Let $x \in \partial\Omega$ be a boundary point of Ω. Then there exist for all $(A_n) \in \mathcal{A}^{\Omega_d}$ the strong limits*

$$s\text{-}\lim V_{-\{nx_d\}}(E_n^F)^{-1} L_\Omega A_n p_\Omega E_n^F P_n^{\Omega_d} V_{\{nx_d\}} =: W_x^{\partial\Omega_d}(A_n)\,. \tag{19}$$

*The mapping $W_x^{\partial\Omega_d} : \mathcal{A}^{\Omega_d} \to \mathcal{L}(im\, P_x)$ is a continuous *-homomorphism which acts on the generating sequences (A1-3) as follows*

$$\begin{aligned} W_x^{\partial\Omega_d}(L_n^{F,G,\Omega_d} S_\Omega|_{S_n^{F,\Omega_d}}) &= P_x T^o(\sigma^{F,G}/\lambda^{F,G})|_{im\, P_x} \\ W_x^{\partial\Omega_d}(L_n^{F,F,\Omega_d} bI|_{S_n^{F,\Omega_d}}) &= b(x)P_x \text{ for all } b \in C(\bar{\Omega}) \text{ and} \\ W_x^{\Omega_d}(J_n) &= 0 \text{ for all } (J_n) \in \mathcal{J}^{\Omega_d}\,. \end{aligned}$$

$W_x^{\partial\Omega_d}(A_n)$ depends on the coset $\phi_x^{\mathcal{J}^{\Omega_d}}(A_n)$ only and $\phi_x^{\mathcal{J}^{\Omega_d}}(A_n)$ is invertible in $\mathcal{A}_x^{\Omega_d}$ if and only if $W_x^{\partial\Omega_d}(A_n)$ is invertible.

Summarizing the results of the propositions 1, 2 and 3 and the local principle assertion (17) we get the stability result.

THEOREM 2 *Let Ω be a bounded convex domain with a smooth boundary $\partial\Omega$. Let (H1)-(H5) to be fulfilled. Then for the stability of an operator sequence $(A_n) \in \mathcal{A}^{\Omega_d}$ it is necessary and sufficient that the following conditions are satisfied:*

1. *the operator $W^1(A_n)$ is invertible in $L^2(\Omega)$,*

2. *the operators $W_x^{\Omega_d}(A_n)$ are invertible in $l^2_{\mathbb{Z}^2}$ for all $x \in \Omega$ and*

3. *the operators $W_x^{\partial\Omega_d}(A_n)$ are invertible in $im\, P_x$ for all $x \in \partial\Omega$.*

REMARK 4
(1) For inner points $x \in \Omega$ the operators $W_x^{\Omega_d}(A_n)$, defined in (18), are discrete convolution operators.

(2) For boundary points $x \in \partial\Omega$ the $W_x^{\Omega_d}(A_n)$, defined in (19), are Toeplitz operators on (rotated) half planes. The rotation depends on the boundary point x.

(3) In case $F = G = \chi_{[0,1)^2}$ ($\lambda^{F,G} \equiv 1$!) one can determine the image if the function $\sigma^{F,G}$ and show that the spectrum of the operator $P_x T^o(\sigma^{F,G}/\lambda^{F,G})|_{im\, P_x}$ is equal to the closed unit disc $\bar{\mathbb{D}} = \{\lambda \in \mathbb{C} : |\lambda| \leq 1\}$ and Theorem 1 follows from the general result.

CONCLUDING REMARKS

The algebraic approach leads in a natural way to a unified study of a whole class of approximation methods for algebras of operators. Morover, one can get a deeper insight into the natur of spline projection methods. Let us only mention that the C*–algebra techniques, developed in [15], apply and allow to determine the asymptotic behaviour of the condition numbers, the ϵ–pseudospectra and others of the sequences in $\mathcal{A}^{\Omega_d}$.

Of course the condition 2. of the Theorem 2 is not very satisfactory. Actually we look for efficient sufficient and necessary stability conditions in terms of the coefficient functions of the operator, and the prespline and the premeasure of the method. Only recently we moved on in establishing invertibility criteria of the Toeplitz operators over half planes for a greater class of spline functions.

REFERENCES

[1] ALLEN,G.R.: *Ideals of vector valued functions.* Proc. London Math. Soc. 18 (1968), 3, 193-216.

[2] BÖTTCHER, A., SILBERMANN, B.: Analysis of Toeplitz Operators. Akademie–Verlag, Berlin 1989 and Springer–Verlag, Berlin, Heidelberg, New York 1990.

[3] DZHURAEV, A., D.: The method of singular integral equations. Nauka, Moscow, 1987 (Russian).

[4] BOITMATOV, K., DZHANGIBEKOV, G.: *On a singular integral operator.* Uspechi mat. Nauk T. 43, 3 (1988), 171-172 (Russian).

[5] GOHBERG,I.,KRUPNIK,N.: One–dimensional singular integral equations. Vol. I and II, Birkhäuser-Verlag Basel-Boston Berlin, 1992.

[6] COSTABEL, M., PENZEL, F., SCHNEIDER, R.: *A collocation method for screen problems in* $\mathbb{R}^3$. in Tagungsberichte "Elliptic operators on singular manifolds", B.G. Teubner Leipzig 1991.

[7] FINCK, T.: *Stability of spline approximation methods for singular integral equations over plane domains.* (to appear)

[8] HAGEN,R., ROCH,S., SILBERMANN,B.: *Stability of Spline Approximation Methods for Multidimensional Pseudodifferential Operators.* IEOT 19 (1994),25–64.

[9] HAGEN,R., ROCH,S., SILBERMANN,B.: Spectral Theory of Approximation Methods for Convolution Equations. Birkhäuser Verlag, Basel-Stuttgart, 1995.

[10] HOPPE,W.: *Stabilität von Splineapproximationsverfahren für Singuläre Integraloperatoren auf kompakten, glatten Mannigfaltigkeiten ohne Rand.* (Diss. TU Chemnitz–Zwickau 1994)

[11] PRÖSSDORF,S., SCHNEIDER,R.: *A Spline collocation method for multidimensional strongly elliptic pseudodifferential operators of order zero.* IEOT 14 (1991), 399-435.

[12] PRÖSSDORF,S., SCHNEIDER,R.: *Spline approximation methods for multidimensional periodic pseudodifferential equations.* IEOT 15 (1992), 626-672.

[13] PRÖSSDORF,S., SILBERMANN,B.: Numerical Analysis for Integral and Related Operator Equations. Birkhäuser Verlag, Basel-Stuttgart and Akademie-Verlag, Berlin 1991.

[14] RATHSFELD,A.: *Piecewise polynomial collocation for the double layer potential equation over polyhedral boundaries, in: Boundary value problems and integral equations in non–smooth domains* (eds.: M. Costabel, M. Dauge and N. Nicaise) (Proceeding of a Conference on Boundary Value Problems and Integral Equations in Non–smooth Domains, Luminy, France) (1994)

[15] ROCH,S., SILBERMANN,B.:*C^*–algebra techniques in numerical analysis.* Preprint TU Chemnitz–Zwickau, 1994.

[16] SCHNEIDER, R.: *Stability of a spline collocation method for strongly elliptic multidimensional singular integral equations.* Numer. Math. 58 (1991), 855–873.

HIERARCHICAL BASIS PRECONDITIONERS FOR COUPLED FEM–BEM EQUATIONS

Stefan A. Funken & Ernst P. Stephan
Institut für Angewandte Mathematik
Universität Hannover
Welfengarten 1
30167 Hannover, Germany

SUMMARY

The purpose of this paper is to present a nearly optimal preconditioned iterative method to solve indefinite linear systems of equations arising from h-adaptive procedures for the symmetric coupling of Finite Elements and Boundary Elements. This solver is nearly optimal in the sense, that its convergence rate grows only logarithmically with the number of unknowns.

The algorithm is based on the conjugate residual method with block-diagonal preconditioning, where no Schur complement construction is required. This method uses different hierarchical basis preconditioners for the positive semi-definite FEM block belonging to an interior Neumann problem and the negative definite boundary element block belonging to the single layer potential.

The efficiency of the hierarchical basis solvers is underlined by a numerical experiment showing fast convergence.

INTRODUCTION

This paper concerns the problem of solving efficiently systems of linear equations which arise from the symmetric coupling of finite elements (FEM) and boundary elements (BEM) when dealing with elliptic transmission problems. These systems are solved iteratively with the preconditioned conjugate residual method (PCR) which uses different HB (hierarchical basis) preconditioners for the seperate FEM and BEM subblocks of the stiffness matrix. This solver is nearly optimal since the number of iterations increases only logarithmically with the number of unknowns. Other solvers for FEM–BEM coupling problems are considered in [3, 4, 5, 6, 8]

We consider the following transmission problem. Let Ω be a bounded Lipschitz domain and $\Omega_c = \mathbb{R}^2 \setminus \overline{\Omega}$ be its complement. The partial differential equation considered in Ω

will be given by the operator $B = (b_{ij})_{i,j=1,2}$. Let the coefficients $b_{ij} = b_{ji} \in L^\infty(\Omega)$ be uniformly bounded in Ω and positive definite.

We consider the following problem involving the prescribed jumps across the interface $\Gamma := \partial\Omega$: *Given $f \in L^2(\Omega)$, $u_0 \in H^{1/2}(\Gamma)$ and $t_0 \in H^{-1/2}(\Gamma)$, find $u_1 \in H^1(\Omega)$, $u_2 \in H^1_{\text{loc}}(\Omega_c)$, such that*

$$\begin{aligned} -\operatorname{div}(B \cdot \operatorname{grad} u_1) &= f \quad \text{in } \Omega \\ -\Delta u_2 &= 0 \quad \text{in } \Omega_c \end{aligned} \tag{1}$$

with

$$u_1 = u_2 + u_0, \quad n.(B \cdot \operatorname{grad} u_1) = \frac{\partial u_2}{\partial n} + t_0 \quad \text{on } \Gamma \tag{2}$$

where n is the normal on Γ pointing from Ω into Ω_c and u_2 satisfies with $a \in \mathbb{R}$

$$u_2(x) = a \log|x| + o(1) \quad \text{for } |x| \to \infty. \tag{3}$$

The interface problem (1)–(3) is equivalent to the following weak formulation which is the symmetric coupling method [2].
Find $(u, \phi) \in H^1(\Omega) \times H^{-1/2}(\Gamma)$, such that

$$\mathcal{B}\left(\begin{pmatrix} u \\ \phi \end{pmatrix}, \begin{pmatrix} v \\ \psi \end{pmatrix}\right) = \mathcal{L}\begin{pmatrix} v \\ \psi \end{pmatrix} \quad \text{for all } \begin{pmatrix} v \\ \psi \end{pmatrix} \in H^1(\Omega) \times H^{-1/2}(\Gamma) \tag{4}$$

with

$$\begin{aligned} \mathcal{B}\left(\begin{pmatrix} u \\ \phi \end{pmatrix}, \begin{pmatrix} v \\ \psi \end{pmatrix}\right) &:= 2 \cdot \int_\Omega (\nabla u)^\top B \nabla v \, dx + \langle Wu + (K' - 1)\phi, v\rangle \\ &\quad + \langle \psi, (K-1)u - V\phi\rangle \end{aligned}$$

and

$$\mathcal{L}\begin{pmatrix} v \\ \psi \end{pmatrix} = 2 \cdot \int_\Omega fv \, dx + \langle Wu_0 + (K' + 1)t_0, v\rangle + \langle \psi, (K-1)u_0 - Vt_0\rangle.$$

Here, the brackets $\langle \cdot, \cdot \rangle$ denote the L^2-duality between the trace space $H^{1/2}(\Gamma)$ and its dual $H^{-1/2}(\Gamma)$, V represents the single-layer and K the double-layer potential operator, K' its adjoint and W denotes the hypersingular integral operator, which is the normal derivative of the double-layer potential operator. These integral operators are defined as follows:

$$\begin{aligned} V\varphi(x) &= -\tfrac{1}{\pi} \textstyle\int_\Gamma \varphi(y) \log|x-y| ds_y, & K\varphi(x) &= -\tfrac{1}{\pi} \textstyle\int_\Gamma \varphi(y) \frac{\partial}{\partial n_y} \log|x-y| ds_y, \\ K'\varphi(x) &= -\tfrac{1}{\pi} \textstyle\int_\Gamma \frac{\partial}{\partial n_x} \varphi(y) \log|x-y| ds_y, & W\varphi(x) &= \tfrac{1}{\pi} \frac{\partial}{\partial n_x} \textstyle\int_\Gamma \varphi(y) \frac{\partial}{\partial n_y} \log|x-y| ds_y. \end{aligned}$$

For the definition of the Sobolev spaces $H^1(\Omega)$, $H^{-1/2}(\Gamma)$ etc. see [7].

In the next section we present the FEM–BEM discretization of the coupling problem 4 which leads to a saddle point problem with an indefinite symmetric system matrix $\mathcal{A}_h$ (see (5)). Then we comment on the behaviour of the eigenvalues of the matrix $\mathcal{P}^{\frac{1}{2}} \mathcal{A}_h \mathcal{P}^{\frac{1}{2}}$ where $\mathcal{P}$ is a block-diagonalmatrix which acts seperately on the FEM and BEM parts of $\mathcal{A}_h$. Then we introduce a hierachical basis method for the weakly singular boundary integral operator V. We show that for its Galerkin matrix $\widehat{V}_J$ in this basis of Haar functions and the corresponding diagonal matrix $\operatorname{diag}(\widehat{V}_J)$ the product $[\operatorname{diag}(\widehat{V}_J)]^{-1}\widehat{V}_J$ has bounded condition number independent of the number of levels J. Since the FEM block when preconditioned with Yserentant's hierarchical basis method ([10]) has a condition number which growth quadratic with J we obtain for thus preconditioned conjugate residual method that the number of iterations grow like $O(|\log(h)|^{3/2})$.

Finally, in the last section, numerical results are presented for the case of a Laplacian transmission problem with an L-shaped boundary Γ. The total number of PCR iterations which is needed in order to achieve an accuracy up to a given tolerance grows moderately with the number of degrees of freedom.

THE DISCRETE PROBLEM

In this section we consider the discretization of problem (4). Therefore we define finite dimensional vector spaces $H_{h,0} \subset H_h \subset H^1(\Omega)$, $H_h^{-\frac{1}{2}} \subset H^{-\frac{1}{2}}(\Gamma)$. Let H_h be the space of continuous piecewise linear functions of a regular triangulation of Ω, which is not assumed to be quasi-uniform and $H_{h,0}$ be the subspace of functions of H_h, which are zero on the boundary Γ. Let $H_h^{-\frac{1}{2}}$ be the space of piecewise constant functions on the discretization of the boundary Γ. Let N_i^Ω, $i = 1, \ldots, n_\Omega$, be basis functions of $H_{h,0}$, i.e. span $_{i=1,n_\Omega}\{N_i^\Omega\} = H_{h,0}$, N_i^Γ, $i = 1, \ldots, n_\Gamma$, be basis functions of $H_h \setminus H_{h,0}$, i.e. span $_{i=1,n_\Gamma}\{N_i^\Gamma\} = H_h \setminus H_{h,0}$ and t_i, $i = 1, \ldots, m$ the basis functions of $H_h^{-\frac{1}{2}}$, i.e. span $_{i=1,m}\{t_i\} = H_h^{-\frac{1}{2}}$. Define $n = n_\Omega + n_\Gamma$. Let $\vec{u} = \{u_i\}_{i=1,\ldots,n} \in \mathbb{R}^n$, $\vec{\phi} = \{\phi_i\}_{i=1,\ldots,m} \in \mathbb{R}^m$ be vectors with $u = \sum_{i=1}^{n_\Omega} u_i N_i^\Omega + \sum_{i=1}^{n_\Gamma} u_{n_\Omega+i} N_i^\Gamma \in H_h$ and $\phi = \sum_{i=1}^{m} \phi_i t_i \in H_h^{-\frac{1}{2}}$, respectively. We represent the discretized forms of the operators W, K, V, etc. by corresponding matrices as follows

$$\begin{aligned}
A_h &:= \left\{2\textstyle\int_\Omega (\nabla N_i^\Omega)^\top \mathbf{B} \nabla N_j^\Omega dx\right\}_{i,j=1,\ldots,n_\Omega} & V_h &:= \{\langle V t_i, t_j\rangle\}_{i,j=1,\ldots,m} \\
B_h &:= \left\{2\textstyle\int_\Omega (\nabla N_i^\Gamma)^\top \mathbf{B} \nabla N_j^\Omega dx\right\}_{i=1,\ldots,n_\Gamma}^{j=1,\ldots,n_\Omega} & K_h &:= \left\{\langle t_j, K\, N_i^\Gamma|_\Gamma\rangle\right\}_{i=1,\ldots,n_\Gamma}^{j=1,\ldots,m} \\
C_h &:= \left\{2\textstyle\int_\Omega (\nabla N_i^\Gamma)^\top \mathbf{B} \nabla N_j^\Gamma dx\right\}_{i,j=1,\ldots,n_\Gamma} & W_h &:= \left\{\langle W\, N_i^\Gamma|_\Gamma, N_j^\Gamma|_\Gamma\rangle\right\}_{i,j=1,\ldots,n_\Gamma}.
\end{aligned}$$

Then, we have for all $u, v \in H_h$, $\phi, \psi \in H_h^{-\frac{1}{2}}$

$$\begin{aligned}
2\int_{\Omega_1} (\nabla u)^\top \mathbf{B} \nabla v d\Omega_1 &= \vec{u}^\top \left(\left(\begin{smallmatrix} A & B^\top \\ B & C \end{smallmatrix}\right)\right) \vec{v}, \\
\langle W u|_\Gamma, v|_\Gamma\rangle &= \{u_i\}_{i=n_\Omega+1,\ldots,n_\Omega+n_\Gamma}^\top W \{v_i\}_{i=n_\Omega+1,\ldots,n_\Omega+n_\Gamma}, \\
\langle \phi, (K-1)u\rangle &= \vec{\phi}^\top (K-1)\{u_i\}_{i=n_\Omega+1,\ldots,n_\Omega+n_\Gamma}, \\
\langle \phi, V\psi\rangle &= \vec{\phi} V \vec{\psi}.
\end{aligned}$$

We define a matrix $\mathcal{A}_h \in \mathbb{R}^{(n+m)\times(n+m)}$, which represents the discretization of the bilinearform $\mathcal{B}((\cdot),(\cdot))$ by functions of $H_h \times H_h^{-1/2}$

$$\mathcal{A}_h = \begin{pmatrix} A_h & B_h^\top & 0 \\ B_h & (C_h + W_h) & (K_h^\top - 1) \\ 0 & (K_h - 1) & -V_h \end{pmatrix}. \tag{5}$$

With $b_h^\top := \left(b_h^\Omega, b_h^\Gamma, b_h^\phi\right)^\top$ where

$$\begin{aligned}
b_h^\Omega &:= \left\{2\int_\Omega f\cdot N_i^\Omega dx\right\}_{i=1,\dots,n_\Omega},\\
b_h^\Gamma &:= \left\{2\int_\Omega f\cdot N_i^\Gamma dx + \langle W u_0 + (K'+1)t_0, N_i^\Gamma|_\Gamma\rangle\right\}_{i=1,\dots,n_\Gamma},\\
b_h^\phi &:= \left\{\langle t_i, (K-1)u_0 - V t_0\rangle\right\}_{i=1,\dots,m}.
\end{aligned}$$

The coupling problem (4) can be rewritten as the linear system of equations

$$\mathcal{A}_h \begin{pmatrix} \vec{u} \\ \vec{\phi} \end{pmatrix} = b_h. \tag{6}$$

with unknown vector $\binom{\vec{u}}{\vec{\phi}} \in \mathbb{R}^{n+m}$.

BLOCKDIAGONAL PRECONDITIONER

As we will see in the following sections, when using iterative solvers, the application of efficient preconditioners plays a crucial role. In the following we use preconditioning matrices, which are symmetric, positive definite and spectrally equivalent to matrices of the form

$$\begin{pmatrix} A_h & B_h^\top & 0 \\ B_h & C_h + W_h + (K_h^\top - 1)V_h^{-1}(K_h - 1) & 0 \\ 0 & 0 & V_h \end{pmatrix}. \tag{7}$$

This matrix has block-diagonal form and it suffices to take a preconditioning matrix $P_A \in \mathbb{R}^{n\times n}$ for the upper left 2×2 subblock and a preconditioning matrix $P_V \in \mathbb{R}^{m\times m}$ for the submatrix V_h.

Then, as preconditioner for the preconditioned conjugate residual (PCR) method we use the block-diagonal matrix

$$\mathcal{P} := \begin{pmatrix} \begin{bmatrix} P_A \end{bmatrix} & \begin{matrix} 0 \\ 0 \end{matrix} \\ 0\ 0 & P_V \end{pmatrix}. \tag{8}$$

In this section we analyse the spectral behaviour of the symmetric and indefinite matrix $\mathcal{P}^{\frac{1}{2}}\mathcal{A}_h\mathcal{P}^{\frac{1}{2}}$ in dependence of h when $\mathcal{P}$ is the preconditioning matrix of the form (8). Note that the matrix $\mathcal{A}_h$ in (5) is symmetric and indefinite, since $\left(\begin{smallmatrix} A_h & B_h^\top \\ B_h & C_h + W_h \end{smallmatrix}\right)$ is positive semi–definite and V_h is positive definite.

With $\beta > 0$ we define mass matrices $M_h^\beta := \{\beta\int_\Omega N_i N_j dx\}_{i,j=1,\dots,n}$ and then $\left[\left(\begin{smallmatrix} A_h & B_h^\top \\ B_h & C_h \end{smallmatrix}\right) + M_h^\beta\right]$ is positive definite.

In [3] it is shown that there exist positive constants Δ, θ and Θ which are independent of the meshsize h such that for all $x \in \mathbb{R}^{n_\Omega + n_\Gamma}$

$$x^\top \begin{pmatrix} A_h & B_h^\top \\ B_h & (C_h + W_h) \end{pmatrix} x \le \Delta\, x^\top \left[\begin{pmatrix} A_h & B_h^\top \\ B_h & C_h \end{pmatrix} + M_h^\beta\right] x \tag{9}$$

$$x^\top \begin{pmatrix} 0 & 0 \\ 0 & (K_h^\top - 1_h)V_h^{-1}(K_h - 1_h) \end{pmatrix} x \le \Theta^2 \, x^\top \left[\begin{pmatrix} A_h & B_h^\top \\ B_h & C_h \end{pmatrix} + M_h^\beta \right] x \tag{10}$$

$$x^\top \left[\begin{pmatrix} A_h & B_h^\top \\ B_h & C_h \end{pmatrix} + M_h^\beta \right] x \ge \theta^2 \, x^\top \begin{pmatrix} A_h & B_h^\top \\ B_h & C_h + W_h + (K_h^\top - 1_h)V_h^{-1}(K_h - 1_h) \end{pmatrix} x \,. \tag{11}$$

The existence of these constants follows from the isomorphism between $\mathbb{R}^n$ and H_h and the mapping properties of the integral operators V, K, K', W and the differential operator in (1). With (9)–(11) one obtains bounds for the eigenvalues of the preconditioned matrix $\mathcal{P}^{\frac{1}{2}}\mathcal{A}_h\mathcal{P}^{\frac{1}{2}}$.

Theorem 1 ([3]) *Let $0 < \lambda_1 \le \ldots \le \lambda_n$ denote the eigenvalues of $P_V \cdot V_h$ and let $0 < \eta_1 \le \ldots \le \eta_m$ denote the eigenvalues of $P_A \left[\left(\begin{smallmatrix} A_h & B_h^\top \\ B_h & C_h \end{smallmatrix} \right) + M_h^\beta \right]$. Then the eigenvalues of the preconditioned matrix $\mathcal{P}^{\frac{1}{2}}\mathcal{A}_h\mathcal{P}^{\frac{1}{2}}$ lie in the union of the intervals*

$$\left[-\frac{1}{2}\left(\lambda_n + \sqrt{\lambda_n^2 + 4\Theta^2\eta_m\lambda_n} \right) \,,\, -\lambda_1 \right] \cup$$
$$\left[\frac{1}{2}\left(-\lambda_1 + \sqrt{\lambda_1^2 + 4\theta^2\eta_1\lambda_1} \right), \frac{1}{2}\left(-(\lambda_1 - \Delta) + \sqrt{(\lambda_1 + \Delta)^2 + 4\Theta^2\eta_m\lambda_n} \right) \right] \tag{12}$$

with constants Δ, θ, and Θ satisfying (9)-(11).

$\mathcal{A}_h$ is a sparse matrix, since the submatrices A_h and B_h, arising from the finite element part, are sparse. Because of these properties (sparsity, large number of unknowns, symmetry) we solve (6) iteratively with PCR. It is proved in [9], that if the eigenvalues of $\mathcal{A}_h$ lie in intervalls of the form

$$[-a, -b\alpha] \cup [c\alpha^2, d]$$

with positive constants a, b, c, d and asymptotically decreasing parameter α, then the asymptotic convergence rate of PCR is

$$\lim_{k\to\infty} \left(\frac{\| r_k \|_2}{\| r_0 \|_2} \right)^{\frac{1}{k}} \le 1 - \alpha^{\frac{3}{2}}\sqrt{bc/ad} + O(\alpha^{\frac{5}{2}})$$

where $\| \cdot \|_2$ denotes the Euclidean norm and $r_k := \mathcal{A}_h x_k - b_h$ is the residual of the kth iterate x_k.

Using standard hat-functions for N_i^Ω, $i = 1, \ldots, n_\Omega$ and N_i^Γ, $i = 1, \ldots, n_\Gamma$, which are one at a certain node and zero elsewhere and basis functions for t_i, $i = 1, \ldots, m$ which are one at a certain boundary element and zero on the other elements, we have $\alpha = h$ by preconditioning with the diagonal of $\mathcal{A}_h$ and so $O(h^{-3/2})$ PCR iterations are required for convergence [8]. This shows, that the convergence of the PCR method depends strongly on the behaviour of the eigenvalues of $\mathcal{A}_h$. A more efficient preconditioner $\mathcal{P}$ than the diagonal of $\mathcal{A}_h$ is obtained by applying the hierarchical basis method as described below. In order to compute good approximations with the Galerkin method we may need refined, non-uniform meshes, which may be produced by adaptive processes. That can be necessary, e.g. when singularities appear.

Yserentant [10] has shown that the hierarchical basis method yields an efficient preconditioner for self-adjoint, elliptic boundary value problems when piecewise linear triangular elements are used even on adaptive meshes. Since the upper left 2×2 block in (7) is spectrally equivalent to the Galerkin matrix for the Neumann problem of the Helmholtz operator $-\Delta + I$ (see [3]) this block can be efficiently preconditioned by the hierarchical basis method. To convert the eigenvalue bounds from Theorem 1 into estimates in terms of the meshsize parameter h we will assume that the hierarchical basis method applied to the finite element block for the Neumann problem of the Laplacian satisfies with a constant $C_3 > 0$

$$C_3 \left(\log \frac{1}{h}\right)^{-2} x^\top x \le x^\top P_A \left[\left(\begin{smallmatrix} A_h & B_h^\top \\ B_h & C_h \end{smallmatrix}\right) + M_h^\beta\right] x \le x^\top x \tag{13}$$

which can always be achieved by scaling (see [3]).

Next we give an efficient preconditioner for V_h with respect to the underlying and not necessarily quasi-uniform meshes and for this purpose we construct a hierarchical basis method for the matrix representation of the single layer potential.

Let $\mathcal{M}$ be an m-dimensional space of piecewise constant functions on a subdivision $\mathcal{T}$ of Γ. We consider an initial subdivision $\mathcal{T}_1$ of Γ and a sequence of refinements

$$\mathcal{T} =: \mathcal{T}_J \supset \mathcal{T}_{J-1} \supset \ldots \supset \mathcal{T}_2 \supset \mathcal{T}_1$$

with corresponding finite dimensional subspaces $\mathcal{M}_k$ of piecewise constant functions with dimension m_k satisfying

$$\mathcal{M} =: \mathcal{M}_J \supset \mathcal{M}_{J-1} \supset \ldots \supset \mathcal{M}_2 \supset \mathcal{M}_1 .$$

The elements of $\mathcal{T}_k$, $k = 1, \ldots, J$, will be denoted by Γ_k^i, $i = 1, \ldots, m_k$, and the refinement of $\mathcal{T}_k$, $k = 2, \ldots, J$, is obtained from $\mathcal{T}_{k-1}$ by halving certain elements of $\mathcal{T}_{k-1}$. For $k = 1, \ldots, J$, $i = 1, \ldots, m_k$ we define standard basis functions ϕ_k^i, by

$$\phi_k^i(x) := \begin{cases} 1 & \text{, if } x \in \Gamma_k^i \\ 0 & \text{, else} \end{cases}$$

and hierarchical basis functions Ψ_k^i as follows. $\Psi_1^i \equiv \phi_1^i$ for $\mathcal{T}_1$. Whenever an element $\Gamma_{k-1}^i \in \mathcal{T}_{k-1}$, $k = 2, \ldots, J$ is bisected yielding the elements Γ_k^i, Γ_k^{i+1} we get a new Haar basis function

$$\Psi_k^i(x) := \begin{cases} 1 & \text{, if } x \in \Gamma_k^i \\ -1 & \text{, if } x \in \Gamma_k^{i+1} \\ 0 & \text{, else} \end{cases} .$$

For the Galerkin matrix $\widehat{V}_J := \{\langle V\Psi_J^i, \Psi_J^j\rangle\}_{i,j=1,\ldots,m_J}$ of the single layer potential computed in this hierarchical basis there holds the following result.

Theorem 2 ([3]) *The minimal and maximal eigenvalues of the diagonally preconditioned Galerkin matrix $[diag(\widehat{V}_J)]^{-1}\widehat{V}_J$ are independent of the number of levels J, where $diag(\widehat{V}_J)$ is the diagonal matrix obtained from $\widehat{V}_J$.*

In order to justify that the diagonal matrix $[\mathrm{diag}(\widehat{V}_J)]^{-1}$ is an efficient preconditioner we only need to show that the transformation to the hierarchical basis can be implemented easily at a low computational cost. Next we give a fast algorithm to compute $\widehat{V}_J x$ with arbitrary $x \in \mathbb{R}^{m_J}$.

Let $T_k \in \mathbb{R}^{m_k \times m_k}$, $k = 2, \ldots, J$, be the matrix which maps the basis $\{\Psi_k^i\}_{i=1,m_k}$ into the basis $\{\phi_k^i\}_{i=1,m_k}$. Then with

$$T_k := \left\{ \frac{(\Psi_k^j, \phi_k^i)}{(\phi_k^i, \phi_k^i)} \right\}_{i,j=1,\ldots,m_k} \tag{14}$$

and the Galerkin matrix in the standard basis

$$V_k := \left\{ \langle V\phi_k^i, \phi_k^j \rangle \right\}_{i,j=1,\ldots,m_k} .$$

there holds

$$T_k^\top V_k T_k = \widehat{V}_k \quad \text{for } k = 2, \ldots, J. \tag{15}$$

Due to the following lemma T_k, $k = 2, \ldots, J$, can be calculated recursively by $m_k \times m_k$-matrices $S_k = \{s_k^{ij}\}$ with entries $s_k^{ij} = (\phi_{k-1}^i, \phi_k^j)/(\phi_k^j, \phi_k^j)$ if $i \leq m_{k-1}$ and $s_k^{ij} = (\Psi_k^i, \phi_k^j)/(\phi_k^j, \phi_k^j)$ if $i > m_{k-1}$.

Lemma 1 ([3]) *With S_k, T_k, $k = 2, \ldots, J$, as above there holds $T_2 = S_2$ and*

$$T_k = S_k^\top \circ \begin{pmatrix} T_{k-1} & 0 \\ 0 & I_{m_{k-1}} \end{pmatrix} \quad , k > 2$$

where $I_{m_{k-1}}$ is the $m_{k-1} \times m_{k-1}$ identity matrix.

Remark 1 *From Theorem 2 and* (15) *we have, that $T_k[\mathrm{diag}(\widehat{V}_k)]^{-1} T_k^\top$ is a symmetric and positive definite matrix, which is spectrally equivalent to V_k^{-1}, $k = 1, \ldots, J$. Hence $T_k[\mathrm{diag}(\widehat{V}_k)]^{-1} T_k^\top$ is an efficient preconditioner for V_k.*

Now, let P_A be the corresponding preconditioner for the FEM part of $\mathcal{A}_h$ in the hierarchical FEM basis (see [10]) and $P_V = T_J[\mathrm{diag}(\widehat{V}_J)]^{-1} T_J^\top$ the corresponding hierarchical basis preconditioner for the single layer potential from Remark 1. The PCR method with this HB (hierarchical basis) preconditioner is a fast solver even for adaptive FEM–BEM coupling problems due to the following result.

Theorem 3 *Then the preconditioned coupled scheme*

$$\mathcal{P}\mathcal{A}_h \binom{\vec{u}}{\vec{\phi}} = \mathcal{P} b_h$$

can be solved iteratively with the PCR method and the error e_k of the kth iteration behaves like

$$\lim_{k \to \infty} e_k^{1/k} = 1 - O(|\log(h)|^{-\frac{3}{2}})$$

and the number of iterations grows like $O(|\log(h)|^{\frac{3}{2}})$.

The proof follows by inserting Theorem 2, (13) in (12).

NUMERICAL EXAMPLE

Let Ω be the L-shaped domain with vertices $(0,0)$, $(s,0)$, (s,s), $(-s,s)$, $(-s,-s)$, $(0,-s)$ with $s = 0.25$. Then the single layer potential operator V is positive definite. On a uniform mesh with triangular elements we use piecewise linear functions in $\bar{\Omega}$ and piecewise constant functions on Γ. The initial mesh is shown in Fig. 1 (left).

Into the right hand side of the linear problem (4) we insert the data f, u_0, t_0, which are computed via (1), (2) from the harmonic functions

$$
\begin{aligned}
u_1(x,y) &= r^{\frac{2}{3}} \sin(\frac{2}{3}\alpha) \\
u_2(x,y) &= \frac{1}{2} \log\left((x+\frac{1}{8}) + (y-\frac{1}{8})\right)
\end{aligned}
$$

(in polar and Cartesian coordinates (r,α) and (x,y), respectively).

Note, that the solution of the transmission problem become singular at the origin. By an adaptive process we get a sequence of 10 meshes which are locally refined towards the origin. The final mesh is shown in Fig.1 (right).

In Tab. 1 and Fig. 2 we present the number of iterations for the PCR method with the above described HB-preconditioner and give an comparison with the conjugate residual method without preconditioning. Here we have to solve the linear problems (6) belonging to these adaptively refined meshes.

This numerical example shows, that the HB-preconditioner together with the PCR method is an efficient and nearly optimal solver for coupled FEM–BEM equations even on non-uniform meshes.

Table 1: Number of iterations with and without preconditioning.

no. mesh	# ndof (n+m)	# iterations without prec.	# iterations with prec.
1	37	20	19
2	46	22	20
3	60	23	20
4	95	29	25
5	153	37	31
6	229	40	33
7	341	45	34
8	539	78	40
9	834	96	41
10	1263	113	44

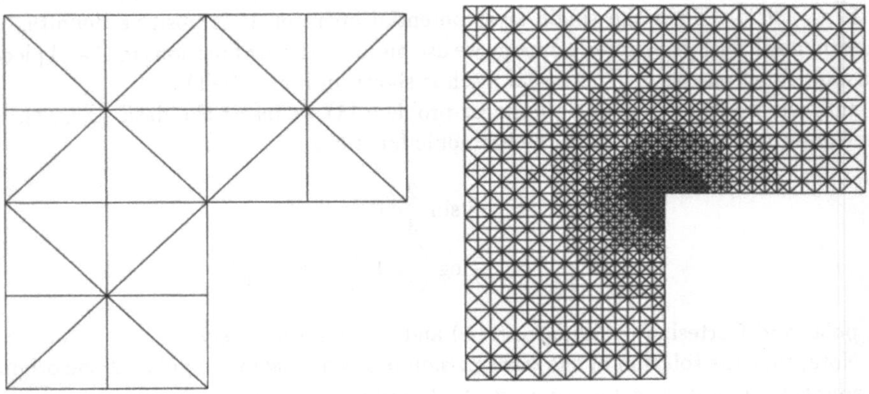

Figure 1: Initial and final mesh.

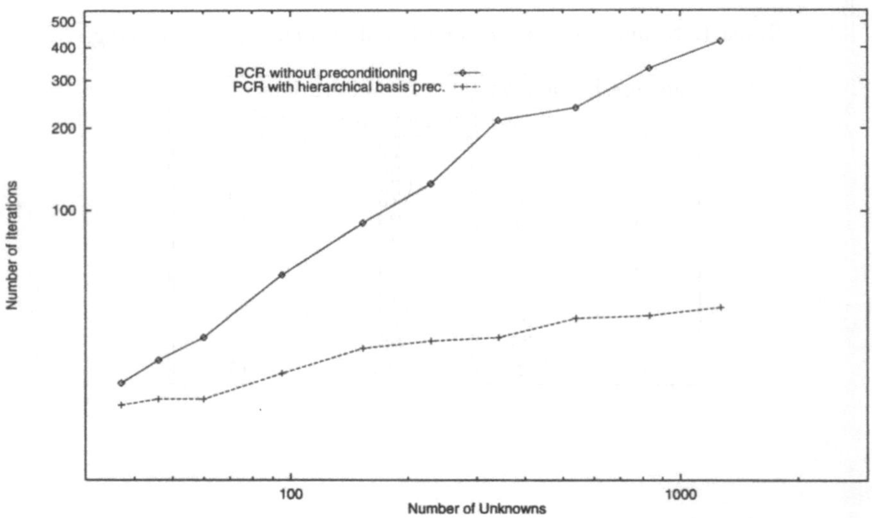

Figure 2: Number of iterations with and without preconditioning.

REFERENCES

[1] COSTABEL, M.: "Boundary Integral Operators on Lipschitz Domains: Element Results", SIAM J. Math. Anal. 19 (1988) 613-626.

[2] COSTABEL, M.: "Symmetric methods for the coupling of finite elements and boundary elements", in: C.A. Brebia et. al. (Eds.), Boundary Elements IX Vol. 1, 411–420, Springer–Verlag, Berlin 1987.

[3] FUNKEN, S.A.: "Schnelle Lösungsverfahren für FEM–BEM Kopplungsgleichungen", Ph.D.-Thesis, Universität Hannover, 1996.

[4] FUNKEN, S.A.: "Fast Solvers with Multigrid Based Preconditioners for Linear FEM–BEM Coupling", Preprint Institut für Angewandte Mathematik, Universität Hannover.

[5] HAHNE, M., STEPHAN, E.P., THIES, W.: "Fast Solvers for Coupled FEM–BEM Equations I" Notes on Numerical Fluid Mechanics 49 (1994) 121–130.

[6] HAHNE, M., MAISCHAK, M., STEPHAN, E.P., WATHEN, A.: "Efficient Preconditioners for Coupled FEM–BEM Equations", Preprint Institut für Angewandte Mathematik, Universität Hannover.

[7] LIONS, J.L., MAGENES, J.: *Non-homogeneous boundary value problems and applications I*, Springer, Berlin 1972.

[8] STEPHAN, E.P., WATHEN, A.: "Convergence of Preconditioned Minimum Residual Iteration for Coupled Finite Element/Boundary Element Computations", Preprint Institut für Angewandte Mathematik, Universität Hannover.

[9] WATHEN, A.J., FISCHER, B., SILVESTER, D.J.: "The convergence rate of the minimum residual method for the Stokes problem", Institut für Angewandte Mathematik, Universität Hamburg, Preprint 76, 1993.

[10] YSERENTANT, H.: "On the multi-level splitting of finite element spaces", Numer. Math. 49 (1986) 379–412.

MULTIGRID AND MULTIPOLE TECHNIQUES IN THE BOUNDARY INTEGRAL EQUATION METHODS

Csaba Gáspár
Szechenyi István College, Department of Mathematics
P.O.Box 701, H–9007 Györ, Hungary
e-mail: csgaspar@d7.szif.hu

SUMMARY

Because of the relatively bad properties of the boundary element matrices (they are generally neither self-adjoint nor sparse) the computational cost of the Boundary Integral Equation Method is often unnecessarily high. Moreover, in case of mixed boundary conditions, the corresponding boundary integral equation is not of the second kind, so that the traditional well-known iterative methods can hardly be applied. In this paper we present a special iterative method which converts the original mixed boundary value problem to a sequence of pure Dirichlet and pure Neumann subproblems converging rapidly to the solution of the original problem. In the solution of these subproblems, standard multigrid tools can be used, so that a significant reduction of the computational cost can be achieved. We also derive a multipole-based technique to evaluate the appearing boundary integrals in an economic way, which can further reduce the overall computational cost.

INTRODUCTION

The Boundary Integral Equation or Boundary Element Method (BEM) converts a *partial differential equation* (defined on an n-dimensional domain) to an *integral equation* (defined on the $(n-1)$-dimensional boundary of the domain). This results in a significant reduction of the number of the unknown parameters in the discretisation and simplifies the discretisation procedure as well. However, the algebraic properties of the discretised problem are not so nice comparing with those of traditional domain-type methods like finite differences or finite elements. The boundary element matrices are generally fully populated (while finite differences result in sparse matrices), non-self-adjoint even if the original problem is self-adjoint (while the finite difference equations often inherit the self-adjointness of the original problem).

Consider, for instance, a 2D elliptic equation. After a usual discretisation (such as the collocation method) with grid size $O(N^{-1})$, where N is the number of boundary nodes, the

BEM produces an algebraic equation with N unknowns, which requires $O(N^3)$ arithmetic operations if a traditional direct solver is applied. An equidistant finite difference method (with the same grid size) results in $O(N^2)$ unknowns: applying a fast multigrid solver, however, the overall computational cost is $O(N^2)$ only.

To improve the computational efficiency of BEM, perhaps the most natural way is to implement the multigrid idea in the boundary element context. For integral equations of the second kind, this is well known and has been applied by many authors, e.g. Schippers [13], Stüben and Trottenberg [14], Hackbusch [9]: it can be also extended for certain types of equations of the first kind, see Hebeker [10]. However, it should be pointed out, that the boundary integral equations of elliptic problems supplied with *mixed* boundary conditions are neither of the second, nor of the first kind, so that the multigrid approach is not obvious. In this paper we present a special iterative technique (see Gáspár [7]), which converts the original mixed problem to a *sequence* of *pure* Dirichlet and Neumann subproblems: the boundary integral equations of these subproblems are of the second kind, so that the multigrid technique can be applied in a standard way. The method can be made even more efficient, if the appearing matrix-vector multiplications (i.e. the evaluations of the discretised boundary integral operators) are performed in a more economic way. The use of *wavelet bases* seems to be such an economic tool as pointed out by Alpert et al [1], since the discrete boundary integral operators are represented by sparse matrices in wavelet bases. Another possibility is to utilize the *multipole series expansion method*, (see Rokhlin [12]), which was also applied to fast summations of point forces (e.g. Greengard and Rokhlin [8]; Carrier et al [2]). In the followings, we show how to use the same idea in the multigrid–boundary integral equation context. The computational cost of the method can be reduced from $O(N^2)$ to $O(N\log N)$, which is close to the "ideal" $O(N)$ (i.e. to the number of unknowns). The multipole expansion method uses a nested cell system known as *quadtree cell system*, which plays a fundamental role also in the *unstructured grid generation*. The cells contained in the quadtree cell system form a non-equidistant, non-uniform grid which becomes fine in the vicinity of the boundary and remains coarse in the middle. It should be pointed out that, using this cell system as a *computational grid* it is possible to define finite volume schemes on this grid as well as a special multigrid technique see Gáspár and Simbierowicz [5]; Gáspár et al [6]. This method can be considered as a special boundary type method, since the grid generation, the discretisation and the solution can be controlled completely by the boundary. The generation of the quadtree cell system needs $O(N\log N)$ operations, while the computational cost of the quadtree-based multigrid method is only $O(N)$.

AN ITERATIVE SOLUTION METHOD

Throughout this paper, the methods are illustrated through the example of the two-dimensional Laplace equation, not because it would be the only problem which could be handled by the presented methods, but it is the simplest one, and therefore, the main steps can be clearly seen. Now we briefly recall the iterative method proposed by Gáspár [7]. Let Ω be a bounded, piecewise smooth domain and introduce the Sobolev space

$$H^1(\Omega, \Delta) := \{U \in H^1(\Omega) : \Delta U \in L_2(\Omega)\}$$

(where Δ means the Laplace operator) supplied with the graph norm. Now consider the following 2D mixed problem:

$$\Delta U = 0, \text{ in } \Omega, \qquad U|_{\Gamma_1} = u_0|_{\Gamma_1}, \qquad \frac{\partial U}{\partial n}|_{\Gamma_2} = v_0|_{\Gamma_2} \,. \tag{1}$$

The boundary Γ is divided into two disjoint parts Γ_1 and Γ_2: along Γ_1 (resp. Γ_2) a Dirichlet (resp. Neumann) boundary condition is prescribed, where the function u_0 is given and belongs to the Sobolev space $H^{1/2}(\Gamma)$, while v_0 is a given functional of $H^{-1/2}(\Gamma)$. The notations $|\Gamma_1$ and $|\Gamma_2$ stand for the restrictions to Γ_1 and Γ_2, respectively. Introducing the following closed subspace of $H^{1/2}(\Gamma)$:

$$W_1 := \{u \in H^{1/2}(\Gamma) : u|_{\Gamma_1} = 0\}$$

the above Dirichlet-type boundary condition means that $(\gamma U - u_0) \in W_1$, while the Neumann-type boundary condition implies that the functional $(\delta U - v_0)$ vanishes on the subspace W_1.

Let $P \in L(H^{1/2}(\Gamma))$ be a projector of the subspace W_1, i.e. assume that $\mathrm{im} P = W_1$ and $P^2 = P$. Defining the projectors $P_1 := I - P$, $P_2 := P^*$ (where I stands for the identity operator of $H^{1/2}(\Gamma)$), it is clear that for every $u \in H^{1/2}(\Gamma)$, $P_1 u$ coincides with u along Γ_1, that is, $P_1 u$ can be considered as an extension from Γ_1 to the whole boundary Γ. On the other hand, the functional $P_2 v$ coincides with v on the subspace W_1 for every $v \in H^{-1/2}(\Gamma)$, which can be interpreted in such a way that they coincide along Γ_2. We also assume that P_1 does not changes the constant functions, i.e. $P_1 w_1 = w_1$ for the function $w_1 = 1$. This immediately implies that $P_2 v$ vanishes on the constant functions, if the functional v does so: $\int_\Gamma P_2 v = 0$ if $\int_\Gamma v = 0$. Using these projectors and the usual trace operators $\gamma U := U|_\Gamma$, $\delta U := \frac{\partial U}{\partial n}$, (1) can be rewritten as:

$$\Delta U = 0, \qquad P_1 \gamma U = P_1 u_0, \qquad P_2 \delta U = P_2 v_0 \tag{2}$$

which has a unique solution in the space $H^1(\Omega, \Delta)$.

We try to solve (2) via the following sequence of subproblems:

$$\begin{aligned} \Delta U_{n+1/2} &= 0, \qquad & \gamma U_{n+1/2} &= u_n + P_1(u_0 - u_n) \\ \Delta U_{n+1} &= 0, \qquad & \delta U_{n+1} &= v_{n+1/2} + P_2(v_0 - v_{n+1/2}) \end{aligned} \tag{3}$$

where $u_n := \gamma U_n$, $v_{n+1/2} := \delta U_{n+1/2}$, and the iteration starts from an arbitrary function $U_1 \in H^1(\Omega, \Delta)$.

This iteration has a natural interpretation. In the first half-step, a Dirichlet subproblem is to be solved, the boundary condition of which coincides with the Dirichlet-type boundary condition of the original problem along Γ_1. Similarly, in the second half-step, a Neumann subproblem is to be solved, in which the Neumann-type boundary condition of the original problem is satisfied along Γ_2. The above boundary conditions are extended to the whole boundary with the help of the projectors P_1 and P_2, respectively. These subproblems can be solved by the Boundary Integral Equation Method without any difficulty: the trick of the iteration is that each subproblem is of *pure* Dirichlet or Neumann type, which allows the use of *fast* solvers for the corresponding boundary integral equations such as the multigrid technique mentioned in the introduction. It is clear that the convergence of the above iteration depends on the proper choice of the projector P.

To simplify the notations, let us introduce the so-called *Dirichlet-to-Neumann* operator $J \in L(H^{1/2}(\Gamma), H^{-1/2}(\Gamma))$ defined by $Ju := \delta U$, where U stands for the unique solution of the Dirichlet problem $\Delta U = 0, \gamma U = u$. J is an isomorphism between two one-codimensional subspaces of $H^{1/2}(\Gamma)$ and $H^{-1/2}(\Gamma)$, respectively. One can easily see that (3) is equivalent to the following iteration:

$$u_{n+1/2} := (I - P_1)u_n + P_1 u_0, \qquad Ju_{n+1} := (I - P_2)Ju_{n+1/2} + P_2 v_0 \tag{4}$$

that is,

$$u_{n+1} = J^{-1}(I - P_2)J(I - P_1)u_n + J^{-1}(I - P_2)JP_1 u_0 + P_2 v_0,$$

from which we immediately obtain that, if u^* denotes the trace of the exact solution of (1), then

$$(u_{n+1} - u^*) = J^{-1}(I - P^*)JP(u_n - u^*) =: A(u_0 - u^*), \tag{5}$$

where obviously $A \in L(H^{1/2}(\Gamma))$, and A maps a one-codimensional subspace of $H^{1/2}(\Gamma)$ into itself (formed by the functions u satisfying $\int_{\Gamma_1} u = 0$). By the proper construction of the projector P, one should assure that A is a contraction in the space $H^{1/2}(\Gamma)$, or, at least that its spectral radius is less than 1.

A trivial choice of P_1 is as follows. For every $u \in H^{1/2}(\Gamma)$, let $P_1 u := \gamma U$, where now U denotes the solution of the mixed problem:

$$\Delta U = 0, \qquad U|_{\Gamma_1} = u|_{\Gamma_1}, \qquad \frac{\partial U}{\partial n}|_{\Gamma_2} = 0\,.$$

It can be easily checked that, in this case, $P_2 v$ is the normal derivative of the solution of another mixed problem, namely:

$$\Delta U = 0, \qquad U|_{\Gamma_1} = 0, \qquad \frac{\partial U}{\partial n}|_{\Gamma_2} = v|_{\Gamma_2}.$$

Furthermore, the corresponding operator A is zero, which implies that the solution of (1) is obtained exactly after the first iteration step. Unfortunately, the definition of the projectors P_1, P_2 is based on the solutions of certain mixed boundary value problems, just like the original problem. However, as pointed out by Gáspár [7], using a multigrid technique for the boundary subproblems of (3), it is sufficient to construct these projectors *on the next coarser grid only*, and the corresponding discretized operator A may still remain contractive. The method was succesfully applied to seepage problems including free surfaces, see Gáspár [4]. It is also possible to incorporate the iteration in the multigrid context in such a way that the iteration starts from the prolongation of the solution obtained on the previous (coarser) grid.

Other projectors can be constructed by using some extension technique from Γ_1 to Γ. To investigate the convergence properties of the corresponding iterations, trivial estimations are insufficient. A more detailed analysis can be carried out on the basis of the concept of *quasi-orthogonality* (which is an efficient tool to investigate a lot of other iterative methods as well).

PROJECTORS AND QUASI–ORTHOGONAL SUBSPACES

Two nontrivial closed subspaces H_1, H_2 of a Hilbert space H are said to be quasi-orthogonal, if the number

$$c(H_1, H_2) := \sup\{|\langle x_1, x_2\rangle| : x_1 \in H_1, x_2 \in H_2, \|x_1\| = \|x_2\| = 1\}$$

is less than 1. In this case $|\langle x_1, x_2\rangle| \leq c(H_1, H_2)\cdot\|x_1\|\cdot\|x_2\|$ for arbitrary $x_1 \in H_1, x_2 \in H_2$, i.e. the strenghtened Cauchy inequality holds.

If H_1, H_2 are quasi-orthogonal then obviously $H_1 \cap H_2 = \{\mathbf{0}\}$. This property is neccesary but in general not sufficient for quasi-orthogonality (only in finite-dimensional cases).

In this subsection, let P_k be the *orthogonal* projector of the subspaces H_k: moreover, denote by $H_k^\perp$ the orthogonal complement of H_k and by $P_k^\perp$ its orthogonal projector, that is: $P_k^\perp := I - P_k$ (k=1,2). By usual arguments, one can easily see that:

Proposition 1: H_1, H_2 are quasi-orthogonal if and only if any of the following two conditions is satisfied: (a) P_1P_2 is a contraction in H: in this case $c(H_1, H_2) = \|P_1P_2\|$; (b) the distance of H_2 and S_1, the unit sphere of H_1 is positive: in this case $c(H_1, H_2)^2 = 1 - \rho(S_1, H_2)^2$.

We will use the following sufficient condition for quasi-orthogonality:

Theorem 2: If there exists a bounded linear operator $B \in L(H)$ which separates the subspaces H_1, H_2 in the sense that im$B \subset H_1$ and im$(I - B) \subset H_2$, then $H_1^\perp$ and $H_2^\perp$ are quasi-orthogonal, moreover, $c(H_1^\perp, H_2^\perp)^2 \leq 1 - \|B\|^{-2}$.

Proof: First of all we note that $\|B\| \geq 1$: otherwise, the operator $I - B$ would be regular, therefore im$(I - B) = H$ would hold. Now let $x_2^\perp \in H_2^\perp$, $\|x_2^\perp\| = 1$ be an arbitrary vector and define $x_1 := Bx_2^\perp/\|Bx_2^\perp\|$, then $x_1 \in H_1$, $\|x_1\| = 1$ ($Bx_2^\perp \neq \mathbf{0}$, otherwise $x_2^\perp - Bx_2^\perp = x_2^\perp$ would hold, whence $x_2^\perp \in$ im$(I - B) \cap H_2^\perp$ follows, that is, $x_2^\perp = \mathbf{0}$). Calculating the distance of the point $x_2^\perp$ and the subspace $H_1^\perp$, we have:

$$\rho(x_2^\perp, H_1^\perp) = \|x_2^\perp - P_1^\perp x_2^\perp\| = \|x_1\| \cdot \|x_2^\perp - P_1^\perp x_2^\perp\| \geq |\langle x_1, x_2^\perp - P_1^\perp x_2^\perp\rangle| = |\langle x_1, x_2^\perp\rangle|,$$

where we have used the Cauchy inequality. By definition, $x_1 = Bx_2^\perp/\|Bx_2^\perp\| = (x_2^\perp - (I - B)x_2^\perp)/\|Bx_2^\perp\|$: from this, since $\|x_2^\perp\| = 1$ and $(I - B)x_2^\perp \in H_2$, it follows that $\rho(x_2^\perp, H_1^\perp) \geq \|Bx_2^\perp\|^{-1} \geq \|B\|^{-1}$. Taking the infimum of the numbers $\rho(x_2^\perp, H_1^\perp)$ for all $x_2^\perp \in H_2^\perp$, $\|x_2^\perp\| = 1$, the theorem follows from Proposition 1.

In particular, if B is a projector and $H_1 =$ im$B$, $H_2 =$ im$(I - B)$, then H_1 and H_2 are quasi-orthogonal as well, and the same estimate holds also for $c(H_1, H_2)$. This follows from Theorem 2. applied to the projector B^* and the subspaces $H_1^\perp$ and $H_2^\perp$.

Returning to our iteration (5), it is clear that $\langle Au, w\rangle = \langle J^{-1}(I - P^*)JPu, w\rangle = [(I - P^*)(JPu)]w = (JPu)[(I - P)w] = \langle Pu, (I - P)w\rangle$ holds for every $u, w \in H^{1/2}(\Gamma)$ (where $\langle .,.\rangle$ denotes the scalar product in $H^{1/2}(\Gamma)$). Setting $H_1 :=$im$P$, $H_2 :=$im$(I - P)$, by the previous theorem we have:

Corollary 3: For the operator A defined by (5), the following estimate holds:

$$\|A\| \leq c(H_1, H_2) \cdot \|P\| \cdot \|I - P\| \qquad (\leq \|P\| \cdot \sqrt{\|P\|^2 - 1})$$

where we have exploited that P is a projector, therefore $\|P\| = \|I - P\|$.

Corollary 3. implies that the iteration is exact if $\|P\| = 1$, that is, P is an *orthogonal* projector, which is the case in the previous trivial example. Corollary 3. also implies that if the projector P is not too "far" from the orthogonal projector of W_1 (with respect to the operator norm), the operator A still remains contractive.

A sharper estimation can be obtained for $r(A)$, the spectral radius of A:

Theorem 4: For the spectral radius of A, the estimation $r(A) \leq c(H_1, H_2)^2 \cdot \|P\|^2$ $(\leq \|P\|^2 - 1)$ is valid.

Proof: The theorem is an immediate consequence of the following lemma: $\|A^{n+1}\| \leq c(H_1, H_2) \cdot \|A\| \cdot \|A^n\|$ for arbitrary integer $n \geq 1$. Let u, w be arbitrary functions of

$H^{1/2}(\Gamma)$, then:

$$\langle A^{n+1}u, w\rangle = \langle A^n u, J^{-1}A^*Jw\rangle = \langle J^{-1}(I-P^*)JPA^{n-1}u, J^{-1}P^*J(I-P)w\rangle$$
$$= \langle (I-P^*)JPA^{n-1}u, P^*J(I-P)w\rangle,$$

where the last scalar product is taken in the space $H^{-1/2}(\Gamma)$. Applying again Theorem 2. for the projector P^*, we have: $\|A^{n+1}\| \le c(H_1, H_2) \cdot \|A^n\| \cdot \|A^*\|$, which proves the lemma.

Consequently, the iteration (3) is still convergent, if $\|P\|^2 < 2$, even if the operator A is not contractive. Numerical experience shows that the estimations of Corollary 3. and Theorem 4. are sharp. Here we cannot detail the possible constructions of the projector P: instead, we restrict ourselves to the problems of the efficient numerical procedure, and suppose that the projector P is defined in such a way that the iteration (3) is convergent.

BOUNDARY INTEGRAL EQUATION FORMULATION

The boundary integral equation of the Laplace equation can be written in the form

$$u + Ku - Rv = 0 \tag{6}$$

where $u := \gamma U$, $v := \delta U$, $U \in H^1(\Omega, \Delta)$ satisfies the Laplace equation in Ω. The boundary integral operators R, K are the single layer and the double layer potentials:

$$(Rv)(x) = \frac{1}{\pi}\int_\Gamma v(y)\log\frac{1}{\|x-y\|}d\Gamma_y, \qquad (Ku)(x) = \frac{1}{\pi}\int_\Gamma u(y)\frac{\langle x-y, n_y\rangle}{\|x-y\|^2}d\Gamma_y, \tag{7}$$

provided that Γ is smooth at the point $x \in \Gamma$. If not, then the coefficient of the first term of the left-hand side of (6) becomes $\alpha(x)/\pi$, where $\alpha(x)$ is the solid angle of Γ at x.

The formula (6) is well known and can be derived by applying the trace operator γ to the third Green formula and taking into account the jump of the double layer potential as well as the continuity of the single layer potential at the boundary. If, instead of γ, the trace operator δ is applied, we obtain the *normal derivative boundary integral equation*:

$$v + K^*v - Qu = 0 \tag{8}$$

where K^* is the adjoint of K, and Q is a hypersingular integral operator which can be obtained by taking the normal derivative of the double layer potential. All the above boundary integral operators are continuous in the corresponding spaces, more precisely: $K \in L(H^{1/2}(\Gamma))$, $R \in L(H^{-1/2}(\Gamma), H^{1/2}(\Gamma))$, $Q \in L(H^{1/2}(\Gamma), H^{-1/2}(\Gamma))$, and $K^* \in L(H^{-1/2}(\Gamma))$. Using these boundary integral operators, the boundary integral equations of the iteration (3) can be written as:

$$(I+K^*)v_{n+1/2} = Q(u_n + P_1(u_0 - u_n)), \qquad (I+K)u_{n+1} = R(v_{n+1/2} + P_2(v_0 - v_{n+1/2})). \tag{9}$$

Both equations are of the second kind, which allows the use of standard multigrid tools. Consequently, if the boundary is discretized by N points, the computational cost is only $O(N^2)$ (up to a prescribed accuracy), since the whole procedure requires a limited number of matrix-vector multiplications i.e. a limited number of evaluations of some boundary integral operators, where this limit does not depend on N. This computational cost is significantly smaller than that of the traditional BEM solvers using some direct method (they require $O(N^3)$ operations). In other words, the efficiency of the multigrid methods can thus be extended to the mixed problems as well. The computational efficiency can be obviously increased further, if the evaluations of the discretized boundary integral operators are performed in a more economic way. In the next section, we show an adaptation of the *fast multipole method* to boundary integral operators.

MULTIPOLE TECHNIQUE IN EVALUATING BOUNDARY INTEGRALS

Using complex notations, let $z_1, z_2, ..., z_N \in \mathbf{C}$ be a finite set of different points. Furthermore, let Φ_j be a complex function, which is analytical on $\mathbf{C}$ except at the point z_j $(j = 1, ..., N)$. For brevity, we call these functions *potentials*. Suppose that each Φ_j has a multipole expansion around z_j:

$$\Phi_j(z) = a_0^{(j)} \log(z - z_j) + \sum_{r=1}^{\infty} \frac{a_r^{(j)}}{(z - z_j)^r} \tag{10}$$

which is convergent outside a circle with radius R_j centered at z_j, moreover, the speed of convergence can be estimated by

$$|\Phi_j(z) - \Phi_j^{(p)}(z)| \leq const. \cdot \left|\frac{R_j}{z}\right|^p$$

where $\Phi_j^{(p)}(z)$ denotes the truncated series

$$\Phi_j(z) = a_0^{(j)} \log(z - z_j) + \sum_{r=1}^{p} \frac{a_r^{(j)}}{(z - z_j)^r}.$$

The fast multipole method makes it possible to compute the matrix-vector products in an economic way, if the matrix entries are of the form $\Phi_j(z_k)$, (at least for $k \neq j$) which is often the case e.g. in computing particle interactions, and also in calculating boundary integral matrices as shown in the followings. To introduce the main notations and concepts, here we briefly recall the fundamental ideas of the multipole method: for details, see Rokhlin [12], Greengard and Rokhlin [8].

THE FAST MULTIPOLE METHOD

Let $u_1, u_2, ...,_u \in \mathbf{C}$ be arbitrary constants. Our goal is to compute the numbers:

$$\Phi(z_k) := \sum_{j \neq k\ j=1}^{N} u_j \Phi_j(z_k - z_j) \tag{11}$$

$(k = 1, 2, ..., N)$. For each point z_k, the above summation is divided into two parts: if z_j lies "near" to z_k, we compute the corresponding term $u_j \Phi_j(z_k - z_j)$ directly. For the points lying "far" from z_k, the sum of the corresponding terms can be (approximately) calculated with the help of the multipole expansions, unifying *several* multipole series to a *single* one which is still convergent at the point z_k. The main trick is that a lot of multipole series can be replaced with a single one: the farther the corresponding z_j's lie from z_k, the greater number of multipole series can be thus unified.

Multipole series can be unified by virtue of the *shifting theorem.* Straightforward calculations show that, if z_0 is the center of the multipole series Φ_0, then it can be expressed as another multipole series around another point $w \in \mathbf{C}$:

$$\Phi_0(z) = a_0 \log(z - z_0) + \sum_{r=1}^{\infty} \frac{a_r}{(z - z_0)^r} = b_0 \log(z - w) + \sum_{r=1}^{\infty} \frac{b_r}{(z - w)^r} \tag{12}$$

where the new coefficients can be calculated as:

$$b_0 = a_0, \qquad b_r = -\frac{a_0(z_0 - w)^r}{r} + \sum_{q=0}^{r-1} \binom{r-1}{q} a_{r-q}(z_0 - w)^q \tag{13}$$

($r = 1, 2, ...$). If the original multipole series is convergent outside the circle centered at z_0 with radius R_0, then the shifted one is convergent outside the circle centered at w with radius $(R_0 + |z_0 - w|)$.

Using truncated series, the determination of the shifted series requires a constant computational cost, depending only on the maximal index p but not on N.

Unifying the multipole series corresponding to the points which lie sufficiently far from z_k, the computational efficiency of evaluation of $\Phi(z_k)$ can be significantly increased.

The only problem is the proper definition of the "far" points in connection with each z_j. This can be made by the so-called *quadtree (QT) cell system*, which was defined originally for completely different purposes (see Jackins and Tanomoto [11]), and also is the basis of the *unstructured grid generation* in numerical grid generation: see e.g. Cheng et al [3]. Let $\Omega_0 \in \mathbf{C}$ a square containing the finite set of points $S_0 := \{z_1, z_2, ..., z_N\}$. The QT cell system generated by S_0 is a nested sequence of subsquares of Ω_0 defined by the following recursive procedure:

1. Initially, set $S := S_0$, $\Omega := \Omega_0$. This is the subdivision level 0.

2. If the number of points of S does not exceed a prescribed number N_0 (N_0 may be even 0 or 1), then the subdivision procedure is finished at the current pair (Ω, S). Otherwise, subdivide Ω into four congruent subsquares $\Omega_1, \Omega_2, \Omega_3, \Omega_4$. For each Ω_j ($j = 1, 2, 3, 4$) determine the corresponding set S_j as the set of points of S which are contained in Ω_j.

3. Invoke the procedure described in 2. for the pairs (Ω_1, S_1), (Ω_2, S_2), (Ω_3, S_3) and (Ω_4, S_4). This will be the subdivision level $(k + 1)$ if the current level is k.

An analogous procedure can be defined in arbitrary finite dimensional space. In particular, the 3D procedure results in the so-called *octree* algorithm.

The resulting cell system can be considered a non-equidistant, non-uniform *grid*, the spatial resolution of which follows the spatial density of the generating points S. Figure

1. shows two typical QT cell systems: the first one is generated by 10 points, while the second is generated by a single point (with $N_0 := 0$ in the latter case).

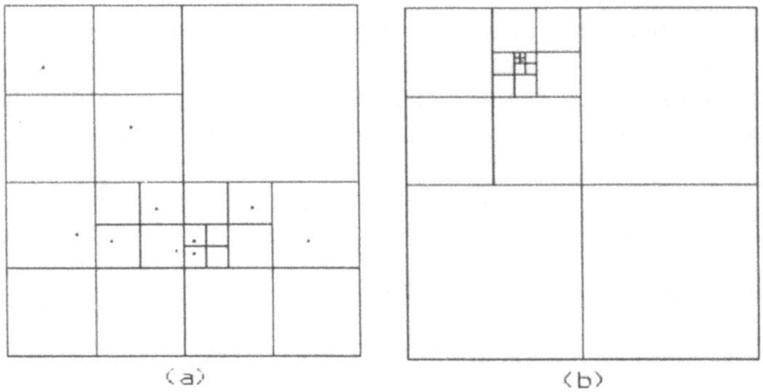

Figure 1. Two typical QT cell systems

In Figure 2. a "boundary-controlled" QT cell system can be seen, which is generated by a discretisation of the boundary of a bounded domain. In this case, the QT-grid is fine in the vicinity of the boundary and remains coarse in the middle. The spatial resolution exponentially increases toward the boundary, consequently, the total number of cells is proportional to N, the number of the boundary discretisation points only.

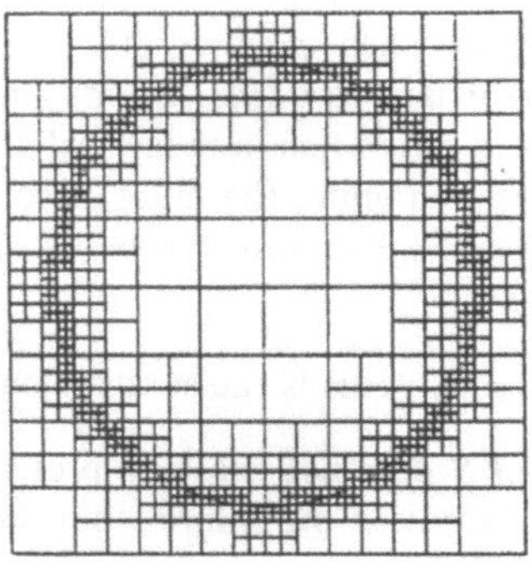

Figure 2. A QT-cell system generated by a circle

Remark: In a lot of practical problems, the resulting QT-grid can be used also as a computational grid for some finite volume schemes. For applications in lake modelling, see e.g. Gáspár et al [6].

The QT cell system makes it possible to define "near" and "far" cells and points. Two cells, Ω_k and Ω_j are said to be "far" cells if the distance of their centers is great enough compared with their sizes, e.g. greater than the cell size of either Ω_k or Ω_j. Otherwise, Ω_k and Ω_j are called "near" cells. Obviously, if Ω_k and Ω_j are far cells, then an arbitrary *child cell* of Ω_k (derived from Ω_k by a subdivision step) is also far cell with respect to Ω_j. The "near" and "far" classification can be determined by a simple recursive algorithm:

details are omitted.

Having constructed the QT cell system, for each z_k, the potentials due to the points lying in the near cells are calculated directly, while the effect of the far cells can be computed by multipole series. It can be easily seen that the number of the far cells of any selected cell is at most $O(\log N)$, therefore the whole procedure requires $O(N\log N)$ operations.

In conclusion, the method consists of the following steps:

1. Generate a QT cell system based on the points $z_1, z_2, ..., z_N$.

2. Using the shifting theorem, calculate the multipole series around the smallest cell centers.

3. Unifying the multipole series of the child cells, calculate the multipole series around the centers of the parent cells. Repeat this procedure until the initial cell Ω_0 is reached.

4. For each point z_k, calculate the potential at z_k: the potentials due to the points of the near and far cells are to be calculated directly and by multipole expansions, respectively.

APPLICATION TO BOUNDARY INTEGRAL OPERATORS

In each step of the iterative-multigrid method defined by (9), one has to calculate the integrals of the type Ku, Rv, Hv, Qu (where H denotes the normal derivative of the single-layer potential). To apply a multipole evaluation method, we have to discretize the boundary integral equations. We restrict ourselves to the simplest collocation method in connection with piecewise constant discretization: other discretization strategies can be handled in a similar way but may require more complicated calculations. Note that the multipole approach is not restricted to the iteration (9): it remains applicable to any iterative technique of boundary integral equations which does not change the boundary element matrices.

Suppose that the boundary is discretised by N points and the boundary itself is approximated by a polygon formed by these points. Consider the piecewise constant trial functions ψ_j $(j = 1, 2, ..., N)$, where ψ_j is identically equal to 1 along Γ_j (the jth side of Γ) and zero on the rest of the boundary. Let the collocation points x_k be the midpoints of the sides Γ_k $(k = 1, 2, ..., N)$. Then the entries of the boundary element matrices have the form $K\psi_j(x_k)$, $R\psi_j(x_k)$, $H\psi_j(x_k)$, $Q\psi_j(x_k)$ $(j, k = 1, 2, ...N)$. Thus, a multipole evaluation of the appearing matrix-vector multiplications is possible if the functions $K\psi_j$, $R\psi_j$, $H\psi_j$, $Q\psi_j$ can be expressed in terms of multipole series.

For each Γ_j define a local coordinate system on the boundary in such a way that Γ_j coincides an interval of the form $[0, a]$. Elementary calculations show that

$$K\psi_j(x,y) = -\omega(x,y), \qquad R\psi_j(x,y) = a + \frac{x-a}{2}\log a_0^2 - \frac{x}{2}\log a_1^2 - y\cdot\omega(x,y) \qquad (14)$$

(see Figure 3. for notations: here x, y denote 1D variables rather than points in $\mathbf{R}^2$.)

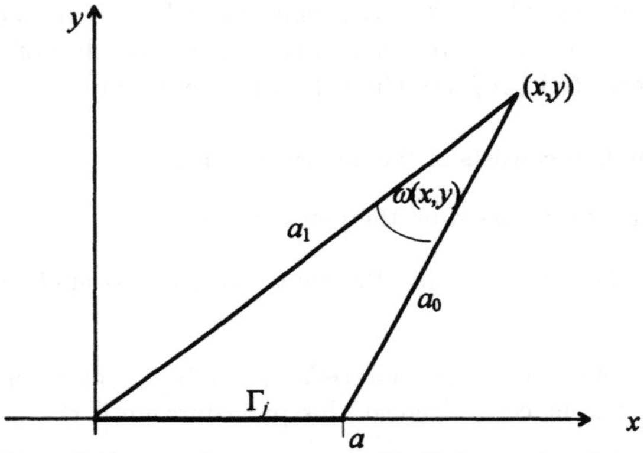

Figure 3. Notations for computations of double layer and single layer potentials

Since Hv and Qu can be expressed by the normal derivative of the functions $K\psi_j$ and $R\psi_j$, it is sufficient to calculate the partial derivatives of the functions defined by (14) and (15), yielding:

$$\frac{\partial K\psi_j}{\partial x}(x,y) = -\frac{y}{a_1^2} + \frac{y}{a_0^2} \qquad \frac{\partial K\psi_j}{\partial y}(x,y) = \frac{x}{a_1^2} - \frac{x-a}{a_0^2}$$
$$\frac{\partial R\psi_j}{\partial x}(x,y) = \frac{1}{2}\log\frac{a_0^2}{a_1^2} \qquad \frac{\partial R\psi_j}{\partial y}(x,y) = -\omega(x,y) \qquad (15)$$

Introducing complex variables, the above expressions can be highly simplified. Identifying $\mathbf{R}^2$ with the complex plane $\mathbf{C}$, let $w_1, w_2, ..., w_N \in \mathbf{C}$ be the vertices of Γ and consider the complex single layer potential:

$$R_j(z) := -e^{-i\omega_j} \cdot \int_{w_{j-1}}^{w_j} \log(z-\zeta)d\zeta,$$

where $e^{i\omega_j} := (w_j - w_{j-1})/|w_j - w_{j-1}|$, it is easy to see that the real part of R_j coincides with the ordinary single layer potential $R\psi_j$. Calculating the integral, we obtain:

$$R_j(z) = -e^{-i\omega_j} \cdot [-(w_j - w_{j-1}) + (z - w_{j-1}) \cdot \log(z - w_{j-1}) - (z - w_j) \cdot \log(z - w_j) \quad (16)$$

Introducing the notations $z_j := (w_j + w_{j-1})/2$, $a_j := w_j - w_{j-1}$, the expression (16) can be easily transformed to a multipole series around z_j. After some algebraic manipulations, using the well-known series for $\log(1+z)$ ($|z| < 1$), we have:

Proposition 5:

$$R_j(z) = \alpha_0 \log(z - z_j) + \sum_{r=1}^{\infty} \frac{\alpha_r}{(z - z_j)^r}$$

where $\alpha_0 = -a_j e^{-i\omega_j}$ and for $r \geq 1$:

$$\alpha_r = a_j e^{-i\omega_j} \cdot \frac{1}{r(r+1)} \left(\frac{a_j}{2}\right)^r$$

if r is even, and $\alpha_r = 0$, if r is odd: the series is convergent for $|z - z_j| > |a_j|/2$.

The complex form of the double layer potential is:

$$K_j(z) := -i \cdot \int_{w_{j-1}}^{w_j} \frac{1}{z-\zeta} d\zeta \,.$$

One can easily verify that the real part of K_j equals to the double layer potential $K\psi_j$. Calculating the integral:

$$K_j(z) = i \cdot \log \frac{z - w_j}{z - w_{j-1}}$$

from which the following multipole series can be derived:

Proposition 6:

$$K_j(z) = \sum_{r=1}^{\infty} \frac{\beta_r}{(z - z_j)^r}$$

where

$$\beta_r = -\frac{2i}{r} \left(\frac{a_j}{2} \right)^r ,$$

if r is odd, and $\beta_r = 0$, if r is even: the series is convergent for $|z - z_j| > |a_j|/2$.

Remark: The complex derivative of R_j equals to K_j (up to a multiplicative constant), more precisely: $R_j'(z) = -ie^{-i\omega_j} K_j(z)$, from which the multipole series of R_j can be obtained in a simpler way.

The evaluation of the vectors Hv and Qu is based on the complex derivatives of the above two multipole series. Note, however, that it is not necessary to store (shift, unify etc.) their multipole series when building the multipole technique. It is sufficient to work with the multipole series of R_j and K_j, and one should apply the derivative series formulations only when evaluating the operators H and Q.

Acknowledgements: The research was partly supported by the Hungarian National Scientific Research Fund under the contract T17323.

REFERENCES

[1] ALPERT, B., BEYLKIN, G., COIFMAN, R., ROKHLIN, V.: "Wavelet-like bases for the fast solution of second-kind integral equations", SIAM J. Sci. Comput. 14, No.1, pp.159-184. 1993.

[2] CARRIER, J., GREENGARD, L., ROKHLIN, V.: "A Fast Adaptive Multipole Algorithm for Particle Simulations", SIAM J. Sci. Stat. Comput. 9, No 4, pp.669-686, 1988.

[3] CHENG, J.H., FINNIGAN, P.M., HATHAWAY, A.F., KELA, A., SCHROEDER, W.J.: "Quadtree/octree meshing with adaptive analysis", Numerical Grid Generation in Computational Fluid Mechanics '88 (ed. by S. Sengupta, J. Häuser, P. R. Eiseman, J. F. Thompson). Pineridge Press, Swansea, 1988.

[4] GÁSPÁR, C.: "Solution of Seepage Problems by Combining the Boundary Integral Equation Method with a Multigrid Technique", Proceedings of the Eighth International Conference on Computational Methods in Water Resources, Venice, Italy, 11-15 June, 1990. Computational Mechanics Publications/Springer-Verlag, 1990.

[5] GÁSPÁR, C., SIMBIEROWICZ, P.: "Difference Schemes in Tree-structured Multigrid Context", Proceedings of the IX. International Conference on Computational Methods in Water Resources, Denver, Colorado, USA, 9-12 June, 1992. Computational Mechanics Publications/Elsevier, 1992.

[6] GÁSPÁR, C., JÓZSA, J., SARKKULA, J.: "Shallow Lake Modelling Using Quadtree-based Grids", Proceedings of the X. International Conference on Computational Methods in Water Resources, Heidelberg, Germany, July 19-22, 1994.

[7] GÁSPÁR, C.: "An Iterative and Multigrid Solution of Boundary Integral Equations", Computers & Mathematics with Applications, 29, No. 7, pp. 89-101, 1995.

[8] GREENGARD, L., ROKHLIN, V.: "A Fast Algorithm for Particle Simulations", Journal of Computational Physics, 76, pp.325-348, 1987.

[9] HACKBUSCH, W.: "Multi-Grid Methods and Applications", Springer-Verlag, Berlin, Heidelberg, New York, Tokyo, 1985.

[10] HEBEKER, F.K.: "On multigrid methods of the first kind for symmetric boundary integral equations of nonnegative order", Roboust Multi-Grid Methods. Proceedings of the Fourth GAMM-Seminar, Kiel, January 22-24, 1988. Friedr. Vieweg & Sohn, Braunschweig/ Wiesbaden, Germany, 1988.

[11] JACKINS, C.L., TANOMOTO, S.L.: "Oct-trees and their use in representing three-dimensional objects", Computer Graphics and Image Processing. 14, No 3, 1980.

[12] ROKHLIN, V.: "Rapid Solution of Integral Equations of Classical Potential Theory", Journal of Computational Physics, 14, pp.187-207, 1985.

[13] SCHIPPERS, H.: "Multiple Grid Methods for Equations of the Second Kind with Applications in Fluid Mechanics" (thesis), Mathematisch Centrum, Amsterdam, 1982.

[14] STÜBEN, K., TROTTENBERG, U.: "Multigrid Methods: Fundamental Algorithms, Model Problem Analysis and Applications", GMD-Studien, Nr. 96. Birlinghoven, Germany, 1984.

RAPID PARALLEL EVALUATION OF INTEGRALS IN POTENTIAL THEORY ON GENERAL THREE DIMENSIONAL REGIONS

Anne Greenbaum[1]
Courant Institute of Mathematical Sciences
51 Mercer Street, New York, NY 10012

Anita Mayo
IBM T.J. Watson Research Center
Yorktown Heights, NY 10598

V. Sonnad
Mail Stop 9233, IBM Corporation
11400 Burnet Road, Austin Texas 78758 USA

SUMMARY

We present a new, high order accurate method for the rapid, parallel evaluation of certain integrals in potential theory on general three dimensional regions. These methods use rapid methods for solving the differential equation which the kernel satisfies, and the number of operations needed to evaluate the integrals is essentially equal to the number of operations needed to solve the differential equation on a regular rectangular grid. In particular, one can evaluate integrals whose kernels are the Greens function for Poissons equation by using Fourier methods on a rectangular grid, or a fast Poisson solver. Thus, these methods avoid the problems associated with using quadrature methods to evaluate an integral with a singular kernel. Numerical results are presented for experiments on a variety of geometries.

INTRODUCTION

New, fast numerical techniques for generating [13] and solving [4], [10] the matrix equations that arise in integral equation methods has lead to their increased use in industrial applications. But, despite the reduction in the cost of solving many integral equations, the methods are still often very expensive. This is because in order to completely solve a

[1]This work was supported by the Applied Mathematical Sciences Program of the U.S. Department of Energy under contract DE-AC02-76ER03077.

problem it is frequently necessary to evaluate a surface integral at many points of a region. Moreover, integral equation formulations for inhomogeneous differential equations usually require the evaluation of a volume integral.

The evaluation of both the volume and surface integrals is very expensive. In particular, we note that $O(n^6)$ operations are needed in order to evaluate a volume integral at every point of a three dimensional $n \times n \times n$ grid, since the evaluation of each integral requires $O(n^3)$ operations.

Aside from the cost of evaluating the integrals, an additional difficulty with integral equation methods is the problem of accurately evaluating the solution at points inside and near the boundary. In integral equation methods for solving elliptic differential equations the solution is expressed as the integral of the product of a density function and a singular kernel. The singularity of the kernel implies that it is difficult to compute the solution accurately by quadrature inside, or near the region of integration.

In this paper we develop fast, parallel three dimensional methods for evaluating volume and surface integrals in potential theory which overcome these two problems. These methods are mesh based, and are generalizations of our previous work (See [5]-[9].). This new work has required a number of new numerical techniques. In particular, it has required the derivation of several new formulas which are not obvious generalizations of the two dimensional formulas, especially when the surface is curved. Specifically, we have had to use the Frenet formulas in order to obtain second order accurate approximations to the integrals. We have also had to use nested triangulations in order to be able to compute volume integrals and integrals of double layer density functions. Furthermore, this is the first time we have used a distributed memory parallel computer to perform these kinds of calculations.

The methods we use to evaluate the integrals rely on fast finite difference methods. In these methods, we embed the irregular region on which we want to evaluate an integral U in a larger rectangular region on which there is a regular grid. If, for example, the kernel of U satisfies Laplace's equation, we compute an approximation to the integral by first computing an approximation to its discrete Laplacian at all points of the regular grid. Then we apply an operator that inverts the discrete Laplacian Δ_h to obtain an approximation to U. To compute an approximation to $\Delta_h U$ at mesh points which have all their neighboring mesh points on the same side of the boundary of the irregular region we use the fact that, up to truncation error, it is equal to ΔU, which is known. However, because the integrals we compute necessarily have discontinuities in some of their derivatives at the the edge of the domain of integration, we must compute special correction terms at certain mesh points near the boundary. It turns out that these correction terms can be computed in terms of the discontinuities in the derivatives of the integral, and how far the mesh points are from the boundary of the region of integration. We show how to compute these discontinuities and how to use them to compute approximations to $\Delta_h U$.

We note that these three dimensional volume integrals also arise in other contexts, in addition to the solution of inhomogeneous differential equations. In particular, they are needed when applying the Biot Savart law to evaluate the magnetic field induced by a conducting wire. This law says that the field B induced by a conducting wire is the curl of the volume integral of the fundamental solution of Poisson's equation multiplied by the current density J:

$$B(x,y,z) = \nabla x \int\int_D\int J * \frac{1}{4\pi R} dx' dy' dz' . \tag{1.1}$$

Therefore, each of the three components of the field is equal to the sum of integrals whose kernels are comprised of products of derivatives of the fundamental solution with components of the current density.

In addition to being fast in serial mode, these methods are easily parallelized. We have, in fact, evaluated the volume and surface integrals on the IBM SP2 parallel computer.

The organization of this paper is as follows. In section 2 we show how compute a second order accurate approximation to volume integrals whose kernels are a fundamental solution of Poisson's equation, and in section 3 we discuss certain extensions of the method to the evaluation of other integrals. In particular, we show how to evaluate surface integrals of single and double layer density functions, and derivatives of these integrals. In section 4 we provide results of numerical experiments, including some on the IBM SP2 distributed memory parallel computer.

EVALUATION OF VOLUME INTEGRALS

As an example of how the method works, we first show how to compute a second order accurate approximation to a volume integral whose kernel satisfies Laplace's equation, that is, an integral of the form

$$U(x,y,z) = \frac{1}{4\pi}\int\int_D\int \frac{1}{R(x,y,z,x',y',z')} f(x',y',z')dx'dy'dz'. \tag{2.1}$$

We embed the region of integration D in a larger rectangular region R with mesh $\{(ih_x, jh_y, kh_z)\}$.

Let $(\Delta_h U)_{i,j,k}$ denote the usual 7 point approximation to the Laplacian:

$$\frac{U(i+1,j,k)+U(i-1,j,k)-2U(i,j,k)}{h_x^2} + \frac{U(i,j+1,k)+U(i,j-1,k)-2U(i,j,k)}{h_y^2}$$
$$\frac{U(i,j,k+1)+U(i,j,k-1)-2U(i,j,k)}{h_z^2}.$$

To obtain our approximation U_h to the integral we compute the approximation to its discrete Laplacian, and then apply a Poisson solver. To compute the discrete Laplacian we use the fact that

$$\Delta U = f \text{ in } D,\ \Delta U = 0 \text{ outside } D. \tag{2.2}$$

Therefore at mesh points inside D that have all their neighbors in D, we set $(\Delta_h U)_{i,k} = f_{ijk}$, and at outside points whose neighbors are also outside D we set $(\Delta_h U)_{ijk} = 0$.

Since U is not a smooth function, we cannot use either formula at points near ∂D. (We can easily see that U isn't smooth at the boundary because the Laplacian of U is discontinuous there.) Therefore, at such points the values of the discrete Laplacian are not well approximated by the values of the exact Laplacian.

Now we show how to compute an approximation to the the discrete Laplacian of U at these other irregular mesh points, that is the set of points B which have one or more of their neighbors on the other side of ∂D.

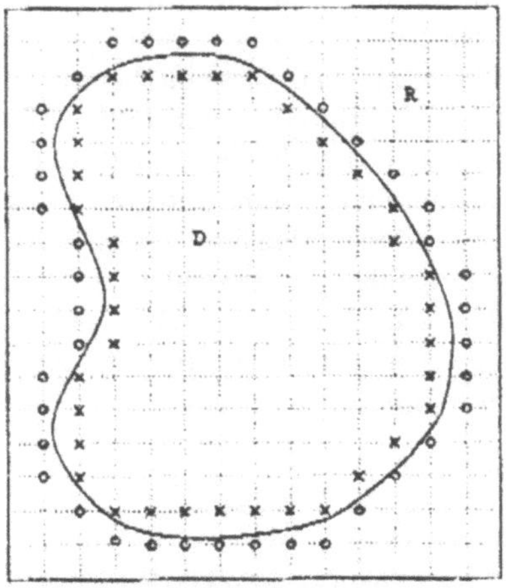

B consists of × and o

FIGURE 1

It turns out that to be able to compute such an approximation it is sufficient to know what the discontinuities in U and its derivatives in the coordinate directions are at the boundary of D. We note that the discrete Laplacian can be derived by using a Taylor series expansion. Such an expansion is not valid when there are discontinuities in derivatives, but if the discontinuities in derivatives are known, then they may be used to compute correction terms to the Taylor series. One can thereby attain an accurate approximation to the difference operator applied to the integral. We first show how to compute the discontinuities in U, and then we show how to use them to evaluate the discrete Laplacian of U.

Suppose the boundary surface is given by $x(s,t), y(s,t), z(s,t)$. where s and t are two real parameters. We assume that $x(s,t), y(s,t)$ and $z(s,t)$ have three continuous derivatives.

For a given function g defined on R which is discontinuous on ∂D let $[g(p)]$ denote the discontinuity in g at a point p on ∂D. An integral of the form (2.1) and it's normal derivative U_n are continuous across ∂D, i.e.

$$[U] = 0, \qquad (2.3)$$

$$[U_n] = 0$$

Therefore, the components of the gradient are also continuous:

$$[U_x] = [U_y] = [U_z] = 0. \qquad (2.4)$$

By (2.2) there must be discontinuities in the second derivatives of U.

To determine discontinuities in the six second derivatives of U we differentiate (2.2) in the normal and tangential directions, and use the discontinuity in the Lapalacian. That is, we use the six equations

$$[U_{ss}] = 0, \ \ [U_{st}] = 0, \ \ [U_{tt}] = 0, \qquad (2.5)$$

$$[U_{ns}] = 0, \ \ [U_{nt}] = 0 \text{ and } [U_{xx}] + [U_{yy}] + [U_{zz}] = f.$$

To be more specific, suppose that the surface D has been decomposed into curved triangular patches with sides

$$a_i = (a_{i1}, a_{i2}, a_{i3}) \quad i = 1, 2, 3 .$$

Suppose also that the curvature is known at each position. To compute the derivatives in (2.5) we differentiate along two of the sides of each patch, i.e. in any trianglular region let s be the direction along a_1, and we let t be the direction along a_2.

Therefore, in any triangular patch

$$U_s = a_1 \bullet \nabla U = a_{11}(s) U_x + a_{21}(s) U_y + a_{31}(s) U_z,$$

and so

$$U_{ss} = a_{11}^2 U_{xx} + a_{11}a_{12}U_{xy} + a_{12}^2 U_{yy} + a_{12}a_{13}U_{xz} + a_{12}a_{13}U_{yz} + a_{13}^2 U_{zz}$$
$$+ \frac{da_1}{ds} \bullet \nabla U .$$

Similarly,

$$U_{ts} = a_{11}a_{21}U_{xx} + a_{11}a_{22}U_{xy} + a_{12}a_{22}U_{yy} + a_{12}a_{13}U_{xz} + a_{12}a_{23}U_{yz} + a_{13}a_{23}U_{zz}$$
$$+ \frac{da_2}{ds} \bullet \nabla U .$$

It follows that in order to use (2.5) to determine discontinuities in derivatives of U, we need to determine quantities such as $\frac{da_1}{ds} a_1(s)$ and $\frac{da_2(t)}{ds}$.

These can be determined in terms of the curvature κ using the Frenet formulas [12] along the curve $a_1(s)$:

$$\frac{dT}{ds} = \kappa N \tag{2.6a}$$

$$\frac{dN}{ds} = -\kappa T + \tau B \tag{2.6b}$$

$$\frac{dB}{ds} = -\tau N \tag{2.6c}$$

where T is the tangent vector along a surface, N is the normal vector, B is the binormal to a_1, and τ is the torsion. By (2.6a)

$$\frac{da_1(s)}{ds} = \kappa N .$$

To determine $\frac{da_2}{ds}$ we note that $a_2 = (a_2, a_1)a_1 + (a_2, B)B$. Since the torsion τ vanishes on the sides of the triangles, this equation and (2.6c) imply that

$$\frac{da_2}{ds} = (a_2, a_1)\kappa N.$$

Thus, the quantities on the right hand side of (2.5) can be determined, and the system solved to find the discontinuities in the six second derivatives of U.

By differentiating the above equations and using higher derivatives of the boundary surface it is possible to derive formulas for discontinuities in higher order derivatives.

We now show how to use these discontinuities to determine the discrete Lapalacian U at the irregular mesh points B.

Let $\tilde{U}(p) = U(p)$ for p in $R - D$. Suppose, for example, p is in D, but it's neighbor to the right p_E, is not. Let p^* be the point on the line between p and p_E which intersects ∂D, let h_1 be the distance between p and p^*, and let $h_2 = h - h_1$.

By manipulating the Taylor series at p and p_E both evaluated at p^*, we can derive the following expression for $\tilde{U}(p_E) - U(p)$ (For details see [7].).

$$\tilde{U}(p_E) - U(p) = [\tilde{U}(p^*) - U(p^*)] + h_2[\tilde{U}_x(p^*) - U_x(p^*)] + \frac{1}{2}h_2^2[\tilde{U}_{xx}(p^*) - U_{xx}(p^*)] \quad (2.7)$$

$$+\frac{1}{6}h_2^3[\tilde{U}_{xxx}(p^*) - U_{xxx}(p^*)] + hU_x(p) + \frac{1}{2}h^2U_{xx}(p) + \frac{1}{6}h^3U_{xxx}(p) + O(h^4).$$

Note that the first four terms depend on the discontinuities in U and in its x derivatives at the boundary. The other terms are the usual Taylor series terms. Therefore, the right hand side of (2.7) is a sum of terms we can evaluate in terms of the discontinuities in U and its derivatives, and terms we would have if there were no boundary between p and p_E.

Now let p_W, p_N, p_S, p_F and p_B be the mesh points to the left of, above, below, in front of and behind p. We get the same type of expressions for the differences between the value of U at p and at its other neighbors, that is $U(p_W) - U(p), U(p_N) - U(p), U(p_S) - U(p), U(p_F) - U(p)$, and $U(p_S) - U(p)$, except that there will not be any boundary terms unless ∂D passes between p and that neighbor. Therefore, we can compute an approximation to the 7 point discrete Laplacian of U, which is just the sum of the above six differences. We can do this at all the irregular points.

More precisely, for mesh points $(x_i, y_j, z_k) \in B$ we define the mesh function m_{ijk} to be the value of the extra terms in the discrete Laplacian we get by our procedure using f and its derivatives.

We define U_h to be the solution of the following equations:

$$\Delta_h U_h = \begin{cases} f_{ijk} & (i,j,k) \in D - B \\ f_{ijk} + m_{ijk} & (i,j,k) \in B \cap D \\ m_{ijk} & (i,j,k) \in B \cap D^c \\ 0 & (i,j,k) \in R - D \cap B \end{cases}.$$

If the values of $m_{i,j,k}$ are third order accurate, then by applying a second order accurate Poisson solver we obtain a second order accurate approximation U_h to U. See [8], [9].

We can clearly also compute higher order accurate solutions by using higher order accurate approximations to the Laplacian.

Before applying a Poisson solver to obtain a solution we must supply appropriate boundary conditions. If we only need a particular solution of Poisson's equation, then there is no difficulty, since the discontinuities in the derivatives of the integral, and therefore the discrete Laplacian, will be the same, independent of which fundamental solution of Laplacian is used as the kernel. The integral we obtain an approximation to is the one associated with the same boundary conditions as the fast Poisson solver we use. For example, if we use a periodic Poisson solver, then we obtain an approximation to the integral whose kernel is the periodic Greens function for the Laplacian. If we need an integral with a specific kernel, then we use the corresponding Poisson solver.

A more difficult problem arises when we need to solve an exterior problem. In that case we can use a method originally developed by Hockney [2] and later improved by James [3] where one calculates the boundary potential by finding a set of correction charges on the boundary of the embedding region, and then convolves them with a suitable Green's function. This method, however, is more expensive than the others since it requires the the application of two Poisson solvers.

EXTENSIONS

In this section we present extensions of the method of the previous section to the evaluation of other integrals. Specifically, we show how to evaluate certain surface integrals in potential theory, and we also show how to evaluate derivatives of the surface and volume integrals.

In order to evaluate the surface integrals we use the same basic method that we use to evaluate the volume integrals. In particular, we can use it to evaluate integrals of single layer density functions:

$$W = 1/4\pi \int_{\partial} \int_{D} \rho(s,t)(1/r)dS \, .$$

As when we evaluate volume integrals, the problem of evaluating this type of integral reduces to evaluating its Laplacian in the two regions bounded by the boundary surface and evaluating the discontinuities in its derivatives at the boundary. We first note that the Laplacian vanishes in the two regions, i.e.

$$\Delta W = \begin{cases} 0 & \text{inside } D \\ 0 & \text{outside } D \, . \end{cases}$$

It is also known [11] that such an integral is continuous in the tangential direction, and has a discontinuity equal to the density in the normal direction:

$$\begin{aligned} W_n^i - W_n^e &= \rho(s,t) \\ W^i - W^e &= 0 \, . \end{aligned}$$

The above two sets of equations and their derivatives determine the discontinuities in the derivatives of W in the coordinate directions. Once determined, these discontinuities are used to determine the discrete Laplacian of W, and the function W itself.

Similarly, we can evaluate integrals of double layer density functions:

$$W = 1/4\pi \int_{\partial} \int_{D} \mu(s,t) \frac{\partial}{\partial n_s}(1/r)dS \, .$$

Just as in the case of an integral of a single layer density function, this type of integral is harmonic in the region D and in the region outside. The only difference is that it is continuous in the normal direction, and has a discontinuity equal to the density in the tangential direction:

$$\begin{aligned} W_n^i - W_n^e &= 0 \\ W^i - W^e &= \mu(s,t) \, . \end{aligned}$$

Once again these equations determine the discontinuities in the derivatives of these integrals in the coordinate directions, and these are used to determine the discrete Laplacian at the irregular mesh points.

We now show how to evaluate derivatives, or a linear combination of the derivatives of a surface or volume integral.

To evaluate discontinuities in derivatives of integrals with differentiated kernels we use the fact that those integrals are derivatives of integrals with undifferentiated kernels. For example, suppose

$$V(p) = \int\int_D\int f(p')G_x(p,p')dV'$$

where $G(p,p')$ is the fundamental solution $\frac{1}{r}$ or its normal derivative.

We note that $V(p) = U_x(p)$ where

$$U(p) = \int\int_D\int f(p')G(p,p')dV'.$$

Since we know how to compute the discontinuities in the derivatives of U, we can, of course, compute the discontinuities in the derivatives of $V = U_x$. As before, once we know the discontinuity terms we can use them to approximate the discrete Laplacian of V.

NUMERICAL EXPERIMENTS

In this section we report on results of numerical experiments.

We first tested this method on problems for which one can evaluate the integrals analytically, both inside and outside the region integration. The test regions D we used were the unit cube, a simplified recording head geometry, and the unit sphere. All the regions were triangulated, and the computations were done in double precision (16 digits) on an IBM 3090 computer. The running time was essentially equal to the time needed to apply a three dimensional Poisson solver. (We used POIS3D from FISHPAK.) Although we have implemented fourth order accurate versions of this method in two dimensions [5],[8], so far we have only implemented a second order accurate method in three dimensions.

We first tested our methods by using using Greens theorem to evaluate harmonic functions. That is, inside a region D we represented a harmonic function U as the sum of a single and a double layer density function:

$$U = 1/4\pi \int_\partial\int_D U\frac{\partial}{\partial n_s}\frac{1}{r} - U_n\frac{1}{r}dS\,.$$

We prescribed the exact values of the function U and its normal derivative at the surface nodes, (i.e. at the vertices of the triangles), and to obtain the values of these functions and their derivatives at other points inside the triangles we used linear interpolation.

In our first test examples we let $U = x + y + z$ on all three regions. As expected, we obtained essentally machine accuracy (14 digits) in computing this function since it is linear. Although the Poisson solver is second order accurate, we did not obtain machine accuracy in computing polynomials of degree 2. This was simply because the linear interpolation of the second degree polynomial values from the vertices to other points in the triangles is not exact. (When we provided the exact values of the U at all points inside the triangle we again obtained essentially machine accuracy.)

In the tables below we present results of other test problems. In these examples the embedding region was the cube $[-.23, 1.3]^3$. The numbers in column 1 are the number of mesh points in each direction in the embedding cube, and the numbers in the second column are the number of surface nodes.

The numbers in column 3 in the following table are the maximum relative errors in computing the function $(x-3)^2 - y^2$ on the cube, and the numbers in column 4 are the maximum relative errors we obtained in computing the function $y((x+4)^2 - z^2) + 3$ on the cube.

Embedded Cube, Max Rel Errors			
1	2	3	4
64	1538	.0009	.0005
32	386	.005	.007
16	26	.009	.02

The numbers in the next table are for computations done on the recording head geometry. This region had equal pole tips of width .1, gap .1, height .8, and uniform depth .8. The numbers in the third and fourth columns are the maximum relative errors in computing the same two functions, $(x-3)^2 - y^2$, and $y((x+4)^2 - z^2) + 3$.

Recording Head, Max Rel Errors			
1	2	3	4
64	1438	.002	.001
32	361	.03	.008
16	96	.09	.08

In the next set of experiments the region of integration D was the unit sphere, and the embedding region was the cube $[-2.3, 2.3]$. We computed the x component of the magnetic field of a unit sphere uniformly magnetized in the x direction. We assumed unit magnetization $M = 1$. This function, which can be represented as the x derivative of a volume integral with density $f(x, y, z) = 1$, is equal to 1 inside the sphere, and equal to $\frac{-3x^2}{r^5} + \frac{1}{r^3}$ outside the sphere. The maximum error we obtained in computing this function when we placed 8190 triangles on the surface was .006, and when there were 2046 surface triangles the maximum error was .04.

The final set of experiments we performed involved parallelization of the code.

For large 3-D problems parallelization is a necessity because of storage requirements. A $1000 \times 1000 \times 1000$ embedding cube contains one billion mesh points and these cannot be stored on a single processor. Although the fast Poisson solver requires only $O(n^3 \log n)$ operations for an $n \times n \times n$ grid, time can also be an issue for large n. Our goal in parallelizing is to be able to solve a problem on a $2n \times 2n \times 2n$ grid using $8p$ processors in about the same amount of time that it takes to solve an $n \times n \times n$ problem using p processors. For this purpose we used the IBM SP2, a distributed memory parallel machine with up to 512 processors.

The part of the code that requires the most time is the Poisson solver. The first step in generating the right-hand side for the Poisson solver is to identify irregular mesh points — those that are inside the region D but have one or more neighbors outside, or vice-versa. This is accomplished by looping through the triangles that make up the boundary of the region and determining which lattice lines intersect each triangle. Each processor can handle its share of the boundary triangles, but a merge of the data is required at the end to check for duplicates, since if a lattice line intersects the boundary at a vertex or edge of a triangle, it might be recorded more than once. Additionally, if a lattice point lies near the surface of the region then it might be reported as outside the region and paired with its neighbor inside, or it might be reported as inside the region and paired with its neighbor outside. If the intersection point is the same, the lattice pairs should be considered the same. It makes little difference whether the point is considered to be just inside or just outside the domain, but it should be recorded only once. Currently, each processor retains the entire list of irregular mesh points.

After determining the irregular mesh points, each processor is assigned certain z-planes in the embedding cube. The processor initializes the right-hand side elements corresponding to its planes to 0 and then loops through the irregular mesh points to determine which ones affect its elements of the right-hand side vector. For the appropriate irregular mesh points, each processor uses information about the boundary intersection point and discontinuities in the integral to determine the right-hand side element corresponding to that point.

Now each processor contains its piece of the right-hand side vector and a parallel 3-D fast Poisson solver can be applied. The fast Poisson solver uses sine transforms in the x, y, and z directions. These are carried out by each processor using ESSL routine SINFT. First the transforms in the x and y directions are computed, and then the data is transposed, so that each processor contains only certain x-planes. Then each processor can compute the sine transforms in the z direction. After this, results are multiplied by appropriate factors, and the inverse sine transforms in the z and y directions are computed. The data is transposed again, so that each processor again contains data for certain z-planes, and inverse transforms in the x direction are computed. This requires two global communication steps. Fortunately, for the problems we have run, this global communication has been a small part of the overall computation time for large problems. The following table shows timings for the fast Poisson solver only, when the amount of data per processor remains fixed. As can be seen, the goal of solving larger problems using more processors, without greatly increasing computation time has been fairly well achieved, although we have yet to perform the largest experiment on 512 processors.

SP-2 Timings (Fast Poisson Solve Only)		
mesh points	processors	time (secs)
128^3	1	9.4
256^3	8	13.0
512^3	64	16.7
1024^3	512	?

REFERENCES

[1] Buneman, O, A compact noniterative Poisson solver, Rep. SUIPR-294, Inst. Plasma Research, Stanford University, Stanford CA, 1969.

[2] Hockney, R., Methods in Computational Physics, vol 9, Academic Press, New York, 1970, pp. 135-211.

[3] James, R., The solution of Poisson's equation for Isolated source distributions, J. Comp. Physics, 25, 71, 1977. pp. 71-93.

[4] Fletcher, R., Conjugate gradient methods for indefinite systems, Numerical Analysis, Dundee 1975, G. A.Watson, Editor, Springer, 1976, pp. 73-89.

[5] Mayo, A. A decomposition finite difference method for the fourth order accurate solution of Poisson's equation on general regions, International Journal of High Speed Computing, vol.3, no. 2, 1991, pp. 89-106.

[6] Mayo, A. and Greenbaum, A., Fast parallel iterative solution of Poisson's and the biharmonic equations on irregular regions, SIAM J. Sci. Stat. Comput. vol. 13, no. 1, 1992.

[7] Mayo, A. , The fast solution of Poisson's and the biharmonic equations on general regions, SIAM J. Numer. Anal, vol. 21, no 2, 1984, pp 285-299.

[8] Mayo, A. Rapid, high order accurate evaluation of volume integrals in potential theory, Journal of Computational Physics, vol. 100, no. 2, 1992, pp. 236-245.

[9] McKenney, A., Greengard, L., and Mayo, A., A fast Poisson solver for complex geometries, J. Comput. Phys, 118, 1995, pp. 348-355.

[10] Meijerink, M. and van der Vorst, H. An iterative solution method for linear systems, Mathematics of Computation, 31, 1977, pp. 148-162.

[11] Mikhlin, S. Integral Equations and their Applications, Pergamon, New York, 1957.

[12] O'Neill, B. Elementary Differential Geometry, Academic Press, London, 1966.

[13] Rokhlin, V. Rapid solution of integral equations of classical potential theory, Journal of Comp. Phy., vol. 60, no. 2, 1985, pp. 187-207.

ADDITIVE SCHWARZ METHODS FOR WEAKLY SINGULAR INTEGRAL EQUATIONS IN $\mathbf{R}^3$ – THE p-VERSION

Norbert Heuer
Institut für Wissenschaftliche Datenverarbeitung
Universität Bremen
Postfach 330 440
28334 Bremen, Germany

SUMMARY

We study additive Schwarz preconditioners with and without overlapping subspaces for the p-version of the boundary element method for solving weakly singular integral equations of the first kind in three dimensions. We prove that the condition numbers of both versions, with and without overlapping subspaces, of the additive Schwarz operator P grow not worse than p^ε where ε is an arbitrary positive constant. Here p is the degree of the polynomials used in the Galerkin boundary element scheme. Numerical experiments underline these theoretical results and demonstrate the applicability of additive Schwarz methods for boundary element schemes in three dimensions.

INTRODUCTION

In this paper we study additive Schwarz methods as preconditioner for the p-version Galerkin scheme for weakly singular integral equations in $\mathbf{R}^3$. We consider the weak formulation

$$a(u,v) := \langle Vu, v\rangle_{L^2(\Gamma)} = \langle g, v\rangle_{L^2(\Gamma)} \quad \text{for all } v \in \tilde{H}^{-1/2}(\Gamma) \tag{1}$$

on a plane open surface piece $\Gamma \subset \mathbf{R}^3$ with the single layer potential operator

$$Vu(x) := \frac{1}{4\pi}\int_\Gamma \frac{u(y)}{|x-y|}\, ds_y, \; x \in \Gamma.$$

V is a continuous and symmetric, positive definite mapping from $\tilde{H}^{-1/2}(\Gamma)$ to $H^{1/2}(\Gamma)$, c f. [7]. Thus we have the norm equivalence

$$a(v,v) \simeq \|v\|^2_{\tilde{H}^{-1/2}(\Gamma)} \quad \text{for any } v \in \tilde{H}^{-1/2}(\Gamma). \tag{2}$$

The Sobolev space $H^{1/2}(\Gamma)$ is defined as

$$H^{1/2}(\Gamma) = \{v|_\Gamma;\ v \in H^{1/2}(\tilde{\Gamma})\}$$

for a closed smooth surface $\tilde{\Gamma}$ containing Γ, and $\tilde{H}^{-1/2}(\Gamma)$ is the dual space of $H^{1/2}(\Gamma)$. For the definition of the Sobolev spaces $\tilde{H}^s(\Gamma)$ and $H^s(\Gamma)$ for arbitrary real s we refer to [2]. Physically the solution u of (1) represents the charge density of a thin, electrified plate Γ loaded with some given potential.

The Galerkin scheme for (1) reads: Find $u_N \in X_N$ such that for every $v \in X_N$

$$a(u_N, v) = \langle g, v\rangle_{L^2(\Gamma)}. \tag{3}$$

Here X_N is a finite dimensional subspace of $\tilde{H}^{-1/2}(\Gamma)$. For the p-version considered in this paper we use a fixed rectangular mesh Γ_h on Γ of width h and use piecewise polynomials of degree p on Γ_h. More precisely we use affine images onto the rectangles of tensor products of scaled Legendre polynomials

$$\sqrt{(2q_1+1)(2q_2+1)}\, L_{q_1}(x_1)\, L_{q_2}(x_2) \tag{4}$$

of degree $q_1 + q_2 \le p$. The scaling by $\sqrt{(2q_1+1)(2q_2+1)}$ normalizes the tensor products with respect to the $L^2(\Gamma)$–norm and already reduces the condition number of the stiffness matrix in (3), c.f. [4] and Figure 1. However, the condition number is still large for high dimensions N of the subspace X_N. Therefore, in order to solve the system (3) efficiently, we need a preconditioner.

Additive Schwarz methods have been studied for several years in the finite element literature, c.f. [1] for a survey and a list of references. For boundary integral equations of the first kind in two dimensions, Schwarz algorithms for the h- and p-versions have recently been investigated, c.f. [3, 4, 6, 8, 9].

Here we present first results for the p-version of the boundary element method in three dimensions. In the next section we briefly recall the additive Schwarz method. We introduce two variants of the additive Schwarz preconditioner and prove for both of them that the condition number grows not worse than p^ε for any $\varepsilon > 0$. In the last section we present numerical results which underline our theoretical estimates and study the dependence, also of several modified versions, on the mesh size h.

Throughout the paper c denotes a generic positive constant which is independent of the degree p.

ADDITIVE SCHWARZ METHODS

We are now giving the general setting for the additive Schwarz method for (3). Let

$$X_N = H_0 + H_1 + \cdots + H_k \tag{5}$$

denote a decomposition of X_N into subspaces H_j. The additive Schwarz method (ASM) consists in solving, by an iterative method, the equation

$$Pu_N := (P_0 + P_1 + \cdots + P_k)u_N = f_N, \tag{6}$$

where the projections $P_j : X_N \to H_j$, $j = 0, \dots, k$, are defined for any $v \in X_N$ by

$$a(P_j v, \varphi) = a(v, \varphi) \quad \text{for any } \varphi \in H_j. \tag{7}$$

The right hand side of (6), $f_N = \sum_{j=0}^{k} P_j u_N$, can be computed without knowing the solution u_N of (3) by

$$a(P_j u_N, \varphi) = \langle g, \varphi \rangle_{L^2(\Gamma)} \quad \text{for any } \varphi \in H_j, \ j = 0, \dots, k. \tag{8}$$

Eq. (6) is the preconditioned linear system and the aim is to estimate its condition number for specific decompositions (5). Ideally this condition number does not depend on the degree p and the mesh width h of the ansatz space X_N of the Galerkin method. In that case the number of iterations of the conjugate gradient method which are neccessary to solve the system (6) up to a given accuracy would not depend on the dimension N of X_N. Such an optimal additive Schwarz preconditioner usually requires a decomposition (5) with a so-called coarse space H_0, which overlaps with each of the subspaces H_j ($j = 1, \dots, k$). However, we consider decompositions of X_N without a coarse space and our numerical experiments show that the influence of this space on the condition number actually is not very large in our cases.

The following lemma is standard in estimating the minimum and maximum eigenvalues of the additive Schwarz operator P.

Lemma 1

(i) If there exists a constant C_1 such that for any $\varphi \in X_N$ there exist $\varphi_j \in H_j$, $j = 0, \dots, k$, satisfying $\varphi = \sum_{j=0}^{k} \varphi_j$ and

$$\sum_{j=0}^{k} a(\varphi_j, \varphi_j) \le C_1^{-1} a(\varphi, \varphi) \tag{9}$$

then

$$\lambda_{\min}(P) \ge C_1.$$

(ii) If there exists a constant C_2 such that for any $\varphi \in X_N$ and $\varphi_j \in H_j$, $j = 0, \dots, k$, satisfying $\varphi = \sum_{j=0}^{k} \varphi_j$ and

$$a(\varphi, \varphi) \le C_2 \sum_{j=0}^{k} a(\varphi_j, \varphi_j) \tag{10}$$

then

$$\lambda_{\max}(P) \le C_2.$$

The next three technical lemmas will be used to prove estimates of the form (9) and (10).

Lemma 2 [10, Lemma 3.2] *Suppose $\Gamma \subset \mathbb{R}^2$ has a Lipschitz boundary and is decomposed into Lipschitz domains $\bar{\Gamma} = \cup_{j=1}^{J} \bar{\Gamma}_j$. Let $s \in \mathbb{R}$. Then for $\tilde{\varphi} \in \tilde{H}^s(\Gamma)$ and $\varphi \in H^s(\Gamma)$ there holds*
(i)

$$\sum_{j=1}^{J} \|\varphi\|^2_{H^s(\Gamma_j)} \le \|\varphi\|^2_{H^s(\Gamma)}$$

and
(ii)

$$\|\tilde{\varphi}\|^2_{\tilde{H}^s(\Gamma)} \le \sum_{j=1}^{J} \|\tilde{\varphi}\|^2_{\tilde{H}^s(\Gamma_j)}.$$

Lemma 3 *Let φ be a polynomial of degree p defined on a rectangle R. Then $\varphi \in \tilde{H}^{-1/2}(R)$ and for any $\varepsilon > 0$ there exists a constant $c > 0$, depending on the size of R, such that*

$$\|\varphi\|_{\tilde{H}^{-1/2}(R)} \le c p^{\varepsilon} \, \|\varphi\|_{H^{-1/2}(R)}. \tag{11}$$

Proof. We have $\varphi \in \tilde{H}^{-1/2}(R)$ since $\varphi \in L^2(R) \subset \tilde{H}^{-1/2}(R)$. The assertion (11) is proved as follows. Let $\varepsilon \in (0,1)$ be given. The space $\tilde{H}^{-1/2+\varepsilon/2}(R)$ is the dual space of $H^{1/2-\varepsilon/2}(R)$. Further, we have the identity

$$\tilde{H}^s(R) = H^s(R) \quad \text{for} \quad s \in [0, 1/2)$$

due to [2, Theorem 1.4.5.2 (b), (c)]. Therefore, we can deduce

$$\|\varphi\|_{\tilde{H}^{-1/2}(R)} \le \|\varphi\|_{\tilde{H}^{-1/2+\varepsilon/2}(R)} = \sup_{\psi \in H^{1/2-\varepsilon/2}(R)} \frac{\langle \varphi, \psi \rangle_{L^2(R)}}{\|\psi\|_{H^{1/2-\varepsilon/2}(R)}}$$

$$= \sup_{\psi \in \tilde{H}^{1/2-\varepsilon/2}(R)} \frac{\langle \varphi, \psi \rangle_{L^2(R)}}{\|\psi\|_{\tilde{H}^{1/2-\varepsilon/2}(R)}} = \|\varphi\|_{H^{-1/2+\varepsilon/2}(R)} \le c h_R^{-\varepsilon/2} p^{\varepsilon} \|\varphi\|_{H^{-1/2}(R)}.$$

For the last estimate we used the inverse property. h_R is the smallest sidelength of R. □

The following lemma is used for estimating the maximum eigenvalue of P. Since it can be applied also to integral operators of orders different from -1 (which is the order of the weakly singular operator) we give a general formulation. The proof is based on a coloring argument as in [11].

Lemma 4 *Let $\{\Gamma_i;\ i = 1, \dots, k\}$ be a finite covering of Γ by subdomains Γ_i with Lipschitz boundary,*

$$\bar{\Gamma} = \cup_{i=1}^{k} \bar{\Gamma}_i,$$

with a covering constant K, i.e. we can color $\{\Gamma_i;\ i = 1, \dots, k\}$ using at most K colors in such a way that subdomains of the same color are disjoint. Let $\varphi = \sum_{i=1}^{k} \varphi_i \in \tilde{H}^s(\Gamma)$ for real s with $\varphi_i \in \tilde{H}^s(\Gamma_i)$, $i = 1, \dots, k$. Then there holds

$$\|\varphi\|^2_{\tilde{H}^s(\Gamma)} \le K \sum_{i=1}^{k} \|\varphi_i\|^2_{\tilde{H}^s(\Gamma_i)}.$$

Proof. Due to the coloring assumption we can split the sum $\sum_{i=1}^{k} \varphi_i$ into at most K sums,

$$\varphi = \sum_{i=1}^{k} \varphi_i = \sum_{j=1}^{K} \sum_{i \in I_j} \varphi_i,$$

where each two functions have at most a subdomain's boundary as a common support:

$$\text{interior } (\text{supp } \varphi_m \cap \text{supp } \varphi_n) = \emptyset \quad \text{for } m, n \in I_j,\ j = 1, \dots, K.$$

Then we can apply Lemma 2(ii) to each of the partial sums,

$$\| \sum_{i \in I_j} \varphi_i \|^2_{\tilde{H}^s(\Gamma)} \leq \sum_{i \in I_j} \|\varphi_i\|^2_{\tilde{H}^s(\Gamma_i)},$$

and obtain by using the triangle inequality

$$\|\varphi\|^2_{\tilde{H}^s(\Gamma)} \leq K \sum_{j=1}^{K} \| \sum_{i \in I_j} \varphi_i \|^2_{\tilde{H}^s(\Gamma)} \leq K \sum_{j=1}^{K} \sum_{i \in I_j} \|\varphi_i\|^2_{\tilde{H}^s(\Gamma_i)} = K \sum_{i=1}^{k} \|\varphi_i\|^2_{\tilde{H}^s(\Gamma_i)}.$$

□

We are now in the position to formulate two specific versions of the additive Schwarz method and we prove that they are nearly optimal for the p-version of the boundary element method.

Non-overlapping subspaces. Let $\Gamma_h = \cup_{j=1}^{J} \Gamma_j$ be a fixed mesh of J rectangles which implicitly define the piecewise polynomials of degree p, see (4). We note that the ansatz functions need not to be continuous. Therefore the simplest choice for the decomposition (5) are the subspaces defined by the restrictions of X_N onto the elements Γ_j. Further we do not use a coarse space, i.e. $H_0 = \emptyset$ in (5). Thus we decompose

$$X_N = H_1 \cup H_2 \cup \cdots \cup H_J \tag{12}$$

where

$$H_j := \{v|_{\Gamma_j};\ v \in X_N\}, \quad j = 1, \ldots, J.$$

Note that (12) is a direct sum decomposition which gives a unique representation for any $v \in X_N$.

Lemma 5 *For any $\varepsilon > 0$ there exists a constant $C_1 > 0$ independent of p such that for the minimum eigenvalue of the additive Schwarz operator P implicitly defined by the decomposition* (12) *there holds*

$$\lambda_{\min}(P) \geq C_1 p^{-\varepsilon}.$$

Proof. Let $v = \sum_{j=1}^{J} v_j$ be the representation of a function $v \in X_N$ according to the decomposition (12). Applying Lemma 3 and 2(i) we obtain

$$\sum_{j=1}^{J} \|v_j\|^2_{\tilde{H}^{-1/2}(\Gamma_j)} \leq c p^{\varepsilon} \sum_{j=1}^{J} \|v_j\|^2_{H^{-1/2}(\Gamma_j)} \leq c p^{\varepsilon} \| \sum_{j=1}^{J} v_j \|^2_{H^{-1/2}(\Gamma)} = c p^{\varepsilon} \|v\|^2_{H^{-1/2}(\Gamma)}.$$

Due to the equivalence (2) and Lemma 1(i) this gives the assertion. □

Lemma 6 *The maximum eigenvalue of the additive Schwarz operator P implicitly defined by the decomposition* (12) *is bounded,*

$$\lambda_{\max}(P) \leq C_2.$$

Proof. Let $v = \sum_{j=1}^{J} v_j$ be the representation of a function $v \in X_N$. We have to show

$$a(v,v) \le C_2 \sum_{j=1}^{J} a(v_j, v_j).$$

Due to the equivalence (2) this is exactly the statement of Lemma 2(ii). □

Combining Lemmas 5 and 6 we have proved

Theorem 1 *For any $\varepsilon > 0$ there exists a constant C independent of p such that there holds*

$$\kappa(P) \le C p^{\varepsilon}$$

for the condition number of the additive Schwarz operator corresponding to the decomposition (12).

Overlapping subspaces. Now we consider an additive Schwarz preconditioner with overlapping subspaces. These overlaps increase the amount of work in numerical applications but, on the other hand, the condition numbers are usually smaller than in the case of non-overlapping methods. Our theoretical results are the same in both cases which means that the bounds for the extremum eigenvalues posess the same asymptotic dependence on the degree p.

We use the same mesh Γ_h as before and choose a decomposition (5) of overlapping subspaces. Let $\{x_i;\ i = 1, \dots, I\}$ denote the set of interior nodes, i.e. the nodes which do not lie on the boundary of Γ. Each of these nodes has four adjacent rectangles which will be denoted by $\Gamma_{i,1}, \Gamma_{i,2}, \Gamma_{i,3}, \Gamma_{i,4}$ $(i = 1, \dots, I)$. Then, each set of rectangles $\bar{\Gamma}'_i := \cup_{j=1}^{4} \bar{\Gamma}_{i,j}$ defines a subspace of X_N, $H_i := \{v|_{\Gamma'_i};\ v \in X_N\}$, which will be used to define the decomposition (5):

$$X_N = H_1 \cup \dots \cup H_I. \tag{13}$$

Lemma 7 *For any $\varepsilon > 0$ there exists a constant $C_1 > 0$ independent of p such that*

$$\lambda_{\min}(P) \ge C_1 p^{-\varepsilon}.$$

Proof. Let $v \in X_N$ be given. We need to find a representation of v according to the decomposition (13) which gives the desired estimate of the minimum eigenvalue based on (9). Since the basis functions of X_N are implicitly defined by the elements Γ_j of the mesh Γ_h the ansatz function v is directly given as $v = \sum_{j=1}^{J} v_j$ where supp $v_j \subset \bar{\Gamma}_j$. This representation can also be used for (13) by assigning each element Γ_j to exactly one domain Γ'_i with $\Gamma_j \subset \Gamma'_i$. This yields a non-overlapping representation $v = \sum_{i=1}^{I} v_i$, i.e. for the interior of the supports $\tilde{\Gamma}'_i$ of v_i there holds

$$\tilde{\Gamma}'_i \cap \tilde{\Gamma}'_j = \emptyset, \quad i,j = 1, \dots, I,\ i \ne j. \tag{14}$$

Applying Lemmas 3 and 2(i) we obtain

$$\sum_{i=1}^{I} \|v_i\|^2_{\tilde{H}^{-1/2}(\tilde{\Gamma}'_i)} \le c p^{\varepsilon} \sum_{i=1}^{I} \|v_i\|^2_{H^{-1/2}(\tilde{\Gamma}'_i)} \le c p^{\varepsilon} \|\sum_{i=1}^{I} v_i\|^2_{H^{-1/2}(\Gamma)} = c p^{\varepsilon} \|v\|^2_{H^{-1/2}(\Gamma)}.$$

For the last inequality we used (14) and the relation $\cup_{i=1}^{I} \tilde{\Gamma}'_i \subset \Gamma$. Due to the equivalence (2) and Lemma 1(i) this gives the assertion. □

Lemma 8 *There exists a constant C_2 independent of p such that*

$$\lambda_{\max}(P) \leq C_2.$$

Proof. Let $v = \sum_{i=1}^{I} v_i$ be any representation of $v \in X_N$. We have to show

$$a(v,v) \leq C_2 \sum_{i=1}^{I} a(v_i, v_i).$$

Due to the equivalence (2) this is exactly the statement of Lemma 4 for $s = -1/2$. □

Combining Lemmas 7 and 8 we have proved

Theorem 2 *For any $\varepsilon > 0$ there exists a constant C independent of p such that there holds*

$$\kappa(P) \leq C\,p^{\varepsilon}$$

for the condition number of the additive Schwarz operator corresponding to the decomposition (13).

NUMERICAL RESULTS

In this section we present numerical results for the Galerkin method for solving the weak formulation (1). We choose the domain Γ to be the square plate $(-1/2, 1/2)^2 \times \{0\}$. The finite-dimensional subspaces X_N of $\tilde{H}^{1/2}(\Gamma)$ are constructed by discontinuous piecewise tensorproducts of scaled or unscaled Legendre polynomials (see (4)) on a partition Γ_h of Γ. For the p-version we use a uniform mesh of 9 rectangles and increase successively the degree p of the polynomials uniformly over the elements.

The entries of the stiffness matrices of the system (3) are implemented by analytical recurrence formulas. For more details we refer to [5]. The condition numbers are computed by reducing the matrices to tridiagonal form and by calculating the minimum and maximum eigenvalues via the QR–algorithm. For this we use standard LAPACK-routines. All the following figures are plotted in a log–log scale.

Figure 1 presents the condition numbers of the unpreconditioned systems and of both additive Schwarz operators covered by Theorems 1 and 2. As mentioned before the scaling of the polynomials reduces the condition number. In the case of unscaled Legendre polynomials this number behaves as p^{α} with α between 2 and 3 and in the case of scaled Legendre polynomials with α between 1 and 2. However, the application of both additive Schwarz preconditioners further reduces considerably the condition number. Both curves have very small slopes and the curve for the overlapping case even appears to be constant at a low level. This behavior is not far off the asymptotic predictions given by Theorems 1 and 2.

In Figure 2 we show the results of numerical experiments using an additional coarse space H_0 in the decomposition (5) and testing diagonal preconditioners. The curves for the additive Schwarz methods without coarse space are the same as in Figure 1.

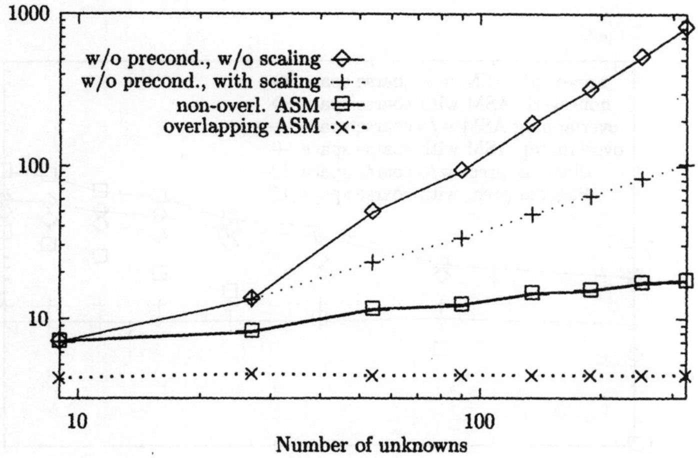

Figure 1: The condition numbers of the original linear systems using unscaled and scaled tensorproducts of Legendre polynomials, and the condition numbers of the additive Schwarz operators defined by the decompositions (12) and (13).

The curves with coarse space base on the same decompositions (12) and (13) with the additional subspace H_0 which consists of the piecewise constant functions on the whole mesh Γ_h. As can be seen in the figure this coarse space does not have a great influence, the respective curves have approximately the same slopes and are close together. In the case of the diagonal preconditioner this influence seems to be greater. However, the condition numbers obtained by applying the diagonal preconditioners increase only slightly faster than those obatined by the non-overlapping additive Schwarz method.

So far we only investigated the dependence of the condition numbers on the degree p of the underlying ansatz space X_N. In Figure 3 we test the dependence of the condition numbers on the mesh width h. In all cases the numbers increase only moderately when decreasing h. This behavior seems not to depend strongly on the presence of a coarse space or an overlap of the subspaces of the additive Schwarz method. Surprisingly, the condition number obtained by the diagonal preconditioner with coarse space even appears to be bounded.

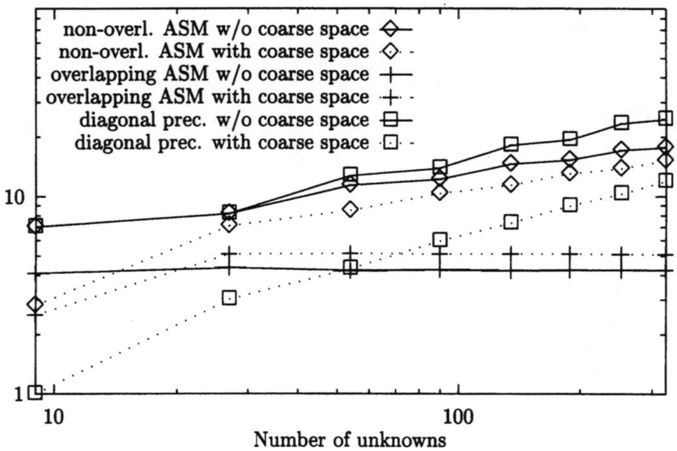

Figure 2: The condition numbers of additive Schwarz operators with and without overlapping subspaces, with and without coarse space, and the diagonal preconditioners with and without coarse space.

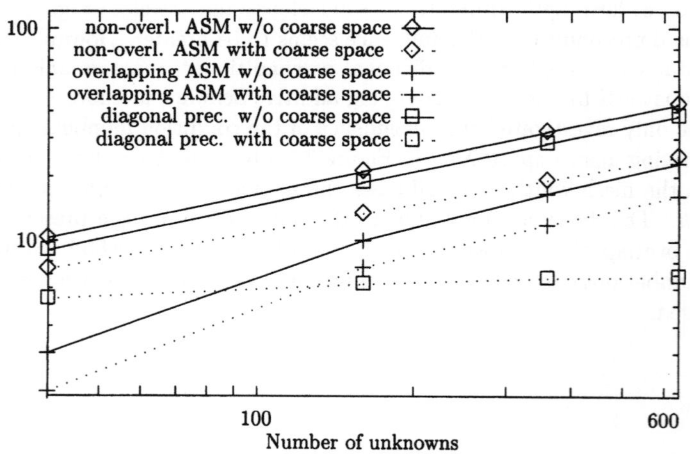

Figure 3: The condition numbers of the different preconditioners for the h-version of the boundary element method. The degree of the ansatz functions is fixed at $p = 3$.

ACKNOWLEDEGMENT. I would like to thank Dr. M. Maischak from the Institut für Angewandte Mathematik at the Universität Hannover for providing me with his boundary element library for calculating the integrals for the stiffness matrices.

REFERENCES

[1] CHAN, T. F. AND MATHEW, T. P.: *Domain decomposition algorithms*, Acta Numerica, 61–143, 1994.

[2] GRISVARD, P.: "Elliptic Problems in nonsmooth domains", Pitman Publishing Inc., Boston, 1985.

[3] HAHNE, M. AND STEPHAN, E. P.: *Schwarz iterations for the efficient solution of screen problems with boundary elements*, Computing (to appear).

[4] HEUER, N.: *Efficient algorithms for the p-version of the boundary element method*, J. Integral Equations Appl. (to appear).

[5] HEUER, N., MAISCHAK, M., AND STEPHAN, E. P.: *The hp-version of the boundary element method for screen problems*, Preprint, Institut für Angewandte Mathematik, Universität Hannover, Germany, 1995.

[6] HEUER, N., STEPHAN, E. P., AND TRAN, T.: *Multilevel additive Schwarz method for the p- and hp-version boundary element method*, Applied Mathematics Report AMR95/37, The University of New South Wales, 1995.

[7] STEPHAN, E. P.: *Boundary integral equations for screen problems in* $\mathbf{R}^3$, Integral Equations Operator Theory 10, 257–263, 1987.

[8] STEPHAN, E. P. AND TRAN, T.: *Additive Schwarz method for the p-version boundary element method*, Applied Mathematics Report AMR95/13, The University of New South Wales, 1995.

[9] TRAN, T. AND STEPHAN, E. P.: *Additive Schwarz method for the h-version boundary element method*, Appl. Anal. (to appear).

[10] VON PETERSDORFF, T.: "Randwertprobleme der Elastizitätstheorie für Polyeder – Singularitäten und Approximation mit Randelementmethoden", PhD thesis, Technische Hochschule Darmstadt, Darmstadt, 1989.

[11] ZHANG, X.: *Multilevel Schwarz methods*, Numer. Math. 63, 521–539, 1992.

ADAPTIVE SCHEMES FOR MULTISCALE DISCRETISATIONS OF BOUNDARY INTEGRAL EQUATIONS

Reinhard Hochmuth[1]
Institut für Mathematik I
Freie Universität Berlin
Arnimallee 2–6
14195 Berlin
email: hochmuth@math.fu-berlin.de

SUMMARY

In this note a-posteriori estimates and adaptive schemes for boundary integral equations in the context of multiscale bases oriented methods are presented. Some basic concepts and ideas from this point of view are applied and related to existing methods in more conventional settings. The local a-posterori estimators arise in a fairly unified fashion essentialy as coefficients of corresponding multiscale expansions. In this way one obtains residual based a-posteriori estimates which are reliable and efficient up to any chosen tolerance. A concrete adaptive strategy based on such estimates can be proved to converge. In principle, the approach applies to an even wider class of elliptic problems (cf. [8]). Those results have been obtained jointly with S. Dahlke, W. Dahmen and R. Schneider.

INTRODUCTION

The understanding of a-posteriori error estimates and adaptive schemes for boundary element methods appears to be far less developed than for finite element methods for partial differential operators (cf. [1,2,4,18]). A discussion of rather general algorithms for boundary integral equations could be found in Rank [16]. Some of the first rigorous results on a-posteriori estimates for boundary element methods for strongly elliptic pseudodifferential operators on smooth boundaries were given by Wendland, Yu in [19]. They introduce a so-called influence index with respect to the underlying set of basis functions, reflecting in some sense the pseudo-local property of strongly elliptic pseudodifferential operators. Their adaptive method is most efficient if the influence index is much smaller than the num. er of basis functions. The involved local error indicators are solutions of

[1]The work of this author was supported by the Graduiertenkolleg "Analyse und Konstruktion in der Mathematik" funded by the Deutsche Forschungsgemeinschaft at the RWTH-Aachen.

local problems. They also transfer their results to those nodal collocation methods which can be interpreted as modified Galerkin methods. Faermann introduced in [13] the approach of local functionals developed by Bank and Rheinboldt for finite element methods for nearly the same class of boundary element methods. In this way she generalizes the results of Wendland, Yu to some extent. Recently Carstensen and Stephan presented in [5,6] a-posteriori estimates within a more general framework. Their approach leads to upper bounds for the global error consisting of terms which can be evaluated locally. Although, they have no proof for the efficiency of their a-posteriori estimates or the convergence of their adaptive algorithms, their numerical examples show the desired behaviour.

The objective of this note is to consider a-posteriori estimates and adaptive schemes in the context of multiscale bases oriented methods. The multiscale approach offers a unified framework within which several important problems could be handled. It differs from usual finite element techniques in that direct use of bases is made which span complements of successive trial spaces. Basic concepts and ideas from this point of view for an even wider class of elliptic operator equations were presented in [8]. In particular, a-posteriori error estimators for Galerkin schemes which are based on stable multiscale bases were derived. These estimators gave upper and lower bounds for the current error by using expressions involving local quantities. Hence they also gave for the first time reliable and efficient a-posterori estimates for operators of negative order. It was shown that the locality of corresponding multiresolution processes combined with certain norm equivalences involving weighted sequence norms lead to adaptive refinement strategies which are guaranteed to converge in a wide range of cases. Multiscale bases seem to be an appropriate tool for taking systematic advantage of the pseudo-local property of boundary integral equations and allow problems like preconditioning, compression and local a-posteriori estimation to be handled in a nearly unified way.

Here we apply this approach to a certain class of boundary integral equations including single layer and hypersingular type operators. We restrict our presentation to boundary integral equations on polygonal closed curves in the plane which are defined by possibly singular kernel functions with typical decay properties and give selfadjoint and positive definite operators with respect to Sobolev spaces. But one should bear in mind that most of the ideas could also be applied to an essential wider class of problems (cf. [8,14]). Especially, most of the results could be directly transferred to integral equations on polygonal geometries including non-closed curves. An extension to integral equations of the second kind would be straightforward.

The note is organized as follows. First, we describe the class of boundary integral equations considered. Then, we introduce multiscale bases and summarize some of their properties which are important in our context. Next we investigate Galerkin schemes with respect to arbitrary finite sets of basis functions. We define local error terms which give via norm-equivalences efficient and reliable a-posteriori error estimators. In a first step we obtain infinite series. Using decay estimates which are also studied in the context of compression strategies for boundary element methods (cf. [10,17]), we yield a-posteriori estimators within a prescribed tolerance by finite sums. The crucial step in this procedure is the appropriate choice of finite index sets. In [8] a general result concerning the existence of appropriate finite index sets with respect to an arbitrary tolerance was given. Finally a convergent adaptive scheme, i.e. a scheme which guarantees an improvement of the solution controlled by the a-posteriori estimates, is described. The convergence result also holds for integral equations which except for a sufficiently small perturbation are symmetric (cf. [8,14]).

For proofs concerning the general aspects and results of the approach we refer the reader to [8]. Proofs and more detailed calculations with respect to the technical terms especially concerning the selection of appropriate index sets can be found in [14].

BOUNDARY INTEGRAL EQUATIONS

We consider singular integral equations of the first kind on polygonal, closed and non-intersecting curves Γ in $\mathbb{R}^2$, i.e. Γ consists of $N \in \mathbb{N}$ straight lines Γ_i, $\Gamma = \bigcup_{i=1}^{N} \overline{\Gamma}_i$, where $\Gamma_i := \gamma_i(0,1)$, $i = 1, \ldots, N$ and $\gamma_i : (0,1) \longrightarrow \mathbb{R}^2$ denote affine functions. The definition of Sobolev spaces for real s is as usal

$$\begin{aligned} H^s(\Gamma) &:= \{u|_\Gamma \mid u \in H^{s+1/2}(\mathbb{R}^2)\}, \quad s > 0, \\ H^0(\Gamma) &:= L^2(\Gamma), \\ H^s(\Gamma) &:= (H^{-s}(\Gamma))', \quad \text{(dual spaces)} \quad s < 0, \\ H^s(\Gamma_i) &:= \{u|_{\Gamma_i} \mid u \in H^s(\Gamma)\}, \quad s \geq 0, i = 1, \ldots, N. \end{aligned}$$

The boundary integral operators are defined with respect to kernel functions $a(\cdot,\cdot) : \Gamma \times \Gamma \to \mathbb{C}$ which are smooth outside the diagonal $\{(x,y) \in \Gamma \times \Gamma \mid x = y\}$ and satisfy the decay estimates

$$|\partial_y^\beta \partial_x^\alpha a(x,y)| \lesssim |x-y|^{-1-\rho-\alpha-\beta}, \quad \alpha, \beta \in \mathbb{N}_0, \ \alpha + \beta > 0, \tag{1}$$

uniformly in $x, y \in \Gamma$ with some $\rho \in [-1, 1]$. The notion ' $\lesssim$ ' means that the left term in (1) can be bounded by a constant multiple of the right side where the constant is independent of any parameters they may depend on. Analogously we use the symbols ' $\gtrsim$ ' and '$\sim$'. The boundary integral operators A are defined by

$$Au(x) := \int_\Gamma a(x,y)\, u(y)\, ds_y. \tag{2}$$

We assume that they are continuous with respect to a certain range of Sobolev spaces, i.e. for certain $s_1 < \rho/2 < s_2$ the operators

$$A : H^s(\Gamma) \longrightarrow H^{s-\rho}(\Gamma), \qquad s \in (s_1, s_2) \tag{3}$$

are continuous. As usual, ρ indicates the order of the operator. The boundary integral equation problem consists in finding for a given $f \in H^{-\rho/2}(\Gamma)$ a function $u \in H^{\rho/2}(\Gamma)$ with

$$Au = f. \tag{4}$$

To simplify matters, we further assume that A is selfadjoint and positive definite with respect to $H^{\rho/2}(\Gamma)$, i.e. $\|u\|^2 := \langle Au, u \rangle$ defines an energy norm, which is equivalent to the $\|\cdot\|_{\rho/2}$-norm. Here $\langle \cdot, \cdot \rangle$ denotes the duality between $H^{-\rho/2}(\Gamma)$ and $H^{\rho/2}(\Gamma)$, such that $\langle u, v \rangle = \int_\Gamma u(x)\overline{v(x)}\, ds_x$ for smooth functions u and v. Our assumptions immediately imply that the operator $A : H^{\rho/2}(\Gamma) \to H^{-\rho/2}(\Gamma)$ is boundedly invertible, i.e. one has

$$\|Au\|_{-\rho/2} \sim \|u\|_{\rho/2}, \quad u \in H^{\rho/2}(\Gamma), \tag{5}$$

and that equation (4) possesses for arbitrary given functions $f \in H^{-\rho/2}(\Gamma)$ a unique solution $u \in H^{\rho/2}(\Gamma)$.
Notice that e.g. the single layer potential is included with $a(x,y) := \ln|x-y|$, $s_1 = -3/2$, $s_2 := 1/2$, $\rho = -1$ if $\text{cap}(\Gamma) \neq 1$, where $\text{cap}(\Gamma)$ denotes the capacity of Γ.

We remark that most of the presented results require only the property (5) and even for the convergence of the adaptive scheme described in the last section the symmetry of A is not really necessary (cf. [8]).

MULTISCALE BASES

Let be given a sequence of nested finite dimensional subspaces

$$S_{j_0} \subset S_{j_0+1} \subset \cdots \subset S_j \subset S_{j+1} \subset \cdots \subset H^{\rho/2}(\Gamma)$$

with $\overline{\cup_{j \geq j_0} S_j} = H^{\rho/2}(\Gamma)$, where $j_0 \in I\!N$ denotes the coarsest level. That $j_0 > 1$ typically appears in the construction of biorthogonal wavelet bases on an interval and secures among other things the nestedness of subspaces. It increases for higher vanishing moments (cf. [11]). We assume that there are suitable complements W_j ($W_{j_0} := S_{j_0}$), i.e.

$$S_j = S_{j-1} \oplus W_j\,, \qquad j \in I\!N, j \geq j_0 + 1,$$

with uniformly stable bases

$$\Psi_j = \{\psi_{j,k} \mid k \in J_j\}, \qquad j \geq j_0,$$

where J_j, $j \geq j_0$, denote appropriate finite index sets, such that for problem relevant norms $\|\cdot\|_s$, especially for $|s| \leq \rho/2$, one has norm–equivalences

$$\|u\|_s \sim \Big(\sum_{j=j_0}^{\infty} \sum_{k \in J_j} 2^{2sj} |\langle u, \psi_{j,k}\rangle|^2\Big)^{1/2}, \qquad u \in H^s(\Gamma). \tag{6}$$

Furthermore we assume, that Ψ_j are local bases, i.e.

$$\text{diam}\,(\text{supp}\,\psi_{j,k}) \sim 2^{-j}\,, \quad j \geq j_0\,. \tag{7}$$

Another important requirement is, that the basis functions $\psi_{j,k}$, $j > j_0$ have vanishing moments of order $d - 1 > -\rho$, $d \in I\!N$, with respect to the straight lines Γ_i, i.e. one has

$$\int_0^1 p(s)\, \psi_{j,k}|_{\Gamma_i} \circ \gamma_i(s)\, ds = 0\,, \quad p \in \Pi_{d-1}(0,1)\,, j \geq j_0 + 1, \tag{8}$$

where $\Pi_{d-1}(0,1)$ denotes the set of polynomials of order less than d on $(0,1)$. We emphasize that the basis functions on the coarsest level, i.e. $\psi_{j_0,k}$, $k \in J_{j_0}$, do not have vanishing moments.

Finally we fix some regularity properties: In local coordinates the basis functions are piecewise polynomials of order less than d, i.e.

$$\psi_{j,k}|_{\Gamma_i} \circ \gamma_i \in \Pi_{d-1,\text{pw}}(0,1)\,, \quad j \geq j_0, k \in J_j, i = 1, \ldots, N. \tag{9}$$

Clearly, the basis functions generally span only a proper subspace of $\Pi_{d-1,\mathrm{pw}}(0,1)$. Furthermore we assume that for $\rho \leq 0$, the basis functions $\psi_{j,k}$ are Hölder continuous with exponent $\sigma \in (0,1]$, i.e.

$$|\psi_{j,k}(x) - \psi_{j,k}(x')| \lesssim 2^{j(1/2+\sigma)}[\mathrm{dist}(x,x')]^{\sigma},$$

while for $\rho > 0$ we require Hölder continuity of order $\sigma > \rho/2$.

We take the existence of such bases for granted and refer the reader for a construction even for the 3D–case to [11]. Basic elements of their approach are Bernstein- and Jackson-estimates, biorthogonality and the concept of stable completions (cf. [7,9]).

Galerkin schemes with respect to suitable "full" spaces $S_j = \oplus_{\ell=j_0}^{j} W_\ell$ are stable and convergent for the described class of boundary integral equations (cf. [10]): If the order ρ of the operator is not zero, it is well-known, that the Galerkin matrices possess growing condition numbers. But the norm-equivalences suggest a diagonal preconditioner which give bounded condition numbers independent of the level j of the discretisation (cf. [10]). Furthermore the Galerkin schemes allow compression techniques leading to so called $\mathcal{O}(N)$–algorithms (cf. [17]).

In the present paper the main concern is not to find possibly sparse representations of the operator A relative to a-priori fixed trial spaces but to find possibly economical trial spaces leading to as small linear systems as possible. This task will be treated in the next sections.

A-POSTERIORI ESTIMATES FOR GALERKIN SCHEMES

Our goal is to study Galerkin schemes with respect to adaptively chosen subspaces. To describe general subspaces we introduce arbitrary finite index sets

$$\Lambda \subset J := \{\lambda = (j,k) \mid k \in J_j,\ j \geq j_0\}$$

and define

$$S_\Lambda := \mathrm{span}\,\{\psi_\lambda \mid \lambda \in \Lambda\}.$$

We assume that $S_{j_0} \subset S_\Lambda$. The corresponding Galerkin solutions $u_\Lambda = \sum_{\lambda' \in \Lambda} u_{\lambda'}\psi_{\lambda'} \in S_\Lambda$ of (4) are the unique solutions of the finite dimensional problems

$$\langle Au_\Lambda, v\rangle = \langle f, v\rangle, \quad v \in S_\Lambda. \tag{10}$$

According to the properties of A, it immediately follows that (10) is uniquely solvable and stable with respect to the subspaces S_Λ. Moreover, $\rho \neq 0$ leads to Galerkin matrices with unbounded condition numbers. But analogously to discretisations with respect to the full spaces S_j the norm-equivalences (6) allow the definition of diagonal preconditioners which yield bounded condition numbers (cf. [8]).

We want to base the adaptive selection of the subspaces S_Λ on a-posteriori estimates. The starting point for their construction are the mapping properties of the operator A

and the norm–equivalences induced by the multiscale basis. More precisely, (5) implies that

$$\|u - u_\Lambda\|_{\rho/2} \sim \|r_\Lambda\|_{-\rho/2} \tag{11}$$

where r_Λ denotes the residuum, i.e. $r_\Lambda := f - Au_\Lambda$. Then the norm–equivalence (6) with $s = -\rho/2$ give

$$\begin{aligned}\|u - u_\Lambda\|_{\rho/2} &\sim \Big(\sum_{\lambda \in J\setminus\Lambda} 2^{-\rho|\lambda|}\,|\langle r_\Lambda, \psi_\lambda\rangle|^2\Big)^{1/2} \\ &= \Big(\sum_{\lambda \in J\setminus\Lambda} \delta_\lambda^2\Big)^{1/2}\end{aligned} \tag{12}$$

where

$$\delta_\lambda := 2^{-\rho|\lambda|/2}|\langle r_\Lambda, \psi_\lambda\rangle| = 2^{-\rho|\lambda|/2}|\langle f, \psi_\lambda\rangle - \sum_{\lambda' \in \Lambda} \langle A\psi_{\lambda'}, \psi_\lambda\rangle u_{\lambda'}| \tag{13}$$

and $|\lambda| := j$ for $\lambda \in J_j$. One can separate the given data f from the Galerkin solution by the triangle inequality and get with $f_\lambda := \langle f, \psi_\lambda\rangle$

$$\delta_\lambda \le 2^{-\rho|\lambda|/2}|f_\lambda| + 2^{-\rho|\lambda|/2}|\sum_{\lambda' \in \Lambda} \langle A\psi_{\lambda'}, \psi_\lambda\rangle u_{\lambda'}|.$$

In (12) and (13) we have used that the Galerkin equations (10) are equivalent to

$$\langle r_\Lambda, \psi_\lambda\rangle = 0\,, \quad \lambda \in \Lambda.$$

Because the basis functions ψ_λ are local (cf. (7)), the error terms δ_λ are also local. Thus (12) is an a-posteriori error estimator existing of local terms which is reliable and efficient. Similar estimates are deduced by Bertoluzza in [3] and Oswald in [15] although in another context of problems.

Furthermore, we observe that the quantities δ_λ can be replaced by

$$\zeta_\lambda := \frac{|\langle r_\Lambda, \psi_\lambda\rangle|}{\|\psi_\lambda\|},$$

which is nothing other than the energy norm $\|\cdot\|$ of

$$w_{\Lambda,\lambda} := \frac{\langle r_\Lambda, \psi_\lambda\rangle}{\|\psi_\lambda\|^2}\psi_\lambda,$$

and $w_{\Lambda,\lambda}$ are the solutions of the local problems

$$\langle Aw_{\Lambda,\lambda}, v\rangle = \langle r_\Lambda, v\rangle, \qquad v \in \operatorname{span}\{\psi_\lambda\}.$$

Because (12) consists of infinitely many terms, the question arises as to get finite sums. A first answer assumes that the saturation property is fullfilled, i.e. there is a canonical finite index set $\widetilde{\Lambda}$, $\Lambda \subset \widetilde{\Lambda} \subset J$, such that for a constant $\kappa < 1$

$$\|u - u_{\widetilde{\Lambda}}\| \le \kappa\,\|u - u_\Lambda\|. \tag{14}$$

Simple calulations show that (12) implies

$$\|u - u_\Lambda\| \sim \Big(\sum_{\lambda \in \widetilde{\Lambda}\setminus\Lambda} \delta_\lambda^2\Big)^{1/2}. \tag{15}$$

The saturation assumption is often used in the context of finite element methods (cf. [2,4]) and seems natural in the sense that a greater subspace should give a better solution. But the constant $\kappa < 1$ could become arbitrarily bad, i.e. $\kappa \sim 1$, even for the selfadjoint and positive definite case, which implies that the equivalence (15) gets lost. A further shortcoming is that the argumentation breaks down if A is not selfadjoint.

In [8] was demonstrated how to obtain a-posteriori estimates for multiscale discretisations of elliptic problems without assuming beforehand the saturation property. For the special case of second order two point boundary value problems a similar result was shown by Bertuluzza in [3]. The crucial idea is to use decay estimates of $|\langle A\psi_{\lambda'}, \psi_\lambda\rangle|$, $\lambda' \in \Lambda$, $\lambda \in J \setminus \Lambda$, for the definition of new a-posteriori estimates which up to any chosen tolerance are still efficient and reliable. Similarly to compression techniques the decay estimates lead to the observation that for almost all $\lambda \in J$ the sums $\sum_{\lambda' \in \Lambda} \langle A\psi_{\lambda'}, \psi_\lambda\rangle u_{\lambda'}$ can actually be neglected.

In the following we distinguish for an arbitrary tolerance $\varepsilon > 0$ and for $\lambda \in J \setminus \Lambda$ index sets $N_\lambda^\varepsilon \subset \Lambda$ such that the terms $2^{-\rho|\lambda|/2} | \sum_{\lambda' \in \Lambda} \langle A\psi_{\lambda'}, \psi_\lambda\rangle u_{\lambda'}|$ will be replaced by

$$a_\lambda := a_\lambda(\Lambda, \varepsilon) := 2^{-\rho|\lambda|/2} \, | \sum_{\lambda' \in N_\lambda^\varepsilon} \langle A\psi_{\lambda'}, \psi_\lambda\rangle \, u_{\lambda'} | , \qquad \lambda \in J \setminus \Lambda. \tag{16}$$

There are only finitely many $a_\lambda \neq 0$ if and only if

$$\sharp N_{\Lambda,\varepsilon} < \infty \qquad \text{for } N_{\Lambda,\varepsilon} := \{\lambda \in J \setminus \Lambda \mid N_\lambda^\varepsilon \neq \emptyset\}. \tag{17}$$

Furthermore, we require an estimate for the remaining terms, i.e. there is for

$$e_\lambda := \sum_{\lambda' \in \Lambda \setminus N_\lambda^\varepsilon} \langle A\psi_{\lambda'}, \psi_\lambda\rangle u_{\lambda'} , \quad \lambda \in J \setminus \Lambda \tag{18}$$

a constant $\beta > 0$ independent of ε and u_Λ with

$$\Big(\sum_{\lambda \in J \setminus \Lambda} 2^{-\rho|\lambda|} |e_\lambda|^2 \Big)^{1/2} \leq \beta\varepsilon \, \|u_\Lambda\|_{\rho/2} \, . \tag{19}$$

Then the following theorem holds (cf. [8]).

Theorem 1 *Under the above assumptions one has*

$$\|u - u_\Lambda\|_{\rho/2} \lesssim \Big(\sum_{\lambda \in N_{\Lambda,\varepsilon}} a_\lambda^2 \Big)^{1/2} + \beta\varepsilon \, \|u_\Lambda\|_{\rho/2} + \Big(\sum_{\lambda \in J \setminus \Lambda} 2^{-\rho|\lambda|} |f_\lambda|^2 \Big)^{1/2} , \tag{20}$$

as well as

$$\Big(\sum_{\lambda \in N_{\Lambda,\varepsilon}} a_\lambda^2 \Big)^{1/2} \lesssim \|u - u_\Lambda\|_{\rho/2} + \beta\varepsilon \, \|u_\Lambda\|_{\rho/2} + \Big(\sum_{\lambda \in J \setminus \Lambda} 2^{-\rho|\lambda|} |f_\lambda|^2 \Big)^{1/2} . \tag{21}$$

Moreover, for $\Lambda \subset \widetilde{\Lambda} \subset J$ we have for the Galerkin solutions u_Λ and $u_{\widetilde{\Lambda}}$

$$\Big(\sum_{\lambda \in \widetilde{\Lambda} \cap N_{\Lambda,\varepsilon}} a_\lambda^2 \Big)^{1/2} \lesssim \|u_{\widetilde{\Lambda}} - u_\Lambda\|_{\rho/2} + \beta\varepsilon \, \|u_\Lambda\|_{\rho/2} + \Big(\sum_{\lambda \in J \setminus \Lambda} 2^{-\rho|\lambda|} |f_\lambda|^2 \Big)^{1/2} . \tag{22}$$

The inequalities (20) and (21) show that $\eta_{\Lambda,\varepsilon} := (\sum_{\lambda\in N_{\Lambda,\varepsilon}} a_\lambda^2)^{1/2}$ defines up to the tolerance $\varepsilon > 0$ an efficient and reliable a-posteriori estimator. The inequality (22) relates two successive Galerkin solutions and is crucial for the convergence of the adaptive scheme presented in the last section. The former considerations indicate that it remains to find index sets N_λ^ε with (17)–(19).

To formulate decay estimates which are decisive for our approach we introduce some notation. For a basis function ψ_λ, $\lambda \in J$, we define $\Omega_\lambda := \text{supp } \psi_\lambda$ and $\Omega_\lambda^s := \text{sing supp } \psi_\lambda$. For two functions $\psi_{\lambda'}, \psi_\lambda$, $\lambda', \lambda \in J$, we denote the distance between their supports by $d_{\lambda',\lambda}$, i.e. $d_{\lambda',\lambda} := \text{dist }(\Omega_{\lambda'}, \Omega_\lambda)$ and the distance between the singular support of $\psi_{\lambda'}$ and the support of ψ_λ by $d_{\lambda',\lambda}^s := \text{dist }(\Omega_{\lambda'}^s, \Omega_\lambda)$. Then one has for any $\lambda' \in \Lambda \setminus J_{j_0}, \lambda \in J \setminus \Lambda$ estimates like (cf. [8,10,17])

$$2^{-\rho\frac{|\lambda|+|\lambda'|}{2}} \, |\langle A\psi_{\lambda'}, \psi_\lambda\rangle| \lesssim \begin{cases} 2^{-||\lambda|-|\lambda'||(1/2+\sigma)}, & d_{\lambda',\lambda} \lesssim 2^{-\min(|\lambda'|,|\lambda|)}, \\ d_{\lambda',\lambda}^{-1-\rho-2d} 2^{-(d+1/2+\rho/2)(|\lambda|+|\lambda'|)}, & d_{\lambda',\lambda} > 0. \end{cases} \tag{23}$$

A more careful analysis leads to the following additional estimate for $\Omega_\lambda \subset \Omega_{\lambda'}, d_{\lambda',\lambda}^s > 0$ (cf.[17])

$$2^{-\rho\frac{|\lambda|+|\lambda'|}{2}} \, |\langle A\psi_{\lambda'}, \psi_\lambda\rangle| \lesssim d_{\lambda',\lambda}^{s\,-\rho-d} 2^{-(d+1/2+\rho/2)|\lambda|} 2^{(1/2-\rho/2)|\lambda'|}.$$

Furthermore for any $\lambda' \in J_{j_0}, \lambda \in J \setminus \Lambda$ and $d_{\lambda',\lambda} > 0$ one has (cf. [14])

$$2^{-\rho\frac{|\lambda|+|\lambda'|}{2}} \, |\langle A\psi_{\lambda'}, \psi_\lambda\rangle| \lesssim d_{\lambda',\lambda}^{-1-\rho-d} 2^{-(d+1/2+\rho/2)|\lambda|} 2^{-(1/2+\rho/2)j_0}. \tag{24}$$

For $d_{\lambda',\lambda} \geq 0$ we assume, that there is some d' with $d' - \rho/2 > 1/2$ and

$$2^{-\rho\frac{|\lambda|+|\lambda'|}{2}} \, |\langle A\psi_{\lambda'}, \psi_\lambda\rangle| \lesssim 2^{-||\lambda|-|\lambda'||(d'-\rho/2)}. \tag{25}$$

For the single layer potential these requirements are satisfied under rather weak assumptions on the multiscale basis if $d' \in (0, 1/2)$. Similar results hold for the hypersingular and the double-layer potential (cf. [14]).

In [8] a principal result concerning index sets N_λ^ε with (17)–(19) is proved by the strategy which was also successfully applied in the context of compression techniques. Here we state a result which is based on (23), (24) and (25).

Theorem 2 *Assume that (23), (24) and (25) hold and let be given $\varepsilon > 0$. Suppose that for $k_1, k_2 \in \mathbb{N}$*

$$2^{-k_1(d+\rho/2)} + 2^{-k_2(d'-\rho/2-1/2)} \leq \varepsilon. \tag{26}$$

Furthermore let $0 < \delta < \sigma$ be fixed and choose positive numbers $\varepsilon_1, \varepsilon_2 > 0$ such that

$$\varepsilon_1^{2d+\rho} + 2^{-\delta/\varepsilon_2} \leq \varepsilon. \tag{27}$$

Then for the following index sets with $\lambda \in J \setminus \Lambda$

$$\begin{aligned} N_{\lambda;1} &:= \{\lambda' \in J_{j_0} \mid d_{\lambda',\lambda} \geq 2^{-j_0}, |\lambda| \leq j_0 + k_1\}, \\ N_{\lambda;2} &:= \{\lambda' \in J_{j_0} \mid d_{\lambda',\lambda} < 2^{-j_1}, |\lambda| \leq j_0 + k_2\}, \\ N_{\lambda;3} &:= \{\lambda' \in \Lambda \setminus J_{j_0} \mid 2^{\min(|\lambda|,|\lambda'|)} d_{\lambda',\lambda} \leq \varepsilon_1^{-1} \textit{ and } ||\lambda| - |\lambda'|| \leq \varepsilon_2^{-1}\} \end{aligned}$$

and

$$N_\lambda^\varepsilon := \bigcup_{i=1}^{3} N_{\lambda;i},$$

there exists a constant $\beta > 0$ such that for

$$e_\lambda := \sum_{\lambda' \in \Lambda \setminus N_\lambda^\varepsilon} \langle A\psi_{\lambda'}, \psi_\lambda \rangle u_{\lambda'}, \quad \lambda \in J \setminus \Lambda,$$

one has

$$\Big(\sum_{\lambda \in J \setminus \Lambda} 2^{-\rho|\lambda|} |e_\lambda|^2 \Big)^{1/2} \leq \beta \varepsilon \, \|u_\Lambda\|_{\rho/2} \,.$$

The condition (26) is derived in [14] and the condition (27) in [10]. Clearly, there are different strategies possible to define index-sets N_λ^ε with (17)–(19). The decision which strategy is "optimal" depends on further investigations concerning the complexity of adaptive multiscale algorithms.

ADAPTIVE SCHEMES

We describe an adaptive feedback procedure, where the algorithm itself decides where to add basis functions in order to improve the Galerkin solution. The strategy is based on the presented a-posteriori estimates and can be proved to converge without any regularity assumptions on the solution (cf. [8]). It is similar in spirit to those by Dörfler [12] who considers adaptive refinements of piecewise linear finite elements for Poisson's equations in two dimensions.

Theorem 3 *There exists a sufficiently small $\mu > 0$ such that for an arbitrary tolerance* eps > 0, *an index set $\Lambda \subset J$ with*

$$\Big(\sum_{\lambda \in J \setminus \Lambda} 2^{-\rho|\lambda|} |f_\lambda|^2 \Big)^{1/2} < \frac{1}{2} \, \mu \, eps$$

and

$$\varepsilon := \frac{\mu \; eps}{2\beta \, \|u_\Lambda\|_{\rho/2}},$$

any finite index set $\widetilde{\Lambda} \subset J$, $\Lambda \subset \widetilde{\Lambda}$, which satisfies

$$\Big(\sum_{\lambda \in \widetilde{\Lambda} \setminus \Lambda} a_\lambda^2 \Big)^{1/2} \geq 1/2 \Big(\sum_{\lambda \in N_{\Lambda,\varepsilon}} a_\lambda^2 \Big)^{1/2}$$

ensures, that for some $\kappa \in (0,1)$ either

$$\|u - u_{\widetilde{\Lambda}}\| \leq \kappa \, \|u - u_\Lambda\|$$

or

$$\Big(\sum_{\lambda \in N_{\Lambda,\varepsilon}} a_\lambda^2 \Big)^{1/2} < eps \,.$$

The appearing constants, including κ, depend only on global constants with respect to the operator A and constants in the applied norm–equivalences.

We remark that the term $\|u_\Lambda\|_{\rho/2}$ could be replaced by $(\sum_{\lambda'\in\Lambda} 2^{\rho|\lambda'|}|u_{\lambda'}|^2)^{1/2}$. The constants then change in an obvious way. Theorem 3 in particular claims, that, if $(\sum_{\lambda\in N_{\Lambda,\varepsilon}} a_\lambda^2)^{1/2} \geq$ eps, the saturation property (14) holds, which implies (15). Thus, the assumptions of Theorem 3 strengthen in some sense the estimates (20) and (21) in the selfadjoint case, which could also be seen directly.

If μ and β are known and an initial accuracy eps_1 and a final accuracy eps are choosen a convergent adaptive algorithm reads as following:

Set $\Lambda_{1,0} := \emptyset$.
Choose for $i = 1, \infty$ index sets $\Lambda_{i,1} \subset J$, $\Lambda_{i,1} \supset \Lambda_{i,0}$ such that

$$(\sum_{\lambda\in J\setminus\Lambda_{i,1}} 2^{-\rho|\lambda|}|f_\lambda|^2)^{1/2} < \frac{1}{2}\mu\,\text{eps}_i.$$

Compute for $j = 1, \infty$ the Galerkin solution $u_{\Lambda_{i,j}}$ with respect to $S_{\Lambda_{i,j}}$.
Set

$$\varepsilon := \frac{\mu\cdot\text{eps}_i}{2\beta\|u_{\Lambda_{i,j}}\|_{\rho/2}}$$

Determine an appropriate index set $N_{\Lambda_{i,j},\varepsilon}$ and

$$\eta_{\Lambda_{i,j};\varepsilon} := (\sum_{\lambda\in N_{\Lambda_{i,j},\varepsilon}} a_\lambda^2)^{1/2}.$$

If $\eta_{\Lambda_{i,j},\varepsilon} < \text{eps}_i$ set $j := \infty$,
else choose $\Lambda_{i,j+1} \supset \Lambda_{i,j}$, $\Lambda_{i,j+1} \setminus \Lambda_{i,j} \subset N_{\Lambda_{i,j},\varepsilon}$ such that

$$(\sum_{\lambda\in\Lambda_{i,j+1}\setminus\Lambda_{i,j}} a_\lambda^2)^{1/2} \geq 1/2\,\eta_{\Lambda_{i,j},\varepsilon}.$$

If $\text{eps}_i < \text{eps}$ go to Stop,
else set $\text{eps}_{i+1} := \text{eps}_i/2$ and $\Lambda_{i+1,0} := \Lambda_{i,j}$.

Stop

REFERENCES

[1] Babuška, I., Miller, A.: " A feedback finite element method with a posteriori error estimation: Part I. The finite element method and some basic properties of the a posteriori error estimator." Comput. Methods Appl. Mech. Engrg. 61, 1987, 1–40.

[2] Bank, R.E., Weiser, A.: "Some a posteriori error estimates for elliptic partial differential equations." Math. Comp. 44, 1985, 283 – 301.

[3] Bertoluzza, S.: "A posteriori error estimates for the wavelet Galerkin method." Appl. Math. Lett. 8-5, 1995, 1–6.

[4] Bornemann, F., Erdmann, B., Kornhuber, R.: "A posteriori error estimates for elliptic problems in two and three space dimensions." Int. J. Numer. Methods Eng. 36, 1993, 3187-3203.

[5] Carstensen, C., Stephan, E.P.: "A posteriori error estimates for boundary element methods." To appear in Math. Comp.

[6] Carstensen, C., Stephan, E.P.: "Adaptive boundary element methods for some first kind integral equations." To appear in SIAM J. Numer. Anal.

[7] Carnicer, J.M., Dahmen, W., Peña, J.M.: "Local decomposition of refinable spaces and wavelets." RWTH-Aachen IGPM Preprint # 112, 1994, to appear in ACHA.

[8] Dahlke, S., Dahmen, W., Hochmuth, R., Schneider, R.: "Stable Multiscale Bases and Local Error Estimation for Elliptic Problems." RWTH-Aachen, IGPM-Preprint 124, 1996.

[9] Dahmen, W.: "Stability of multiscale transformations." RWTH-Aachen, IGPM-Preprint 109, 1994.

[10] Dahmen, W., Prössdorf, S., Schneider, R.: "Multiscale methods for pseudo-differential equations on smooth manifolds." in: Proceedings of the International Conference on Wavelets: Theory, Algorithms, and Applications, C.K. Chui, L. Montefusco, L. Puccio (eds.), Academic Press, 385-424, 1994.

[11] Dahmen, W., Schneider R.: "Multiscale methods for boundary integral equations I: Biorthogonal wavelets on 2D-manifolds in $\mathbb{R}^3$." In preparation.

[12] Dörfler, W.: "A convergent adaptive algorithm for Poisson's equation." To appear in SIAM J. Numer. Anal.

[13] Faermann, B.: "Lokale a-posteriori-Fehlerschätzer bei der Diskretisierung von Randintegralgleichungen." Dissertation, Christian-Albrechts-Universität Kiel, Germany, 1993 (in German).

[14] Hochmuth, R.: "A-posteriori estimates and adaptive schemes for transmission problems." In preparation.

[15] Oswald, P.: "Multilevel finite element approximation." Teubner Skripten zur Numerik, Stuttgart, 1994.

[16] Rank, E.: "Adaptive h-, p- and hp-versions for boundary element methods." Int. J. Numer. Meth. Engrg. 28, 1989, 1335 – 1349.

[17] Schneider, R.: "Multiskalen- und Wavelet-Matrixkompression: Analysisbasierte Methoden zur Lösung großer vollbesetzter Gleichungssysteme." Habilitationsschrift, TH Darmstadt, Germany, 1995 (in German).

[18] Verführt, R.: "A posteriori error estimation and adaptive mesh refinement techniques." J. Comput. and Appl. Math. 50, 1994, 67–83.

[19] Wendland, W.L., Yu, D.: "Adaptive BEM for strongly elliptic integral equations." Numer. Math. 53, 1988, 539 – 558.

DOMAIN DECOMPOSITION BOUNDARY ELEMENT METHODS: PREPROCESSING AND PARALLEL SOLUTION

Michael Kuhn
Institute of Mathematics, Johannes Kepler University Linz,
Altenbergerstr. 69, A-4040 Linz, Austria.
e-mail: kuhn@miraculix.numa.uni-linz.ac.at

SUMMARY

The Domain Decomposition Method (DD) is a powerful tool for deriving boundary element equations approximating boundary value problems with piecewise homogeneous material properties and for constructing the corresponding parallel solvers. Although the method allows the coupling of different discretization techniques, i.e., Boundary Element Methods (BEM) and Finite Element Methods (FEM), as it is desired in various applications, we discuss pure BEM formulations in this paper. We introduce the **A**daptive **D**omain **D**ecomposition **Pre**processor **ADDPre** which realizes an automatic decomposition of the domain under consideration into p subdomains, where p is the number of processors to be used. Furthermore, the parallel algorithm and the preconditioners being involved are discussed. Numerical examples, including potential and linear elasticity problems, which demonstrate the high efficiency of the algorithm are presented.

INTRODUCTION

Nowadays, the BEM is well established as a powerful tool for the numerical approximation to solutions of partial differential equations. Recent research has underlined that the BEM is very competitive compared to other discretization techniques as, e.g., FEM and that it depends on the actual application which discretization method is of advantage for solving the problem numerically. Even the coupling of BEM and FEM, using the advantages of both methods, via DD-formulations [9, 11] has been realized very efficiently. Nevertheless, pure DD–BEM models are of great interest. The application of DD ideas in Boundary Element Methods gives rise to algorithms for solving BEM–equations which are well suited to run on Multiple Instruction Multiple Data (MIMD) parallel computers. The aim of parallelization is to split up the work including, e.g., matrix generation, matrix-by-vector operations and preconditioning, to numerous processors.

In this paper we consider a model problem where Ω is a bounded domain and we have a weight function $a \in L^\infty(\Omega)$ which is piecewise constant.

*Problem 1 (**Strong Formulation**). Given $g \in H^{1/2}(\Gamma)$ find $u \in H^1(\Omega)$ such that*

$$- \operatorname{div} a\nabla u = 0 \quad \text{in } \Omega \quad \text{and} \quad u = g \quad \text{on } \Gamma := \partial\Omega. \tag{1}$$

The combination of DD methods and BEM allows us naturally to model those problems with piecewise constant coefficients $a(.)$. That is, the coefficients define an a-priori decomposition of the domain Ω in the following way:

$$\bar{\Omega} = \bigcup_{j=1}^{P_M} \bar{\hat{\Omega}}_j, \quad \text{with} \quad a(.) = a_j \text{ in } \hat{\Omega}_j \quad \text{and} \quad \hat{\Omega}_i \cap \hat{\Omega}_j = \emptyset \quad \forall i \neq j. \tag{2}$$

Thus, applying BEM we are faced with the following interface problem.

Problem 2. Given $g \in H^{1/2}(\Gamma)$ find $(u_1, \dots, u_{P_M}) \in H^1(\Omega_1) \times \cdots \times H^1(\Omega_{P_M})$ such that

$$\begin{aligned} - \operatorname{div} a_i \nabla u_i &= 0 \quad \text{in } \Omega_i, \quad u_i = g \quad \text{on } \Gamma \cap \partial\Omega_i & (3)\\ u_i &= u_j \quad \text{on } \partial\Omega_i \cap \partial\Omega_j & (4)\\ \partial u_i / \partial n_i &= -\partial u_j / \partial n_j \quad \text{on } \partial\Omega_i \cap \partial\Omega_j & (5) \end{aligned}$$

for all $i, j \in \{1, \dots, P_M\}$ (n_i (n_j) is the unit outer normal on $\partial\Omega_i$, ($\partial\Omega_j$)).

The outline of this paper is as follows. In Section 2 we are going to define a second interface problem based on a further decomposition of the domains $\hat{\Omega}_j$ ($j = 1, \dots, P_M$) into overall $p \geq P_M$ subdomains, where p is also the number of processors to be used. We introduce the preprocessor ADDPre which realizes the task of such a decomposition. Section 3 is devoted to both, the DD-based weak formulation and the resulting Galerkin discretization. The parallel iterative solution of the linear system and the definition of the preconditioners being involved are discussed in Section 4. Numerical results are presented in Section 5 and, finally, we give some concluding remarks in Section 6.

PREPROCESSING

Having in mind to make use of a multiprocessor computer for solving DD-equations the most natural idea is to use one processor per subdomain. In our case we assume that the number of processors (p) to be used is (much) larger than the number of a-priori given material-defined domains, i.e., we have $p > P_M$. Thus, we wish to decompose the domains $\hat{\Omega}_j$ determined by the materials further, which leads finally to

$$\bar{\Omega} = \bigcup_{i \in \mathcal{I}} \bar{\Omega}_i, \quad \text{with} \quad \bar{\hat{\Omega}}_j = \bigcup_{i \in \mathcal{I}_j} \bar{\Omega}_i \quad \forall j = 1, \dots, P_M, \tag{6}$$

where the Ω_i's are pairwise disjoint, i.e., $\Omega_i \cap \Omega_k = \emptyset \ \forall i \neq k \in \mathcal{I}$, and the sets of indices are given by

$$\mathcal{I}_j \subset \mathcal{I} := \{1, \dots, p\}, \quad \bigcup_{j=1}^{P_M} \mathcal{I}_j = \mathcal{I}, \quad \mathcal{I}_j \cap \mathcal{I}_k = \emptyset \quad \forall j \neq k.$$

We assume that there exist open balls $B_{\underline{r}_i}$ and $B_{\overline{r}_i}$ $(i \in \mathcal{I})$ with positive radii $0 < \underline{r}_i, \overline{r}_i < 1$, such that $B_{\underline{r}_i} \subset \Omega_i \subset B_{\overline{r}_i}$ and $0 < \underline{c} \leq \overline{r}_i / \underline{r}_i \leq \overline{c}$ $\forall i \in \mathcal{I}$ with fixed (i-independent) constants $\underline{c}$ and $\overline{c}$. Based on this decomposition we can now formulate the new interface problem.

Problem 3. Given $g \in H^{1/2}(\Gamma)$ find $(u_1, \dots, u_p) \in H^1(\Omega_1) \times \cdots \times H^1(\Omega_p)$ such that Equations (3)–(5) are satisfied for all $i, j \in \{1, \dots, p\}$.

The task of decomposing Ω into p subdomains starting with a fixed discretization of the interfaces $\Gamma_M := \cup_{j=1}^{P_M} \partial\hat{\Omega}_j$ aiming at a load-balanced decomposition is not trivial. Actually we seek such a decomposition for an a-priori defined discretization of the boundary Γ_M which minimizes the imbalance between the processors, i.e.,

$$\max_{i=1,\dots,p} \{N_i\} \Big/ \min_{i=1,\dots,p} \{N_i\} \longrightarrow \text{Minimum} \tag{7}$$

where N_i $(i = 1, \dots, p)$ is the number of local boundary elements arising after the decomposition. In the following we describe both, the preprocessor ADDPre [7], which performs the decomposition, and its application to DD–BEM problems.

DATA FORMATS

Throughout the preprocessing procedure we are concerned with two types of describing data: dd-data and tri-data. Both of them describe the geometry of the problem (shape of the domain, material interfaces) as well as the discretization (density and grading of the mesh). Each of them consists of certain classes of objects which are defined hierarchically, see Table 1. The main difference between the two data types is how the discretization is being represented. The dd-data describe the refinement only on the interfaces (as it is sufficient for a BEM discretization), whereas the tri-data describe the full 2d-discretization (as it is required for a FEM discretizations). Thus, tri-data are, in some sense, the 2d-realization of the refinement information contained in the dd-data. The way the refinement on the geometrical edges is actually being described (number of elements on that edge, distributed uniformly or graded towards the crosspoints) follows the definitions in [6, 7].

Table 1: The data formats, describing objects and their hierarchical definition.

type	object	definition
dd–data	crosspoints	coordinates
	geometrical edges	two crosspoints, (midpoint), refinement parameter, (boundary condition)
	subdomains	geometrical edges, material pointer
tri–data	points	coordinates
	edges	two points, (midpoint), (boundary condition)
	triangles	edges, material pointer

Remark 1. All edges may be straight lines or arcs of a circle. In the latter case an additional point (midpoint) defines this edge. This information would then be inherited throughout all stages of further refinement.

THE CODES

At this place, we are going to explain the codes and the main ideas they are based on. More information and technical details can be found in the complete documentation of the preprocessor [7]. The code ADDPre consists of three parts: Decomp, Tri2DD and AdapMesh which may be described as follows.

AdapMesh creates a mesh (Fig. 1b) from given dd-data (Fig. 1a) using, optionally, adaptivity information. AdapMesh applies an improved version of the mesh generator described in [6].

Decomp determines $\hat{p}_1, \ldots, \hat{p}_{P_M}$ with $\sum_{j=1}^{P_M} \hat{p}_j = p$ and decomposes $\hat{\Omega}_j$ into $\hat{p}_j$ ($j = 1, \ldots, P_M$) subdomains applying a modification of the recursive spectral bisection method (rsb) [15]. As a result, each triangle is assigned to one of the subdomains (gray shades in Fig. 1c).

Tri2DD converts triangular-based data (Fig. 1c) into the dd-data format (Fig. 1d). Interfaces between different materials will be maintained as they were given originally, whereas artificially created boundaries within one material are smoothed.

Since the rsb requires a full 2d-mesh it is always, i.e., also for DD–BEM as considered here, necessary to generate such a mesh ! At a first glance this may seem disadvantageous, especially if one already starts with a dd-description of the problem. On the other hand a slightly improved decomposition may give major improvements if time-dependent or optimization problems involving repeated calls of the linear solver are to be solved. Furthermore, the full 2d-mesh can also be used for the BEM postprocessing.

THE PREPROCESSOR ADDPRE

The complete **algorithm** can now be formulated as follows.

0. Create (manually) a dd-description of the problem and state a-priori refinement information on the boundaries (see Fig. 1a).
1. Use AdapMesh to convert the dd-data into tri-data. (see Fig. 1b).
2. Use Decomp to decompose the domain, i.e., to assign each triangle to one of the subdomains (see Fig. 1c).
3. Use Tri2DD to convert the tri-data back into dd-data (see Fig. 1d).
4. Stop or modify the refinement information in the dd-file and goto (1.).

Starting off with a dd-description the preprocessor has to maintain the original information throughout all stages ending up with a dd-file which describes finally the decomposition of our domain into subdomains according to the a-priori given refinement information. The preprocessor performs a quasi-static decomposition. It is possible to

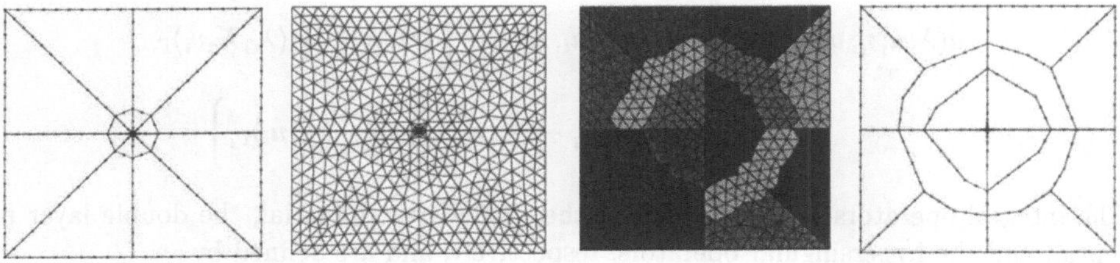

Figure 1: The 4 stages (a–d from left to right) of preprocessing. The dots in (a,d) represent the discretization on the interfaces.

restart the preprocessing optionally in step (4.). This may be necessary if the refinement information (grading or density of the mesh) have been modified, e.g., according to the behaviour of the solution obtained from a coarse grid calculation. The a-posteriori change of the refinement information, i.e., of the discretization, destroys in general the load-balance, that is the present decomposition is no longer a solution to (7).

Remark 2. The preprocessor allows us to start off with any data format and to decompose Ω into any number $p \geq P_M$ of subdomains. The only data which are fixed throughout the preprocessing are the definition of Γ_M and the a-priori given refinement information on Γ_M !

THE DD–BEM FORMULATION

We recall that $\Omega_1, \dots, \Omega_p$ are p pairwise disjoint bounded Lipschitz domains in the plane which separate Ω, $\overline{\Omega} = \cup_{j=1}^{p} \overline{\Omega}_j$ and $a_1, \dots, a_p$ are positive numbers. Then, the weak form is straightforward.

Problem 4 ***(Variational Formulation).*** *Given* $g \in H^{1/2}(\Gamma)$ *find* $u \in \mathbf{W}_g := \{v \in H^1(\Omega) : v = g \text{ on } \Gamma\}$ *such that for all* $v \in \mathbf{W}_0 := \{v \in H^1(\Omega) : v = 0 \text{ on } \Gamma\}$

$$\sum_{i=1}^{p} a_i \int_{\Omega_i} \nabla^T u \nabla v \, dx = 0. \tag{8}$$

Problem 4 can be rewritten as a system of boundary integral equations as in [12] with

$$\mathbf{\Lambda} := H^{-1/2}(\Gamma_1) \times \cdots \times H^{-1/2}(\Gamma_p) \tag{9}$$

$$\mathbf{U}_g := \{u \in L^2(\Gamma) : u|_\Gamma = g \text{ and for all } i = 1, \dots, p : u|_{\Gamma_i} \in H^{1/2}(\Gamma_i)\} \tag{10}$$

and $\langle .,. \rangle_{\Gamma_i}$ defines the duality product between $H^{-1/2}(\Gamma_i)$ and $H^{1/2}(\Gamma_i)$ extended from the scalar product in $L^2(\Gamma_i)$. Here, $\Gamma_i := \partial\Omega_i$ and $u_i = u|_{\Gamma_i}$, $\lambda_i = \partial u / \partial n|_{\Gamma_i}$ for $i \in \{1, \dots, p\}$.

Problem 5. Given $g \in H^{1/2}(\Gamma)$ find $(\lambda, u) \in \mathbf{\Lambda} \times \mathbf{U}_g$ that satisfies

$$a(\lambda, u; \eta, v) = 0 \qquad ((\eta, v) \in \mathbf{\Lambda} \times \mathbf{U}_0). \tag{11}$$

The bilinear form $a(.;.)$ is defined by

$$\begin{aligned} a(\lambda, u; \eta, v) \;:=\; & \sum_{i=1}^{p} a_i \Big\{ \langle \mathcal{D}_i u_i, v_i \rangle_{\Gamma_i} + \frac{1}{2} \langle \lambda_i, v_i \rangle_{\Gamma_i} + \langle \lambda_i, \mathcal{K}_i v_i \rangle_{\Gamma_i} \\ & + \langle \eta_i, \mathcal{V}_i \lambda_i \rangle_{\Gamma_i} - \langle \eta_i, \mathcal{K}_i u_i \rangle_{\Gamma_i} - \frac{1}{2} \langle \eta_i, u_i \rangle_{\Gamma_i} \Big\}. \end{aligned}$$

The integral operators $\mathcal{V}_i$, $\mathcal{K}_i$, $\mathcal{D}_i$ denote the single layer potential, the double layer potential and the hypersingular operators, respectively, and are defined by

$$\begin{aligned} \mathcal{V}_i \lambda_i(x) &:= \int_{\Gamma_i} \mathcal{E}(x,y) \lambda_i(y)\, ds_y, & \mathcal{V}_i &\in \mathcal{L}(H^{-1/2}(\Gamma_i); H^{1/2}(\Gamma_i)), \\ \mathcal{K}_i v_i(x) &:= \int_{\Gamma_i} \partial_y \mathcal{E}(x,y) v_i(y)\, ds_y, & \mathcal{K}_i &\in \mathcal{L}(H^{1/2}(\Gamma_i); H^{1/2}(\Gamma_i)), \\ \mathcal{D}_i u_i(x) &:= -\partial_x \int_{\Gamma_i} \partial_y \mathcal{E}(x,y) u_i(y)\, ds_y, & \mathcal{D}_i &\in \mathcal{L}(H^{1/2}(\Gamma_i); H^{-1/2}(\Gamma_i)). \end{aligned} \tag{12}$$

Here, $\mathcal{E}(x,y)$ is the kernel of the fundamental solution of the differential operator (e.g., $-\frac{1}{2\pi}\log|x-y|$ for the 2d-Laplacian) and where, for Banach spaces $X$and Y, $\mathcal{L}(X;Y)$ denotes the Banach space of linear and bounded operators from X to Y. The mapping properties of the boundary integral operators (12) on Sobolev spaces are well-known [5]. The space $\mathbf{V} := \mathbf{\Lambda} \times \mathbf{U}_0$ is endowed with the norm

$$\|(\lambda, u)\|_{\mathbf{V}} := \left(\|\lambda\|_{\mathbf{\Lambda}}^2 + \|u\|_{\mathbf{U}_0}^2 \right)^{1/2} \quad \text{with} \tag{13}$$

$$\|\lambda\|_{\mathbf{\Lambda}}^2 := \sum_{i=1}^{p} \|\lambda_i\|_{H^{-1/2}(\Gamma_i)}^2 \quad \text{and} \quad \|u\|_{\mathbf{U}_0}^2 := \sum_{i=1}^{p} \|u_i\|_{H^{1/2}(\Gamma_i)}^2.$$

Problems 4 and 5 are equivalent. The bilinear form $a(.;.)$ is $\mathbf{V}$-elliptic and both Problems, 4 and 5, have unique solutions [12].

To describe a Galerkin discretization of Problem 5, we choose finite dimensional subspaces

$$\begin{aligned} \mathbf{\Lambda}_h &:= \text{span}\,\{\psi_1, \psi_2, \ldots, \psi_{N_\Lambda}\}, \\ \mathbf{U}_h &:= \text{span}\,\{\phi_1, \ldots, \phi_{N_C}\} \end{aligned}$$

of $\mathbf{\Lambda}$ and $\mathbf{U}_0$, respectively, and set

$$\mathbf{V}_h := \mathbf{\Lambda}_h \times \mathbf{U}_h.$$

Problem 6. Given $f \in \mathbf{V}_h^*$ find $(\lambda_h, u_h) \in \mathbf{V}_h$ that satisfies

$$a(\lambda_h, u_h; \eta_h, v_h) = f(\eta_h, v_h) \qquad ((\eta_h, v_h) \in \mathbf{V}_h). \tag{14}$$

Remark 3. In Equation (14), $f(\eta_h, v_h) := a(0, g; \eta_h, v_h)$ and $g \in \mathbf{U}_g$ is arbitrarily extended from the given data $g|_\Gamma$. Thus, the final approximation is $(\lambda_h, g + u_h) \in \mathbf{\Lambda} \times \mathbf{U}_g$.

To reflect the domain decomposition, we assume a certain order in the basis function. Then, the nodal BE basis consists of piecewise polynomial trial functions

$$\Phi := \Psi_\Lambda \times \Phi_C = \text{span}\ \{\psi_1, \psi_2, \dots, \psi_{N_\Lambda}, \phi_1, \dots, \phi_{N_C}\}$$

with $N_\Lambda = \sum_{i=1}^{p} N_{\Lambda,i}$ and

$$\begin{aligned} \Psi_\Lambda &:= \Psi_{\Lambda_1} \times \cdots \times \Psi_{\Lambda_p} = \text{span}\ \{\psi_1, \dots, \psi_{N_{\Lambda_1}}\} \times \cdots \times \text{span}\ \{\psi_{N_\Lambda - N_{\Lambda_p}+1}, \dots, \psi_{N_\Lambda}\}, \\ \Phi_C &:= \text{span}\ \{\phi_1, \dots, \phi_{N_C}\}. \end{aligned}$$

Utilizing the isomorphism $\Phi : \mathbb{R}^{N_\Lambda + N_C} \to \mathbf{V}_h$ we are lead to the linear system of equations

$$\begin{pmatrix} \mathbf{K}_\Lambda & -\mathbf{K}_{\Lambda C} \\ \mathbf{K}_{C\Lambda} & \mathbf{K}_C \end{pmatrix} \begin{pmatrix} \underline{u}_\Lambda \\ \underline{u}_C \end{pmatrix} = \begin{pmatrix} \underline{f}_\Lambda \\ \underline{f}_C \end{pmatrix}. \tag{15}$$

The block entries are given by

$$\begin{aligned} (\mathbf{K}_\Lambda \underline{u}_\Lambda, \underline{v}_\Lambda) &:= \sum_{i=1}^{p} a_i \langle \eta_i, \mathcal{V}_i \lambda_i \rangle_{\Gamma_i} \quad \text{with } \lambda_i = \Psi_i \underline{u}_{\Lambda i},\ \eta_i = \Psi_i \underline{v}_{\Lambda i}, \\ (\mathbf{K}_{C\Lambda} \underline{u}_\Lambda, \underline{v}_C) &:= \sum_{i=1}^{p} a_i \{\langle \lambda_i, \mathcal{K}_i v_i \rangle_{\Gamma_i} + \frac{1}{2} \langle \lambda_i, v_i \rangle_{\Gamma_i}\} \\ \mathbf{K}_{\Lambda C} &:= \mathbf{K}_{C\Lambda}^T \\ (\mathbf{K}_C \underline{u}_C, \underline{v}_C) &:= \sum_{i=1}^{p} a_i \langle \mathcal{D}_i u_i, v_i \rangle_{\Gamma_i}, \quad \text{with } u_i = \Phi_i \underline{u}_{Ci},\ v_i = \Phi_i \underline{v}_{Ci}. \end{aligned}$$

Remark 4. Problem 6 is equivalent to (15).

PARALLEL SOLUTION AND PRECONDITIONING

According to the Bramble-Pasciak transformation [2], (15) is equivalent to

$$\mathbf{M} \underline{u} = \underline{q} \tag{16}$$

with the symmetric coefficient matrix

$$\mathbf{M} := \begin{pmatrix} \mathbf{K}_\Lambda \mathbf{C}_\Lambda^{-1} \mathbf{K}_\Lambda - \mathbf{K}_\Lambda & -(\mathbf{K}_\Lambda - \mathbf{C}_\Lambda)\, \mathbf{C}_\Lambda^{-1} \mathbf{K}_{\Lambda C} \\ -\mathbf{K}_{C\Lambda} \mathbf{C}_\Lambda^{-1} (\mathbf{K}_\Lambda - \mathbf{C}_\Lambda) & \mathbf{K}_{C\Lambda} \mathbf{C}_\Lambda^{-1} \mathbf{K}_{\Lambda C} + \mathbf{K}_C \end{pmatrix}$$

and the vectors

$$\underline{q} := \begin{pmatrix} (\mathbf{K}_\Lambda - \mathbf{C}_\Lambda) \cdot \mathbf{C}_\Lambda^{-1} \underline{f}_\Lambda \\ -\mathbf{K}_{C\Lambda} \mathbf{C}_\Lambda^{-1} \underline{f}_\Lambda + \underline{f}_C \end{pmatrix} \quad \text{and} \quad \underline{u} := \begin{pmatrix} \underline{u}_\Lambda \\ \underline{u}_C \end{pmatrix}.$$

Here, $\mathbf{C}_\Lambda \in \mathbb{R}^{N_\Lambda \times N_\Lambda}$ denotes some suitable scaled symmetric positive definite (spd) matrix which satisfies the following spectral equivalence inequalities

$$\underline{\gamma}_\Lambda \mathbf{C}_\Lambda \le \mathbf{K}_\Lambda \le \overline{\gamma}_\Lambda \mathbf{C}_\Lambda \quad \text{with } \underline{\gamma}_\Lambda, \overline{\gamma}_\Lambda > 1. \tag{17}$$

Theorem 1 stated below is equivalent to the results in [2] and gives sufficient conditions for the spectral equivalence of $\mathbf{M}$ with the preconditioner

$$\mathbf{C} := \begin{pmatrix} \mathbf{K}_\Lambda - \mathbf{C}_\Lambda & 0 \\ 0 & \mathbf{C}_C \end{pmatrix}. \tag{18}$$

Theorem 1. *Let $\mathbf{C}_C$ be a symmetric positive definite preconditioner for the BE Schur-complement $\mathbf{K}_C + \mathbf{K}_{C\Lambda}\mathbf{K}_\Lambda^{-1}\mathbf{K}_{\Lambda C}$, i.e., there exist positive constants $\underline{\gamma}_C$ and $\overline{\gamma}_C$ with*

$$\underline{\gamma}_C \cdot \mathbf{C}_C \le \mathbf{K}_C + \mathbf{K}_{C\Lambda}\mathbf{K}_\Lambda^{-1}\mathbf{K}_{\Lambda C} \le \overline{\gamma}_C \cdot \mathbf{C}_C. \tag{19}$$

Then, as defined in (18), the spd-preconditioner $\mathbf{C}$ is spectral equivalent to the matrix $\mathbf{M}$ of the spd system (16), i.e.,

$$\underline{\gamma} \cdot \mathbf{C} \le \mathbf{M} \le \overline{\gamma} \cdot \mathbf{C} \tag{20}$$

where $\alpha := 1 - (1/\overline{\gamma}_\Lambda)$,

$$\underline{\gamma} := \left(1 + \frac{\alpha}{2} + \sqrt{\alpha + \frac{\alpha^2}{4}}\right)^{-1} \cdot \min\{1, \underline{\gamma}_C\}, \quad \overline{\gamma} := \frac{1+\sqrt{\alpha}}{1-\alpha} \cdot \max\{1, \overline{\gamma}_C\}. \tag{21}$$

For solving (16) we utilize a conjugate gradient (cg) method preconditioned by $\mathbf{C}$, i.e., we apply the cg-iteration to the linear system of equations

$$\mathbf{C}^{-1}\mathbf{M}\underline{u} = \mathbf{C}^{-1}\underline{q}. \tag{22}$$

Thus, we have to find preconditioners $\mathbf{C}_C$ and $\mathbf{C}_\Lambda$ which satisfy (17) and (19), respectively, independently of the discretization parameter h.

We propose the following two realizations for $\mathbf{C}_C$:

$$\mathbf{C}_C^{(1)} \;:=\; \text{S–BPX}, \qquad \mathbf{C}_C^{(2)} \;:=\; \mathbf{K}_C\,(\mathbf{I}_C - \mathbf{M}_C)^{-1}. \tag{23}$$

The Schur–complement BPX (S–BPX) and its implementation is well known [17]. $\mathbf{C}_C^{(2)}$ is defined via a standard symmetric multigrid iteration applied to the global operator (assembled over all subdomains) $\mathbf{K}_C$ which is the discretization of a pseudo-differential operator of order one (see [18]). $\mathbf{K}_C$ becomes positive definite after implementing the Dirichlet boundary conditions. For either choice of $\mathbf{C}_C$ there exist h-independent constants $\underline{\gamma}_C$ and $\overline{\gamma}_C$ such that (19) is valid [4].

The matrix $\mathbf{K}_\Lambda$ is block-diagonal with blocks $\mathbf{K}_{\Lambda,i}$ $(i = 1, \dots, p)$. Thus, in order to construct a preconditioner for $\mathbf{K}_\Lambda$ it is sufficient to find preconditioners for each $\mathbf{K}_{\Lambda,i}$. Following [1, 18] we propose as realization for $\mathbf{C}_\Lambda$:

$$\mathbf{C}_\Lambda \;:=\; \delta \cdot \text{diag}\,[\mathbf{K}_{\Lambda,i}\,(\mathbf{I}_{\Lambda,i} - \mathbf{M}_{\Lambda,i})^{-1}]_{i=1,\dots,p}. \tag{24}$$

In (24), $\mathbf{M}_{\Lambda,i}$ is the multigrid operator satisfying the conditions formulated in [13]. From the cited papers we conclude that there exist h-independent constants $\underline{\gamma}_\Lambda$ and $\overline{\gamma}_\Lambda$ such that (17) is valid.

Remark 5. In order to ensure $\underline{\gamma}_\Lambda > 1$ in (17) the scaling factor δ in (24) has to be chosen such that $\delta < 1 - \eta$ where η is the multigrid convergence rate. In practice $\delta = 0.4$ is a good choice.

NUMERICAL RESULTS

As an academic test problem we consider Problem 1 with boundary conditions $g := r^{1/2} \sin \phi/2$, where (r, ϕ) are the polar coordinates with respect to the origin, and the domain Ω being the square $(-1, 1) \times (-1, 1)$ with a slit $([0, 1], 0)$. As preconditioner

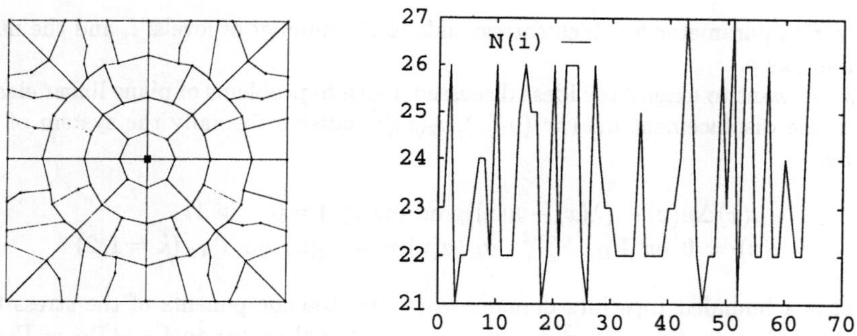

Figure 2: The decomposition (D3) into 64 subdomains (left) and the corresponding local problem size N_i $(i = 1, \ldots, 64)$ for the first level $(l = 1)$.

$\mathbf{C}_C$ we choose $\mathbf{C}_C^{(1)}$ (S–BPX). The BE–matrices were computed fully analytically using piecewise linear functions for the Φ_i's (potential) and piecewise constant functions for the Ψ_i's (normal derivative) [16]. We discuss the results for three different decompositions: a manual decomposition (D1) into 16 subdomains (see Fig. 1a), the resulting automatic decomposition (D2) into 16 subdomains (see Fig. 1d) and the automatic decomposition (D3) into 64 subdomains (see Fig. 2). (D1) and (D2) have an identical discretization of $\Gamma = \partial\Omega$, whereas, in the case of (D3), the discretization on Γ is as double as fine as for (D2). The result of the preprocessing for (D3) can be seen in Fig. 2 on the right. The local problem size varies between 21 and 27 which is quite satisfying taking into account the strong grading $(h_{max}/h_{min} > 100)$ of the grid on Γ.

Comparing (D1) and (D2) in Table 2, we observe that (D2) can be solved much faster than (D1) which is not only a result of lower iteration numbers that occur in the table. Additionally the maximum local problem size has been reduced from 32 (D1) to 30 (D2) as a result of the preprocessing. The latter alone implies a reduction of the CPU time for generating the matrices, which is a purely local task, from 17.4sec (D1) to 14.7sec (D2) on the 5th level $(l = 5)$. Looking at the results for (D2) and (D3) we can compute directly the scaled efficiency, which is about 0.89, according to the CPU time for $l = 5$. This value also profits from a lower maximum local problem size (27) for (D3) compared to (D2) (30). The performance of the preconditioners turns out to be independent of both, the

Table 2: Number of unknowns (N), iteration count ($I(\epsilon)$, $\epsilon = 10^{-6}$), CPU time (system generation and solution) in seconds. The experiments were carried out on a GC–PowerPlus using one processor per subdomain.

p	(D1) 16 manually			(D2) 16 automatically			(D3) 64 automatically		
l	N	I(ε)	CPU	N	I(ε)	CPU	N	I(ε)	CPU
1	541	34	1.3	579	28	1.2	2114	29	4.4
2	1085	35	1.8	1159	29	1.9	4249	30	5.4
3	2173	36	3.5	2319	29	2.8	8519	30	6.4
4	4349	37	9.6	4639	30	7.7	17059	30	10.8
5	8701	39	34.3	9279	30	26.8	34139	30	27.6

discretization parameter h, which corresponds to the number of levels l, and the number of subdomains p.

Now, we want to extend the ideas discussed above to problems of plane linear elasticity in which the displacement $u(x) = (u_1(x), u_2(x))^T$ satisfies formally the system of Lamé equations

$$\begin{array}{l} -\mu(x)\Delta u(x) - (\lambda(x) + \mu(x))\text{grad div}\, u(x) = 0 \quad \text{in } \Omega \\ u(x) = 0 \text{ on } \Gamma_D, \quad \sum_{l=1}^{2} \sigma_{kl}(u(x))n_l = g_k(x) \text{ on } \Gamma_N, \ (k = 1, 2) \end{array} \tag{25}$$

where Ω is a bounded Lipschitz domain, $\sigma_{kl}(u)$ are the components of the stress tensor $\sigma(u)$ and $n(x) = (n_1(x), n_2(x))^T$ is the outward normal vector to $\Gamma_D \cup \Gamma_N = \Gamma := \partial\Omega$ ($\Gamma_D \neq \emptyset$) and λ and μ, $\lambda, \mu > 0$, are the Lamé coefficients of the elastic material. In (25), $g = (g_1, g_2)^T$ is the vector of boundary tractions. The extension of the theory presented above for potential problems to linear elasticity is straightforward. The equations and definitions can be found, e.g., in [9].

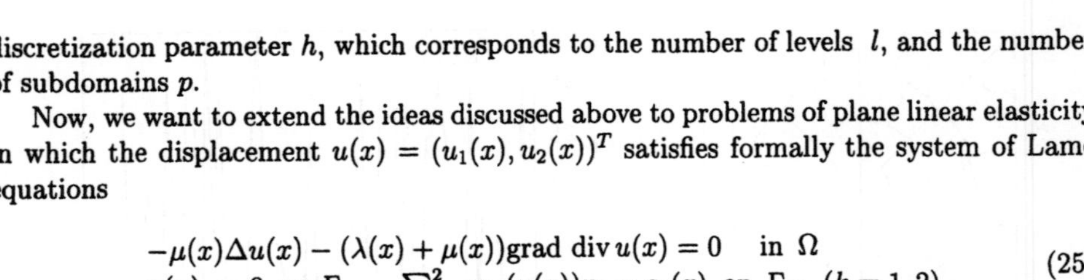

Figure 3: The dam (left) and its decomposition into 16 subdomains (D4) and 64 subdomains (D5). Dots in (D4,D5) represent crosspoints.

As a test problem we consider a dam filled with water as sketched in Fig. 3. As indicated there, Dirichlet boundary conditions (b.c.) are given on Γ_D (zero displacement) and Neumann b.c. on Γ_N (the tractions are equal to zero, or they are chosen according to the water pressure). The Lamé constants are given for rock by $\mu_r = 7.4e5$MPa, $\lambda_r = 1.7e6$MPa and for concrete by $\mu_c = 1.2e6$MPa, $\lambda_c = 1.2e6$MPa. We consider two automatic decompositions: one into 16 subdomains (D4) and one into 64 subdomains (D5),

Table 3: Number of unknowns (N), iteration count ($I(\epsilon)$, $\epsilon = 10^{-6}$), CPU time (system generation and solution) in seconds. The experiments were carried out on a GC–PowerPlus using one processor per subdomain.

p	(D4) 16 automatically			(D5) 64 automatically		
l	N	I(ϵ)	CPU	N	I(ϵ)	CPU
1	714	27	3.6	3099	33	29.9
2	1430	31	5.2	6271	35	30.8
3	2862	30	9.4	12615	36	38.1
4	5726	31	20.5	25313	37	52.9
5	11454	31	69.5	50679	37	96.1

where for the latter the discretization on Γ is as double as fine for (D5) than for (D4). Both decompositions are the result of the automatic preprocessing started with an a–priori graded grid towards the regions where high stresses are expected.

For the results presented in Table 3, the preconditioner $\mathbf{C}_C$ was realized using $\mathbf{C}_C^{(2)}$ (multigrid). The BE–matrices were computed fully analytically using piecewise linear functions for the displacements and piecewise constant functions for the tractions [16]. From the CPU time for the 5th level we compute the scaled efficiency 0.80.

CONCLUSIONS

The generalization of the approach to unbounded domains and to the coupling of BEM and FEM is straightforward [5, 8, 11, 14] Moreover, the potential problem can then be non-linear in the finite element subdomains [10]. Linear elasticity problems and local plasticity problems can be treated in the same fashion [3]. With the exception of plasticity problems, the code FEM⊗BEM [8] can solve these problems on massively parallel computers and workstation clusters (PVM).

REFERENCES

[1] J. H. Bramble, Z. Leyk, and J. E. Pasciak. The analysis of multigrid algorithms for pseudodifferential operators of order minus one. *Math. Comp.*, 63:461–478, 1994.

[2] J. H. Bramble and J. E. Pasciak. A preconditioning technique for indefinite systems resulting from mixed approximations of elliptic problems. *Math. Comp.*, 50(181):1–17, 1988.

[3] C. Carstensen. *Nonlinear interface problems in solid mechanics - Finite element and boundary element coupling.* Habilitation thesis, University of Hannover, 1992.

[4] C. Carstensen, M. Kuhn, and U. Langer. Fast parallel solvers for symmetric boundary element domain decomposition equations. Report 50x, Institute of Mathematics, Johannes Kepler University Linz, 1996.

[5] M. Costabel. Boundary integral operators on Lipschitz domains: Elementary results. *SIAM J. Math. Anal.*, 19:613–626, 1988.

[6] G. Globisch. PARMESH - a parallel mesh generator. *Parallel Computing*, 21(3):509–524, 1995.

[7] M. Goppold, G. Haase, B. Heise, and M. Kuhn. Preprocessing in BE/FE domain decomposition methods. Technical Report 96-2, Institute of Mathematics, Johannes Kepler University Linz, 1996.

[8] G. Haase, B. Heise, M. Jung, and M. Kuhn. FEM⊗BEM - a parallel solver for linear and nonlinear coupled FE/BE-equations. Priority Research Programme "Boundary Element Methods", Report 94-16, University of Stuttgart, 1994.

[9] G. Haase, B. Heise, M. Kuhn, and U. Langer. Adaptive domain decomposition methods for finite and boundary element equations. In *Reports from the Final Conference of the Priority Research Programme Boundary Element Methods 1989–1995, (W. Wendland ed.), Stuttgart, October 1995*, Berlin, 1996. Springer Verlag. Also Technical Report 95–2, Institute of Mathematics, Johannes Kepler University Linz, 1995.

[10] B. Heise. Nonlinear field calculations with multigrid-Newton methods. *IMPACT of Computing in Science and Engineering*, 5:75–110, 1993.

[11] B. Heise and M. Kuhn. Parallel solvers for linear and nonlinear exterior magnetic field problems based upon coupled FE/BE formulations. *Computing*, 56(3), 1996. To appear.

[12] G. C. Hsiao and W. L. Wendland. Domain decomposition in boundary element methods. In *Proc. of IV Int. Symposium on Domain Decomposition Methods, (R. Glowinski, Y. A. Kuznetsov, G. Meurant, J. Périaux, O. B. Widlund eds.), Moscow, May 1990*, pages 41–49, Philadelphia, 1991. SIAM Publ.

[13] M. Jung, U. Langer, A. Meyer, W. Queck, and M. Schneider. Multigrid preconditioners and their applications. In G. Telschow, editor, *Third Multigrid Seminar, Biesenthal 1988*, pages 11–52, Berlin, 1989. Karl–Weierstrass–Institut. Report R–MATH–03/89.

[14] U. Langer. Parallel iterative solution of symmetric coupled FE/BE- equations via domain decomposition. *Contemporary Mathematics*, 157:335–344, 1994.

[15] H. D. Simon. Partitioning of unstructured problems for parallel processing. *Comput. System in Eng.*, 2:135–148, 1991.

[16] O. Steinbach. Gebietsdekompositionsmethoden in der BEM. Priority Research Programme "Boundary Element Methods", Report 92-17, University of Stuttgart, 1992.

[17] C. H. Tong, T. F. Chan, and C. J. Kuo. A domain decomposition preconditioner based on a change to a multilevel nodal basis. *SIAM J. Sci. Stat. Comput.*, 12(6):1486–1495, 1991.

[18] T. von Petersdorff and E. P. Stephan. On the convergence of the multigrid method for a hypersingular integral equation of the first kind. *Num. Math.*, 57:379–391, 1990.

[19] T. von Petersdorff and E. P. Stephan. Multigrid solvers and preconditioners for first kind integral equations. *Num. Methods for Partial Differential Equations*, 8:443–450, 1992.

OBJECT ORIENTED DESIGN ASPECTS FOR BOUNDARY ELEMENT METHODS

Christian Lage
Seminar für Angewandte Mathematik
ETH Zürich
Rämistrasse 101
CH-8092 Zürich, Switzerland

SUMMARY

In this paper we give a description of an object-oriented approach to boundary element methods which has been elaborated as a C++ class library. It is shown that this kind of development results in a flexible and expansible software.

INTRODUCTION

The software design for boundary element has to take a great variety of tasks into account. Among others, there are different discretization schemes, different kind of integral operators and quite a lot of quadrature techniques to consider. Moreover, advanced methods, such as panel clustering (cf. [3], [4]) or wavelet based methods (cf. [2]), require particular informations which have to be provided by the underlying data structures. A design strategie that isolates the common parts and 'focus upon the essential characteristics' [1] to build the basic abstract data types, so-called key abstractions, would be very useful. This property is given by object-oriented methods, known to be a suitable tool to manage complex systems. For a comprehensive introduction to object-oriented methods see [1].

The essential tool used in the following design of a class library is the concept of an abstract base class. It serves to describe the common behaviour of an entity and captures the outside view of the entity in form of an interface. With concrete classes derived from the abstract base class several instances of the entity can be realized.

In order to identify the key abstractions of the system we have to investigate the mathematical formulation of the problem. It turns out that the concepts of this formulation, i.e. the boundary element method, could basically be adopted to the design.

We consider the boundary integral equation reformulation of a boundary value problem in $\Omega \subset \mathbb{R}^d$:

$$(\lambda I + K)\, u = f \qquad \text{on } \Gamma := \partial\Omega,\ \lambda \in \mathbb{R} \tag{1a}$$

$$\text{with } (Ku)(x) := \text{p.f.} \int_\Gamma k(x, y)\, u(y)\, dy. \tag{1b}$$

In the strong singular case the kernel function k in (1b) is not Lebesgue-integrable. Therefore we have to take the finite part (p.f. = part finit) of the integral (see [5], [4]).

We focus on the weak formulation of the problem (1a):

For given right-hand side $f \in X$ find $u \in X$ such that

$$\langle \lambda u, v\rangle + \langle Ku, v\rangle = \langle f, v\rangle \qquad \forall v \in X', \tag{2}$$

where X denotes a suitable function space, X' its dual space and $\langle \cdot, \cdot \rangle_{X \times X'}$ the dual form onto $X \times X'$.

To obtain a discretization of (2) by the Petrov-Galerkin method the trial space X and test space X' are to be replaced with finite dimensional subspaces $X_n := \text{span}\{\psi_i\}_{i<n} \subset X$ and $X'_n := \text{span}\{\phi_i\}_{i<n} \subset X'$, $n \in \mathbb{N}$. Inserting the basis functions of X_n and X'_n we arrive at the desired linear system

$$(\lambda M_n + K_n) u_n = f_n \tag{3}$$

with $n \times n$ matrices $M_n := (\langle \psi_j, \phi_i\rangle)_{i,j<n}$ and $K_n := (\langle K\psi_j, \phi_i\rangle)_{i,j<n}$ and the vector $f_n := (\langle f, \phi_i\rangle)_{i<n}$. Note that this formulation is equivalent to the discretization by means of projection methods defined for example in [3].

Example 1 *The collocation method is characterized by piecewise continuous functions as trial functions ψ_j and Dirac delta functions δ_{ξ_i} concentrated at suitable points $\xi_i \in \Gamma$ as test functions ϕ_i. Then the evaluation of the dual form in (3) is given by*

$$\langle K\psi_j, \phi_i\rangle = \text{p.f.} \int_\Gamma k(\xi_i, y) \psi_j(y)\, dy. \tag{4a}$$

Identifing X with X' (trial functions equal to test functions), one obtains the Galerkin method. In this case the the dual form is identical to the scalar product in X leading to

$$\langle K\psi_j, \phi_i\rangle = \int_\Gamma \text{p.f.} \int_\Gamma k(x, y)\, \psi_j(y)\, \phi_i(x)\, dy\, dx. \tag{4b}$$

GEOMETRY

The foundation of the discretization is the choice of the subspaces X_n and X'_n. A basic information necessary to construct these spaces is a description of the boundary Γ. Here, two levels of abstraction are used:

- the *analytical* description that denotes a splitting of the boundary in elementary parts, so-called *patches*, which can be associated with problem-oriented attributes such as boundary conditions of the initial boundary value problem, and
- the *numerical* description that denotes an approximation of the boundary by means of geometric objects, called *panels*, which can be handled by the numerical algorithms for instance quadrature subroutines.

The second level defines a geometric discretization, i.e. for instance, a polyhedron of plane triangles with corners located on a given surface (cf. Figure 1).

The two levels of boundary description are represented by two classes called Boundary and BoundaryApx in the library:

```
class Boundary {                    class BoundaryApx {
  public:                             public:
    Boundary(char* fname);              BoundaryApx(const Boundary& bd, int level);
    :                                   :
};                                  };
```

The information used to construct the geometric disretisation, i.e. the instantiation of objects of BoundaryApx, is provided by an instance of the class Boundary as indicated by the first parameter of the constructor of BoundaryApx. The second parameter called level controls the resolution of the approximation, so that we are able to form a hierarchy of panels with different instances of BoundaryApx. This hierarchy, for example, can be used with multilevel methods. In Figure 1 three instances of BoundaryApx according to the choices level equal to 0, 2, and 3 are depicted. All approximations are initiated by the same analytical boundary description. Note that we are not restricted to plane panels. It is also possible to generate curved panels with the supplied information of the analytical description.

Besides the flexible generation of multiple approximations of one boundary, the separation of the abstraction of boundary information in two distinct classes serves to specify an interface to the numerical treatment of the problem. To say, for many practical three-dimensional problems the boundary Γ is not defined by an analytical description. Only discretely measured surface points are given, so that a reasonable interpolation of the surface is required. This interpolation could be hidden in a variation of the class Boundary, whereas the class BoundaryApx considered as interface to the numerical algorithms remains unchanged.

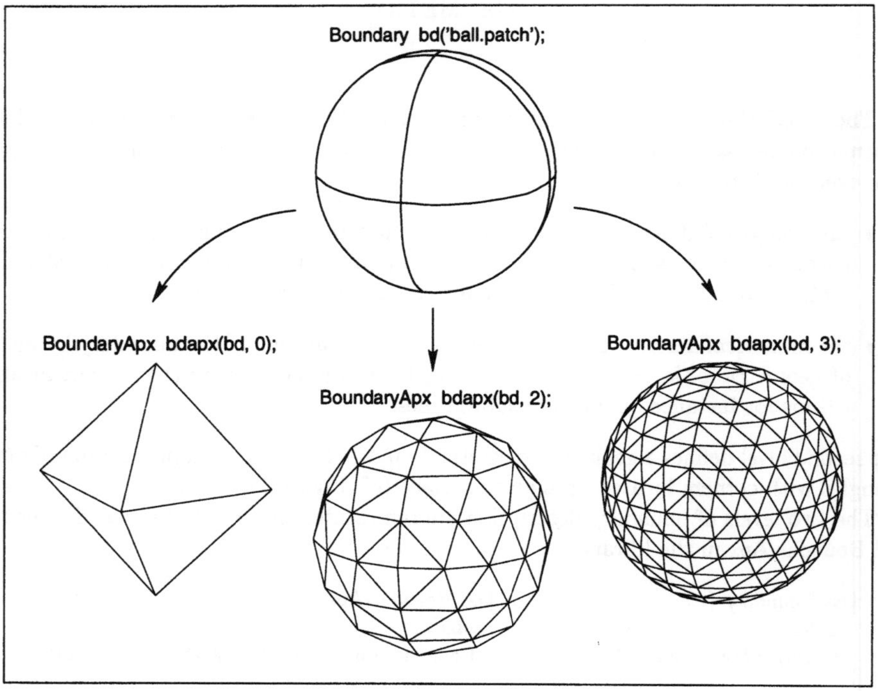

Figure 1: Description of geometry by means of the classes Boundary and BoundaryApx.

BASIS FUNCTIONS AND SUBSPACES

The next step in the classification of boundary element methods deals with the abstraction of the subspaces in particular the abstraction of bases and basis functions. To characterize a single basis function, two well-known approaches are available. The first one, we call it the *vertex-oriented* version, relates each basis function to its support in conjunction with an associated vertex (cf. Figure 2). This kind of abstraction corresponds most to the given mathematical formulation, because every function has its own representation. In contrast to this the second, *panel-oriented* approach. For every panel the non-vanishing parts of the basis functions are collected. A linear combination of these restricted functions forms a *panel function* and the possibly conditioned combination of all panel functions forms the subspace. Of course, this approach allows us to formulate a basis function only in an indirect way, but the time consuming calculation of the integrals – a panel-oriented task – can be organized much more efficiently.

Following this way we are looking for an abstraction of panel functions. With the facilities of object-oriented methods, it is possible to isolate the structure and behaviour that are common for diffenent types of panel functions in an interface class. Subclasses that inherit from this interface class form concrete panel functions such as constant or linear functions on varied shapes of panels. Even Dirac delta functions match this characterization. Figure 3 and the following class definitions illustrate this construction:

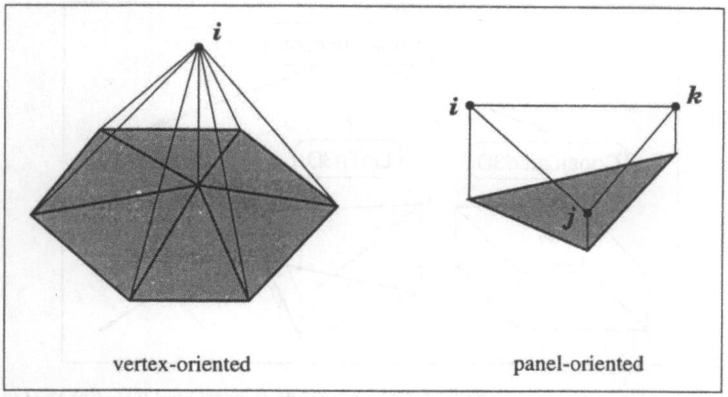

Figure 2: Two characterizations of basis functions

```
class PanelFunction {
  public:
    virtual const Panel&   support() const = 0;
    virtual int            dimension() const = 0;
    virtual const int*     index() const = 0;
    ⋮
};

class DiracVtx3D:  public PanelFunction {
    const Vertex3D*   vtx;
    int               idx;

  public:
    const Panel&   support() const { return *vtx; }
    int            dimension() const { return 1; }
    const int*     index() const { return &idx; }
    ⋮
};
class LinTri3D:  public PanelFunction {
    const TriFace*   fce;
    int              idx[3];

  public:
    const Panel&   support() const { return *fce; }
    int            dimension() const { return 3; }
    const int*     index() const { return idx; }

    ⋮
};
```

The interface or abstract base class PanelFunction defines several virtual functions to specify the common properties of panel functions:

- the connection to the underlying panel (member function support()),
- the number of non-vanishing basis functions (member function dimension()) and

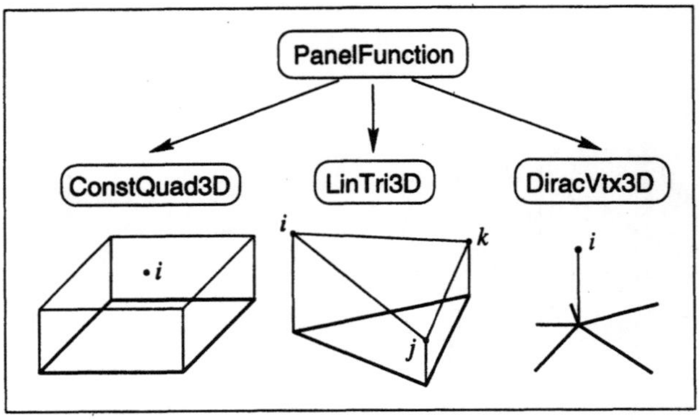

Figure 3: Spezifications of class **PanelFunction**: constant (**ConstQuad3D**), linear (**LinTri3D**) and Dirac delta function (**DiracVtx3D**).

- some information abount indices associated to these basis functions (member function **index()**).

The latter serves to relate a vector of coefficients to panel functions for the matter of specifying a concrete function of the subspace. As indicated before, we construct by inheritance specializations of the base class. These classes properly define the abstract virtual functions of the base class depending on the desired type of panel function. For linear functions on a triangle the member function **LinTri3D::support()** returns a reference to the triangle. Compatible with this there are three basis functions associated with the triangle as indicated by the member function **LinTri3D::dimension()**. In the same way the definition of a Dirac delta function concentrated in a vertex follows.

For the instantiation and maintenance of panel functions we have to establish a further class called **Space**. This provider is responsible for the correct arrangement of the panel functions, such as the organization of conforming indices in the case of continuous trial functions. To realize various subspaces for the Petrov-Galerkin scheme, several specializations of the base class **Space** are used. Corresponding to the subclasses of **PanelFunction** shown before we could define classes representing, for example, the subspace of piecewise constant, piecewise linear functions or the subspace of Dirac delta functions concentrated at the vertices.

DUAL FORMS AND INTEGRATION

To formulate an abstraction of integral operators, we previously have to deal with the dual forms of equation (2). For reasons of compatibility with the representation of spaces discussed in the preceding section a dual form is characterized by a function that assigns to each pair of panel functions an element matrix known from the context of finite element methods. In terms of a class definition this concept results in:

```
class DualForm {
  public:
    virtual const real* operator()(const PanelFunction& trialfnc,
                                   const PanelFunction& testfnc) const;
    ⋮
};
```

Again, the class DualForm is used to specify an interface for dual forms. Derived classes of DualForm realize concrete forms, as we can see in the following example illustrating the usage of this construction:

```
⋮
DiracVtx3D&   f(...);
LinTri3D&     g(...);
DlpDF         df(...);
real*         matrix;
⋮
matrix = df(f, g);
⋮
```

The derived class DlpDF represents the calculation of the form $\langle Kf, g\rangle$ with K the double layer potential and f, g two panel functions.

Before the assignment of an element matrix can be carried out, an appropriate integration or quadrature rule dependent on the particular type of panel functions has to be chosen at run-time (in the example above a rule for a Dirac delta function and a linear function). This choice is controled by two parameters, the trial and test function. Thus, we have to use two nested virtual function calls to solve this kind of multi method problem (see [6]). The first call determines the type of the test function, the second the type of the trial function. The drawback of this technique is that a class definition of a panel function (such as ConstQuad3D or LinTri3D) needs informations about other classes of panel functions, i.e., for the definition of a new panel function a change of existing classes is necessary (cf. [14, 13.8]). Moreover, using virtual member functions of the class PanelFunction to call the quadrature means a rather strong connectivity between the abstraction of the panel function and the quadrature rule. An exchange of the rules is not possible without modifying existing code.

For these reasons we recommend another approach that uses the concept of run-time type information (for pro and cons using RTTI see [6]). This mechanism offers the possibility to determine the type of an object during run-time in an efficient way. In the mean time, RTTI is a supported feature of the programming language C++.

The code used to determine the type of a panel function is similar to a case-statement and is best located in the member function

```
operator()(const PanelFunction& trialfnc, const PanelFunction& testfnc)
```

which is called when references to concrete types of panel functions are missing (note that PanelFunction is the base class). For the class DlpDF we have the following design:

```
class DlpDF   public DualForm {
  public:
   const real* operator()(const PanelFunction& trialfnc, const PanelFunction& testfnc) const;
   const real* operator()(const LinTri3D& trialfnc, const DiracVtx3D& testfnc) const;
   const real* operator()(const LinTri3D& trialfnc, const LinTri3D& testfnc) const;
     ⋮
};
```

where the member functions with concrete types,

```
operator()(const LinTri3D& trialfnc, const DiracVtx3D& testfnc)    and
operator()(const LinTri3D& trialfnc, const LinTri3D& testfnc),
```

evaluate the corresponding integrals, i.e., the collocation integral (4a) or the galerkin integral (4b) respectively. The casting of types is given by

```
const real* DlpDF::operator()(const PanelFunction& trialfnc,
                              const PanelFunction& testfnc) const {
   const LinTri3D* trialA;
   if (castUp(&trialfnc, &trialA)) {
    const DiracVtx3D* testA;
    if (castUp(&testfnc, &testA)) return operator()(*trailA, *testA);
    const LinTri3D* testB;
    if (castUp(&testfnc, &testB)) return operator()(*trailA, *testB);
   }
   return (real*)NULL;
}
```

This approach is not as elegant as the first one using nested virtual function calls but offers much flexibility when maintaining quadrature rules for several integral operators and avoids the unnatural mixing of panel functions and quadrature rules.

FUNCTIONS AND OPERATORS

As already mentioned, an arbitrary function of a subspace is characterized by the associated coefficient vector corresponding to the chosen basis. This representation is used for the abstraction of functions by the class **Function** which essentially includes a reference to the subspace and an array of coefficients. In addition, operations such as linear combinations and inner products are provided.

To assemble the local element matrices generated by the dual forms in a global matrix the class **MatrixOp** is introduced in the design. It is a specialization of the base class **Operator** which first of all is responsible for the abstraction of integral operators. This is done by defining a virtual member function that declares the mapping of the considered operator:

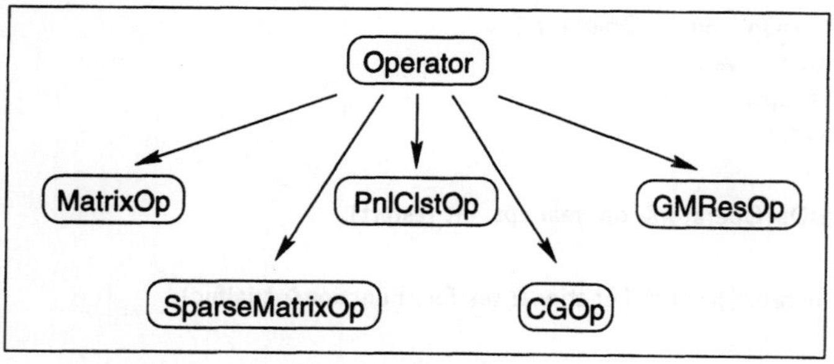

Figure 4: Specializations of the class Operator.

```
class Operator {
  public:
    Operator(const Space& test, const Space& trial);
      :
    virtual void operator()(const Function& trialfnc, Function& testfnc);
      :
};
```

In the special case of the class MatrixOp, this function is to be overloaded by a simple matrix-vector product.

```
class MatrixOp  public Operator {
    real*  matrix;
  public:
    MatrixOp(const Space& test, const Space& trial, const DualForm& df);
    ~MatrixOp(){ delete[] matrix; }
      :
    void operator()(const Function& trialfnc, Function& testfnc);
      :
};
```

The initialization of an instance of MatrixOp, i.e. calling the constructor, is equivalent to the setup of the matrix K_n in (3). In advance, it is necessary to define a trial and test space as well as a dual form. With these parameters the projection method and the integral operator under consideration are specified.

There are cases in which the representation of the operator by means of a two dimensional array is not adequate. When storing for example the mass matrix of the Galerkin method one should use hash tables or lists instead. Also in the case of matrix condensation by panel clustering or wavelet based methods special structures must be applied. To deal with these advanced techniques, new classes can be derived having the behaviour of an operator in common (see Figure 4]).

Another application of the interface class **Operator** is given by solving the discrete system (3), because this task can be read as applying the inverse of an operator. Best suited are interative solvers such as *cg* or *gmres* which only use matrix-vector products. In these cases, we can overload the mapping characterizing an operator by the iteration process. The following definition illustrates the usage of the class **Operator** to formulate an abstraction of the gmres algorithm:

```
class GMResOp   public Operator {
    Operator& inverseOp;
    real maxEps;
    int maxIt;
  public:
    GMResOp(Operator& op, real eps, int restart);
      :
    void operator()(const Function& testfnc, Function& trialfnc);
      :
};
```

When calling the constructor of GMResOp, the restart value, the break condition and the operator that is to be inverted must be specified. For the latter we can use every description derived from the base class Operator. In particular, for the solver to work it is of no consequence whether the discrete integral operator of equation (3) is represented by the full matrix using the class MatrixOp, by a panel clustering approximation using PnlClstOp or something else.

AN EXAMPLE

To summarize the rough introduction of our class library we present a short application:

```
int main() {
    // initialize the geometry of file "cube"
    Boundary        bd("cube");
    BoundaryApx   bdapx(bd, 2);

    // generate test and trial space of the collocation method
    DiracSpace   test(bdapx);
    LinSpace        trial(bdapx);

    // assemble the system matrix
    MatrixOp      op(test, trial, df);

    // declare and define two functions
    Function f(trial); f = "(sin(x*y))";
    Function g(test);
```

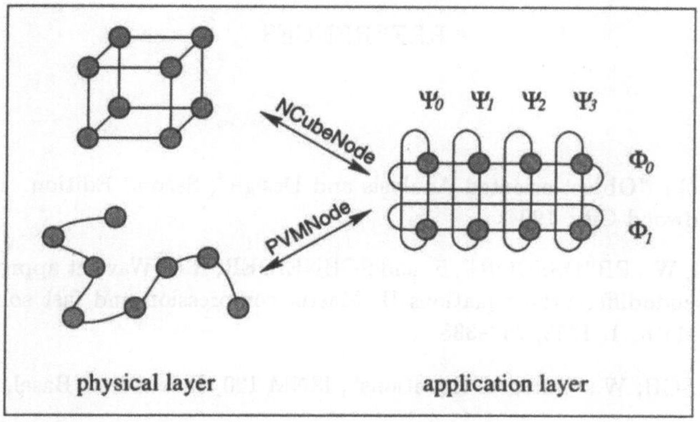

Figure 5: Virtual machine abstraction.

```
    // map f by means of op and store it in g
    op(f, g);

    // apply the inverse of op
    GMResOp invOp(op, 1e-8, 100);
    invOp(g, f);
}
```

Chosing **test** equal to **trial** in the example above would change the discretization to Galerkin's method.

PARALLEL ARCHITECTURES

When dealing with parallel architectures, especially MIMD-architectures, we have to organize different topologies of processor grids. Having a parallelization approach to boundary elements by means of data decomposition in mind, i.e. splitting the trial and test spaces, we would prefer a two dimensional torus to match the block structure of the system matrix (cf. [4]). To formulate the classes discussed so far independently of the given architecture, we introduce a class **ProcessNode**, which defines the basic communication operations in the terminology of a torus, for example, a broadcast along a *row* or sending messages to the *upper, lower, left or right* neighbour.

Specializations of this class, as **PVMNode** for workstation cluster or **NCubeNode** for the nCube, establish the connection to the underlying architecture designated in Figure 5 as *physical layer*. This kind of usage of object-oriented methods is known as *virtual machine abstraction.*

REFERENCES

[1] BOOCH, G.: "Object-oriented Analysis and Design", Second Edition. Benjamin/Cummings, Redwood City, 1994.

[2] DAHMEN, W., PR"OSSDORF, S. and SCHNEIDER, R.: "Wavelet approximation methods for pseudodifferential equations II: Matrix compression and fast solution. Adv. in Comput. Math. 1, 1993, 259–335.

[3] HACKBUSCH, W.: "Integral Equations", ISNM 120, Birkhäuser, Basel, 1995.

[4] LAGE, C.: "Softwareentwicklung zur Randelementmethode: Analyse und Entwurf effizienter Techniken". Dissertation, Universität Kiel, 1996.

[5] SCHWAB, C. and WENDLAND, W. L.: "On numerical cubatures of singular surface integrals in boundary element methods". Numer. Math. 62, 1992, 343–369.

[6] STROUSTRUP, B.: "The Design and Evolution of C++". Addison-Wesley, Reading, 1994.

PARALLEL SETUP OF GALERKIN EQUATION SYSTEM FOR A GEODETIC BOUNDARY VALUE PROBLEM

Rüdiger Lehmann, Geodetic Institute, University of Karlsruhe, Englerstr. 7, D-76128 Karlsruhe (Germany)
Roland Klees, Delft University of Technology, Faculty of Geodetic Engineering, Thijsseweg 11, NL-2629 JA Delft (The Netherlands)

SUMMARY

A particular problem in setting up large Galerkin equation systems in the 3D-BEM is the huge number of offdiagonal elements, which are mainly regular integrals over pairs of boundary elements. The numerical effort of cubature depends strongly on the distance between these elements. Therefore, the polynomial degree of exactness of the cubature formula may be chosen based on an estimate of the distance. In a straightforward implementation on a parallel computer, this may lead to severe imbalances of workload, which considerably reduces the speedup. To overcome this difficulty, a dynamic load balancing scheme is applied. It is shown how this works in the solution of the classical oblique boundary value problem (BVP) of potential theory (which results from the linearization of the fixed BVP of Physical Geodesy) on a IBM 9076 SP/2.

INTRODUCTION

Let $\Omega \subset \mathbb{R}^3$ denotes a smooth bounded domain, and $\Omega^c = \mathbb{R}^3 \setminus \Omega$. We consider the boundary value problem

$$\begin{aligned} \Delta u(x) &= 0, && x \in \Omega^c \\ D_l u(x) &= f(x), && x \in \Gamma \\ u(x) &\to 0, && |x| \to \infty\ . \end{aligned} \tag{1}$$

Here, Δ is the Laplacian; $l = l(x) : \Gamma \to \mathbb{R}^3$ is a given smooth unit vector field on the boundary Γ which is oblique to the surface normal vector field n; D_l is the directional derivative into direction l, and $f(x)$ is a sufficiently smooth given function on Γ. It has been proved by GIRAUD [4] that this problem is well-posed provided that l is nowhere tangential to Γ.

This boundary value problem arises in Physical Geodesy in relation with the determination of the Earth's external gravity field. Then, Ω^c denotes the external space of the Earth, u denotes the so-called disturbing potential, defined as the difference between the actual gravity potential and a suitable approximation, Γ the Earth's surface, and f represents the measurements taken on Γ.

There are several possibilities to represent u in the domain Ω^c, e.g. a single layer representation

$$u(x) = \frac{1}{4\pi} \int_{y \in \Gamma} \frac{\mu(y)}{|y-x|} d\Gamma(y), \quad x \in \Omega^c \tag{2}$$

or a double layer representation

$$u(x) = \frac{1}{4\pi} \int_{y \in \Gamma} \frac{\cos[n(y), y-x]}{|y-x|^2} \nu(y) d\Gamma(y), \quad x \in \Omega^c. \tag{3}$$

For alternative representations, see [2],[3]; [8]. When taking the boundary values in the representation formulae, we obtain the boundary integral equations

$$f(x) = -\frac{1}{2}\mu(x)\cos[l(x), n(x)] + \frac{1}{4\pi} p.v. \int_{y \in \Gamma} \mu(y) k_S(x,y) d\Gamma(y), \quad x \in \Gamma \tag{4}$$

and

$$f(x) = \frac{1}{2}\langle l(x), \nabla\nu(x)\rangle + \frac{1}{4\pi} p.f. \int_{y \in \Gamma} \nu(y) k_D(x,y) d\Gamma(y), \quad x \in \Gamma. \tag{5}$$

respectively, with kernel functions

$$k_S(x,y) := \frac{\cos[l(x), y-x]}{|y-x|^2} \tag{6}$$

$$k_D(x,y) := \frac{\cos[n(y), l(x)] - 3\cos[n(y), y-x]\cos[l(x), y-x]}{|y-x|^3}. \tag{7}$$

$\langle l(x), \nabla\nu(x)\rangle$ denotes the directional derivative of ν in the direction of the tangential component of l. If $l = n$ at $x \in \Gamma$ then $\langle l(x), \nabla\nu(x)\rangle = 0$ holds.

The first integral equation is a Cauchy-singular boundary integral equation for the unknown single layer density μ; the second one is a hyper-singular boundary integral equation for the unknown double layer density ν.

GALERKIN METHOD

It is known that (4) and (5) define classical pseudo-differential operators of order 0 and 1, respectively (cf. [13]). Moreover, if l is nowhere tangential to Γ, the corresponding boundary integral operators are strongly elliptic [14]. Thus, a Garding inequality is fulfilled which implies with the Lax-Milgram theorem the invertibility of the integral operators in some Sobolev spaces. Hence, the classical Fredholm alternative holds. If the boundary integral operators are injective the Galerkin-Bubnov method is stable and converges quasi-optimally. Therefore, we decided to use the Galerkin-Bubnov method to discretize the boundary integral equations. Let

$$\tilde{\mu}(x) = \sum_{i=1}^{n} \alpha_{S,i}\mu_i(x) \tag{8}$$

$$\tilde{\nu}(x) = \sum_{i=1}^{n} \alpha_{D,i}\nu_i(x) \tag{9}$$

denote the approximate solutions of the boundary integral equations (4) and (5), respectively with basis functions $\{\mu_i, i = 1\ldots, n\}$ and $\{\nu_i, i = 1, \ldots, n\}$. Then, for the Galerkin-Bubnov method we multiply the integral equations with the corresponding basis functions μ_k and ν_k, respectively, integrate over Γ and equate the integrals. We obtain:

$$\begin{aligned} a_{S,k,j} &= -\frac{1}{2}\int_{x\in S_{\mu,k,j}} \mu_k(x)\mu_j(x)\cos[l(x), n(x)]d\Gamma(x) \\ &+\frac{1}{4\pi}\int_{x\in S_{\mu,k}} \mu_k(x)\, p.v.\int_{y\in S_{\mu,j}} \mu_j(y)k_S(x,y)d\Gamma(y)d\Gamma(x) \end{aligned} \tag{10}$$

$$b_{S,k} = \int_{x\in S_{\mu,k}} \mu_k(x)f(x)d\Gamma(x), \tag{11}$$

$$\begin{aligned} a_{D,k,j} &= \frac{1}{2}\int_{x\in S_{\nu,k,j}} \nu_k(x)\langle l(x), \nabla\nu_j(x)\rangle d\Gamma(x) \\ &+\frac{1}{4\pi}\int_{x\in S_{\nu,k}} \nu_k(x)\, p.f.\int_{y\in S_{\nu,j}} \nu_j(y)k_D(x,y)d\Gamma(y)d\Gamma(x). \end{aligned} \tag{12}$$

$$b_{D,k} = \int_{x\in S_{\nu,k}} \nu_k(x)f(x)d\Gamma(x), \tag{13}$$

$$k, j = 1, \ldots, n.$$

Here, $S_{\mu,k}$ and $S_{\nu,k}$ denotes the support of the basis functions μ_k and ν_k and $S_{\mu,k,j} = S_{\mu,k} \cap S_{\mu,j}$ and $S_{\nu,k,j} = S_{\nu,k} \cap S_{\nu,j}$. The support of the basis functions μ_k and ν_k consists in general of the union of a certain number of panels Γ_i:

$$\Gamma = \bigcup_{i=1}^{N} \Gamma_i, \quad |\Gamma_i \cap \Gamma_j| = 0 \quad \text{if } i \neq j \tag{14}$$

$$h := \max\{\operatorname{diam}(\Gamma_i), i = 1, \ldots, N\}. \tag{15}$$

Therefore, the following integrals have to be calculated numerically:

$$a^i_{S,k,j} := -\frac{1}{2}\int_{x\in\Gamma_i} \mu^i_k(x)\mu^i_j(x)\cos[l(x), n(x)]d\Gamma_i(x) \tag{16}$$

$$a^{i,m}_{S,k,j} := \frac{1}{4\pi}\int_{x\in\Gamma_i} \mu^i_k(x)\, p.v.\int_{y\in\Gamma_m} \mu^m_j(y)k_S(x,y)d\Gamma_m(y)d\Gamma_i(x) \tag{17}$$

$$b^i_{S,k} := \int_{x\in\Gamma_i} \mu^i_k(x)f(x)d\Gamma_i(x) \tag{18}$$

$$i, m = 1, \ldots, N, \quad k, j = 1, \ldots, n$$

for the discretized Cauchy singular integral equation, and

$$a^i_{D,k,j} := \frac{1}{2}\int_{x\in\Gamma_i} \nu^i_k(x)\langle l(x), \nabla\nu^i_j(x)\rangle d\Gamma_i(x) \tag{19}$$

$$a^{i,m}_{D,k,j} := \frac{1}{4\pi}\int_{x\in\Gamma_i} \nu^i_k(x)\, p.f.\int_{y\in\Gamma_m} \nu^m_j(y)k_D(x,y)d\Gamma_m(y)d\Gamma_i(x) \tag{20}$$

$$b^i_{D,k} := \int_{x\in\Gamma_i} \nu^i_k(x)f(x)d\Gamma_i(x) \tag{21}$$

$$i, m = 1, \ldots, N, \quad k, j = 1, \ldots, n$$

for the discretized hyper-singular integral equation. Here, μ^i_k, ν^i_k denote the restrictions of μ_k, ν_k, respectively, to the panel Γ_i.

The numerical costs for setting-up the linear system of equations depends on the number of basis functions n. Actual models of the Earth's gravity field, which uses spherical harmonics as basis functions, have been developed up to degree and order 360. They contain $361^2 = 130321$ coefficients. The corresponding spatial resolution is about 56 km at the equator. The boundary data used to determine the coefficients are mostly available as block mean values over blocks of side-length $180°/360 = 0.5°$ in spherical latitude and longitude. When using piecewise constant trial functions, a discretization of the Earth's surface into about $N = n = 165000$ triangular elements will give a comparable resolution. For higher order trial functions and the same resolution, the number of elements reduces correspondingly, so that the total number of unknowns will almost remain the same. Thus, the number of basis functions for state-of-the-art geodetic applications is of the order $O(10^6)$. The number of singular and nearly singular integrals is of the same order. However, about $O(10^{12})$ regular integrals have to be evaluated which forms the real bottleneck for every application of boundary element methods to geodetic boundary value problems.

KLEES [8] has investigated the boundary integral equation (4) for piecewise constant trial functions with 10000 boundary elements. The regular integrals have been calculated using Gauß-Legendre product formulae. The number of nodes has been chosen independently on the distance between outer and inner domain of integration such that the error of numerical integration remained uniformly bounded by some ϵ. This choice has been made to avoid IF-statements in the innermost DO-loops what would have reduced the efficiency on SIMD machines considerably. More than 97% of the total CPU-time for setting-up the linear system of equations falled to the share of the regular integrals.

NUMERICAL EVALUATION OF THE INTEGRALS

The computational cost for setting up the Galerkin system of linear equations is mainly determined by numerical evaluation of the regular integrals of type (17),(20). The complexity of this task is $O(N^2)$. For the singular and nearly singular integrals we need only $O(N \log^s N)$ operations for some positive s [7]. Further regular integrals are (16),(19) and the right hand sides (18),(21). The computational cost of their evaluation is only $O(N)$.

From the mathematical point of view, the *numerical evaluation of the regular integrals* does not pose a problem. Standard cubature formulae Q_d with polynomial degrees of exactness d can be applied:

$$\int_{\Gamma_i} \int_{\Gamma_j} f_{i,j}(x,y) d\Gamma_j(y) d\Gamma_i(x) \approx Q_{d_{i,j}}(f_{i,j}) := \sum_{k=1}^{K} w_k f_{i,j}(x_k, y_k). \tag{22}$$

Our aim is to achieve a given polynomial degree of exactness $d_{i,j}$ with a minimum number K of integration nodes. Unfortunately, minimum nodes formulae in more than one dimensions are known only for a few standard regions of integration, e.g. a rectangle or a triangle. But it is possible to transform each of the integrals (17),(20) into an integral over a product of triangles $\Delta_i \times \Delta_j$ and then to apply such formulae both for the inner and outer integration (product formulae). In the following we will use products of symmetric minimum nodes formulae for the triangle of degree $d_{i,j} \in \{1,3,5,7,9\}$ given in [10] (cf. table 1).

Table 1: Runtimes on a 66 MHz POWER2 processor for the numerical evaluation of all regular integrals corresponding to a pair of boundary elements

degree of exactness		1	3	5	7	9
number of nodes (single formula)		1	4	7	13	19
single layer, constant trial	(μs)	3.5	21	61	202	430
single layer, linear trial	(μs)	4.5	24	66	218	460
double layer, linear trial	(μs)	5.8	32	86	278	580

Only one problem is still left: How to choose the polynomial degree of exactness $d_{i,j}$? Different approaches are known. In [12] it is stated that $d_{i,j}$ has to be chosen such that the error of numerical integration does not increase the asymptotic discretization error for $h \to 0$. Local estimates of the cubature error are obtained by the authors, which give rise to consistency requirements in terms of d. The result depends a) on the singularity of the kernel function, b) on the polynomial degree of the trial functions, and c) on the (Sobolev-)Norm, in which we want to measure the error. We think that the situation with the geodetic BVPs is different in the sense that the boundary data f are never given in continuous form. Therefore, any chosen discretization always has to reflect our data situation, and no asymptotic refinement is possible. Our interest is then not so much the asymptotic behaviour for $h \to 0$, but rather the accuracy which can be obtained for a given h.

It is clear that the accuracy of a particular numerical integration formula $Q_{d_{i,j}}$ in (22) depends strongly on the distance of the elements $\mathrm{dist}(\Gamma_i, \Gamma_j)$. Therefore, we determine a relationship of the form

$$d_{i,j} := d(\mathrm{dist}(\Gamma_i, \Gamma_j)) \tag{23}$$

such that a uniform accuracy, i.e. the error is below some $\epsilon > 0$, for all entries of the Galerkin system of equations is guaranteed. $d(\cdot)$ is a decreasing integer-valued function.

We want to use formulae of very low degrees of exactness. In this case, the computation of $\mathrm{dist}(\cdot,\cdot)$ in a strict geometrical sense would increase the computational costs considerably. But it is possible to compute an approximate distance $\widetilde{\mathrm{dist}}(\cdot,\cdot)$ satisfying

$$\widetilde{\mathrm{dist}}(\Gamma_i, \Gamma_j) \leq \mathrm{dist}(\Gamma_i, \Gamma_j). \tag{24}$$

Such a lower bound of a distance can be obtained as follows: Let $\chi_i(\xi,\eta) : \Delta_0 \to \Gamma_i, i = 1,\dots,N$ be a representation of the boundary element Γ_i as the image of the standard triangle $\Delta_0 = \{0 \leq \xi, 0 \leq \eta \leq 1-\xi\}$. Let

$$\widetilde{\mathrm{rad}}(\Gamma_i) := \max_{\xi,\eta\in\Delta_0} \left|\chi_i\left(\frac{1}{3},\frac{1}{3}\right) - \chi_i(\xi,\eta)\right|. \tag{25}$$

$\widetilde{\mathrm{rad}}$ depends only on the geometry of each boundary element, such that the complexity of this task is only $O(N)$. For elements of low curvature it is possible to compute the maximum only over the vertices $(0,0),(0,1),(1,0)$; as will be done here.

Now, let

$$\widetilde{\mathrm{dist}}(\Gamma_i, \Gamma_j) := \max\left(0, \left|\chi_i\left(\frac{1}{3},\frac{1}{3}\right) - \chi_j\left(\frac{1}{3},\frac{1}{3}\right)\right| - \widetilde{\mathrm{rad}}(\Gamma_i) - \widetilde{\mathrm{rad}}(\Gamma_j)\right). \tag{26}$$

This means, that our approximate distance is the distance between two spheres enclosing the boundary elements completely, with centers in the barycenters of the elements and

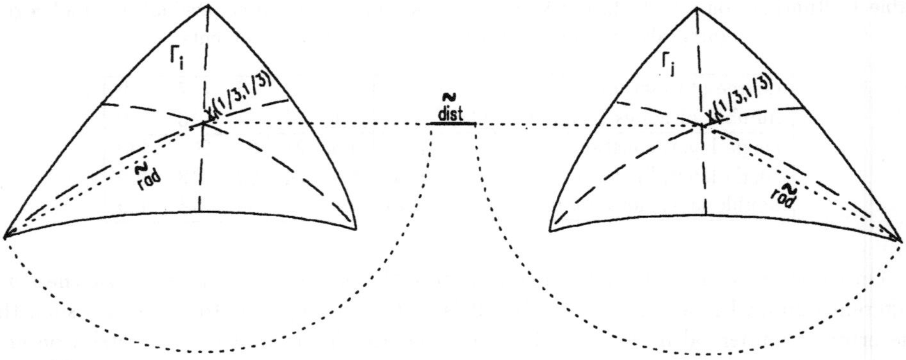

Figure 1: Definition of an approximate distance (26)

minimal radius (cf. figure 1). Then, (24) is immediately fulfilled and implies

$$d(\widetilde{\text{dist}}(\Gamma_i, \Gamma_j)) \geq d(\text{dist}(\Gamma_i, \Gamma_j)). \tag{27}$$

Mostly we get the same degree with both dist and $\widetilde{\text{dist}}$, only sometimes using $\widetilde{\text{dist}}$ may give a higher degree, thus more function evaluations. This extra effort can be neglected. The evaluation of (26) *requires no additional computation time*, because the barycenters are nodes in the numerical integration formulae used, and the distance is part of the kernel functions involved.

As far as the parallelization is concerned, the *numerical evaluation of the nearly singular and singular integrals* does not play an important role. We use the program GBEMINT [1] which performs the necessary transformations suggested in [11].

PARALLELIZATION ON MIMD-COMPUTERS

MIMD-computer are equipped with multiple processing nodes, each consisting of a processor and a *local* memory. The nodes can communicate only via network. If we want to parallelize a certain numerical algorithm on such a computer, we are usually faced with two problems:

1. *Minimizing the communication overhead.* Fortunately, as long as all the geometry data are stored in each local memory, the evaluations of the integrals are independent tasks and must not lead to any communication at all. As has been pointed out by several authors (e.g. [5]), this multiple storage seems to be possible because it requires memory only of the order $O(N)$, while the system matrix requires $O(N^2)$ bytes. However, if this redundancy can be avoided without introducing much communication overhead, then we may use the saved main storage efficiently for a different purpose: We are perhaps able to store all the values corresponding to the nodes of all numerical integration formulae to be used (1+4+7+13+19=44 nodes in our case, cf. table 1), instead of recomputing them for each regular integral. We followed the latter approach.

2. *Balancing the workload among the processing nodes.* The workload $w_{i,j}$ for the regular integration is nonuniformly distributed. For (22) holds

$$w_{i,j} := w(d_{i,j}) = w(d(\widetilde{\text{dist}}(\Gamma_i, \Gamma_j))). \tag{28}$$

For $1 \leq d_{i,j} \leq 9$, the ratio $\max w_{i,j} / \min w_{i,j}$ can be more than 100 (cf. table 1). This imbalance can become even worse for heterogeneous discretizations of the boundary surface or heterogeneous computing environments. Parallelization of the setup of the Galerkin system of equations is therefore mainly concerned with load balancing. In existing parallel BEM implementations, e.g. [6], this problem is mostly avoided by using the same degree of exactness for each regular integral. Applications of load balancing techniques are known only for collocation discretizations, e.g. in [5].

The speedup s_p is defined as

$$s_p := \frac{t_1}{t_p} \tag{29}$$

where t_p is the total runtime of a parallel program using p processing nodes. The corresponding efficiency is defined as

$$e_p := \frac{s_p}{p}. \tag{30}$$

Fortunately, unlike in the case of communication overhead, the speedup does not decrease due to load imbalance. But the efficiency can become very low for $p \to \infty$, which is an improper use of a parallel computer.

TASK BUILDING

The first step towards parallelization is the definition of tasks to be distributed among the processing nodes. Here we follow two principles:

1. Each integral cannot be an extra task. Otherwise, since we consider $N > 10^4$ boundary elements, this would certainly lead to a considerable additional effort of task scheduling.

2. All integrals over one pair of elements should belong to the same task, because their kernel functions are either identical or very similar such that we can save a lot of the most time consuming numerical operations. Transfer of those kernel function values via network must be excluded because it would take too much time.

These principles require that the set of boundary elements is devided into disjoint subsets in a certain way. All integrals (17),(20) over an element Γ_i in the k-th subset and Γ_j in the l-th subset ($l \leq k$) belong to the task $T_{k,l}$. The workload in connection with task $T_{k,l}$ will be denoted $W_{k,l}$.

Such subsets of elements can be formed differently (cf. figure 2). First, they can be *contiguous* surface pieces. This has the advantage that all nearly singular and singular integrals belong to tasks $T_{k,k}$. Here $W_{k,k}$ is large, which makes load balancing easier in some cases. However, the values $W_{k,l}, k > l$ will be nonuniform. Second, we can think of subsets of *scattered* elements in a way that the minimum distance of two different elements in each subset is as large as possible (or maximum). In this case we achieve the opposite effect: $W_{k,k}$ is small, but all $W_{k,l}, k > l$ will be of approximately the same magnitude, which simplifies load balancing in a different way.

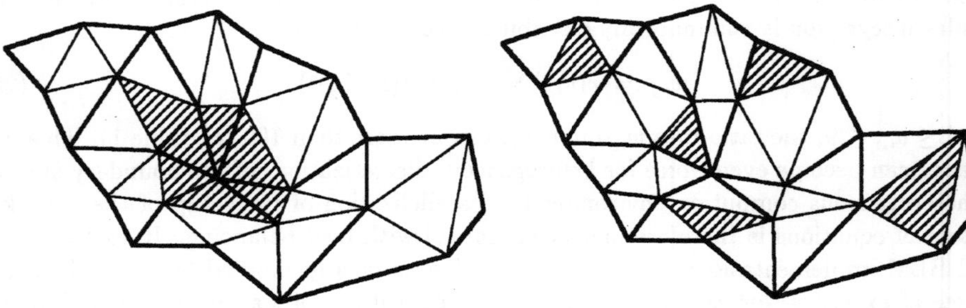

Figure 2: Task building by forming subsets of boundary elements. left: contiguous elements, right: scattered elements

A STRAIGHTFORWARD APPROACH

We want to present a model calculation of the expected speedup and efficiency values under the following ideal circumstances:

1. no communication delay, homogeneous computing environment
2. limited task granularity: 10 tasks per processing node
3. load metric $w_{i,j} = W_1/\widetilde{\text{dist}}(\Gamma_i, \Gamma_j)$, cf. (28)
4. spherical boundary
5. homogeneous discretization of the boundary
6. contiguous subsets of boundary elements .

1 is due to the independence of the tasks for multiple storage of the geometry data, and to keep matters simple. 2 is necessary because the number of tasks for p processing nodes must be of the order $O(p)$. Otherwise we can expect a growing overhead due to task scheduling. 3-5 almost reflect the case for the global geodetic bvp. 6 is chosen because it is easier to implement.

Simple geometrical calculations show that the frequency distribution of distances δ of elements implied by 5 for a spherical boundary of radius R is

$$\frac{3\delta}{4R}\sqrt{R^2 - \frac{\delta^2}{4}}. \tag{31}$$

According to this law, a set of tasks is generated at $\delta_i = i\Delta\delta$, $i = 1, \ldots, 2R/\Delta\delta$. Two very simple models of a distribution of such tasks are compared:

1. Random distribution. Here we only adhere to assumption 2 above.
2. Cyclic distribution. Sort the tasks with respect to their workload, then assign task number $i \in \{1 \ldots 10p\}$ to node number $((i-1) \bmod p) + 1$.

Denoting by t_i the total runtime of all tasks assigned to a node $i \in \{1, \ldots, p\}$, then the speedup (29) becomes

$$s_p = \frac{\sum_{i=1}^{p} t_i}{\max_{i=1}^{p} t_i} \tag{32}$$

and the efficiency follows with (30). The results are presented in figure 3. The speedup is constantly increasing, but not as fast as $O(p)$, which makes the efficiency declining rapidly.

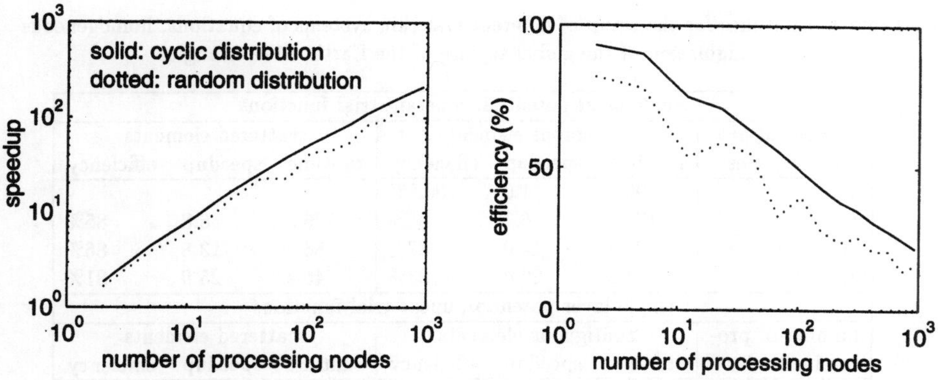

Figure 3: Results of the straightforward approach

This simple calculations show that straightforward approaches may lead to *non-scalable* algorithms (i.e. $e_p \to 0$).

A LOAD DISTRIBUTION SCHEME FOR SETUP OF THE GALERKIN SYSTEM

Methods for load distribution in numerical integration algorithms are discussed lengthy in [9].

In *static* load distribution, we assign tasks $T_{k,l}$ to nodes before the computation starts. Here we need a good estimate of the corresponding workloads $W_{k,l}$. For our problem, this seems to be very difficult. If we have both a homogeneous discretization of the boundary surface and a homogeneous computing environment, then we can perhaps use the *scattered* elements subset method (cf. section 5).

In *dynamic* load distribution, tasks are assigned to nodes partly at runtime, depending on the progress of the nodes. This is much more flexible, and we can expect good results also in heterogeneous situations (which we cannot show here). Unfortunately, the exchange of runtime information among the nodes requires communication, but for reasonable task granularities the corresponding loss of performance is usually very small.

We want to suggest a *task pool method* for the setup of the Galerkin system of equations: The task pool

$$T := \{T_{k,l},\ k = 1, \ldots, p,\ l = 1, \ldots, k\} \tag{33}$$

is the total set of tasks as defined in section 5. Every processing node $P_i, i = 1, \ldots, p$ inquires the presence of the following ordered sequence of tasks in T:

$$\{T_{i+1,i}, \ldots, T_{p,i}, T_{i,1}, T_{i,i-1}, T_{i,i}\}\,. \tag{34}$$

If a task is found in T, it is removed from T and processed. Then, we continue with the next inquiry until the end of the sequence is reached. The latencies of node P_i due to communication can be hidden if in the meantime we already process parts of task $T_{i,i}$,

Table 2: Runtimes for the setup of different Galerkin systems of equations, homogeneous triangulation of the global surface of the Earth, 8192 triangles

number of processing nodes	contiguous elements runtime	speedup	efficiency	scattered elements runtime	speedup	efficiency
single layer potential, constant trial functions						
1	1190 s	1.0	100.0%	-	-	-
8	177 s	6.7	84%	176 s	6.8	85%
16	85 s	14.0	87%	86 s	13.8	86%
32	45 s	26.4	83%	46 s	25.9	81%
single layer potential, linear trial functions						
1	1390 s	1.0	100.0%	-	-	-
8	212 s	6.6	82%	214 s	6.5	81%
16	103 s	13.5	84%	105 s	13.2	83%
32	55 s	25.3	79%	55 s	24.3	79%
double layer potential, linear trial functions						
1	1870 s	1.0	100.0%	-	-	-
8	292 s	6.4	80%	287 s	6.5	81%
16	144 s	13.0	81%	144 s	13.0	81%
32	72 s	26.0	81%	73 s	25.6	80%

which was for this purpose shifted to the end of the sequence (34). The status of a task $T_{i,j}$ must only be known at P_i and P_j, and no synchronization between the nodes takes place. Therefore we call this a *multiple asychronous* task pool method.

The advantage of such a scheme is the following: Assume that the latency due to a transfer of geometry data belonging to one subset of elements per task can be hidden in the way explained above (as it can in our case). Then we may also implement a distributed (non-redundant) storage of the geometry data: each node P_i manages only the data belonging to the i-th subset of elements.

RESULTS

We implemented the method described in the preceeding section on a IBM 9076 SP2 with 34 processing nodes. The results for the different BEM-solutions of the global fixed gravimetric bvp are presented in table 2. The computations aim at a global resolution of the gravity field of about 250 km at the surface of the Earth. We see that the efficiency is usually better than 80%, and does not decline rapidly as e.g. in figure 3. This proves the method to be powerful in our case. It seems to work best for the single layer solution with constant trial functions, which can be explained by the least variability of the load metric (28) in this case. No significant differences between the subset strategies are observed.

Acknowledgment: The first author greatfully acknowledges the financial support from the Deutsche Forschungsgemeinschaft (DFG) under contract Le 962/1-1, and the permission to use the supercomputing facilities of the Gesellschaft für Mathematik und Datenverarbeitung (GMD).

REFERENCES

[1] Andrä, H. (1995): GBEMINT Version 1.03g. Kurzanleitung. Inst. f. Mechanik und Festigkeitslehre, Universität Karlsruhe (TH).

[2] Brovar, V.V. (1963): Solutions of the Molodenskij boundary problem. Geodesy and Aerophotography, 237–240.

[3] Brovar, V.V. (1964): Fundamental harmonic functions with a singularity on a segment and solution of outer boundary problems. Geodesy and Aerophotography, 150–155.

[4] Giraud, G. (1934): Equations integrales principales. Annales Scientifiques de l'Ecole Normale Superieure, Troisieme Serie, 51: 251–372.

[5] Herzig, Th. (1994): Ein paralleles 3D-Randelementprogramm der Elektrostatik für heterogene Rechnerarchitekturen. Dissertation. Universität Karlsruhe (TH).

[6] Ingber, M.S., D.E. Womble, L.A. Mondy (1994): A Parallel Boundary Element Formulation for Determining Effective Properties of Heterogeneous Media. Int. J. Num. Meth. Eng. **37**, 3905–3920.

[7] Johnson, C.G.L., L.R. Scott (1989): An Analysis of Quadrature Errors in Second Kind Boundary Integral Equations. SIAM, J. of Numer. Anal. **16** 1356–1382.

[8] Klees, R. (1992): Lösung des fixen geodätischen Randwertproblems mit Hilfe der Randelementmethode. Veröffentl. d. Dt. Geod. Kommiss. b. d. Bayer. Akad. d. Wiss., Reihe C, Heft Nr. 382, München.

[9] Krommer, A.R., C.W. Ueberhuber (1994): Numerical Integration on Advanced Computer Systems. Lecture Notes in Computer Sciences, Vol. 848. Springer Verlag Berlin Heidelberg.

[10] Lyness, J.N., D. Jesperson (1975): Moderate degree symmetric quadrature rules for the triangle. J. Inst. Math. Appl. **15**, 19–32.

[11] Sauter, S.A. (1992): Über die effiziente Verwendung des Galerkin-Verfahrens zur Lösung Fredholmscher Integralgleichungen. Dissertation. Christian-Albrechts-Universität Kiel.

[12] Sauter, S.A., A. Krapp (1995): On the Effect of Numerical Integration in the Galerkin Boundary Element Method. Inst. f. Informatik und Praktische Mathematik, Christian-Albrechts-Universität Kiel, Report No. 9504.

[13] Schwab, Chr., Wendland, W.L. (1992): Kernel properties and representations of boundary integral operators. Math. Nachr. 156: 187–218.

[14] Wendland, W.L. (1990): Boundary element methods for elliptic problems. In: A.H. Schatz, V. Thomee, W.L. Wendland, "Mathematical Theory of Finite and Boundary Element Methods", DMV Seminar Vol. 15, Birkhaeuser: 223–276.

ON THE IMPLEMENTATION OF THE h-p BOUNDARY ELEMENT METHOD ON CURVED SURFACES

Patrick Mund
Institut für Angewandte Mathematik
Universität Hannover
Welfengarten 1
30167 Hannover, Germany

SUMMARY

We consider the *h-p* Galerkin boundary element method applied to a weakly singular integral equation of the first kind on an open curved surface. We focus on the effective computation of the Galerkin matrix, i.e. we introduce a numerical qudrature rule which is shown (for a model case) to converge exponentially fast with the number of kernel evaluations. Numerical experiments show that the application of this quadrature rule preserves the exponential convergence of our *h-p* Galerkin scheme.

INTRODUCTION

The combination of geometric mesh refinement and *h-p* approximation with boundary element techniques gives a powerful tool for the approximate solution of boundary integral equations. In [3] an *h-p* Galerkin scheme for weakly singular and hypersingular integral equations on plane screens in $\mathbb{R}^3$ was analyzed and in [4] exponential convergence could be proved.

Although the singular integrals in [3] can be evaluated analytically the assembly of the Galerkin matrix is extremely expensive. This becomes even worse if curved surfaces are considered and the entries of the Galerkin matrix have to be computed by a numerical quadrature rule. Here, an application of the *h-p* boundary element method has the advantage that the Galerkin error decays exponentially fast with the size of the Galerkin matrix, i.e. the number of Galerkin entries is kept low.

In this paper we focus on the weakly singular integral equation on an open curved surface G which corresponds to the direct single layer potential formulation of the Dirichlet problem for the homogeneous Laplace equation in $\mathbb{R}^3 \setminus G$. Our aim is now, on one hand, to show numerically that the results in [3] extend to curved surfaces and, on the other hand, to define a quadrature rule which approximates the Galerkin entries exponentially fast with the number n of kernel evaluations. By increasing n at each h-p refinement step we may, hence, expect to preserve the exponential convergence of the Galerkin scheme while keeping the computational costs low.

The quadrature rules which we use are basically applications of Schwab's [8] graded quadrature rules for singular integrals to the inner *and* outer integrals in our Galerkin matrix. Schwab's rule can be applied directly to assemble collocation matrices or the inner integrals of the Galerkin entries. Based on the h-p approximation results in [2] and the interpolation property of e.g. Gaussian quadrature formulae exponential convergence could be proved [8]. For the outer integrals we need a similar rule which is designed to approximate the singularities of the single layer potential.

The outline of the paper is as follows: In the first section we define our quadrature rules. We focus on the critical cases where the supports of the test and trial functions intersect. In Section 2 we prove exponential convergence of the quadrature rule for a simple example and we give numerical results for more general cases. In Section 3 we comment on further computational aspects. In particular we describe briefly how to approximate the curved surface. We conclude with numerical results in Section 4 for the h, p and h-p methods applied to a model problem on an open curved surface G.

1 DEFINITION OF THE QUADRATURE FORMULAE

Let $G \subset \mathbb{R}^3$ be an open curved surface with parameter region $\Gamma = [-1,1]^2$ and parameter function $\gamma : \Gamma \to G$. We assume that G satisfies a Lipschitz condition and that $\gamma(\partial\Gamma) = \partial G$. Let $V : \widetilde{H}^{-1/2}(G) \to H^{1/2}(G)$ be the single layer potential operator defined as

$$V\psi(x) = \frac{1}{4\pi}\int_\Gamma \frac{\psi(y)}{\|x-y\|}\, d\sigma_y\,.$$

For the definition of the Sobolev spaces $H^s(G)$ and $\widetilde{H}^s(G)$ see [5]. We consider the following weakly singular integral equation: Given $f \in H^{1/2}(G)$ find $\psi \in \widetilde{H}^{-1/2}(G)$ such that

$$V\psi = f\,. \tag{1.1}$$

For the Galerkin scheme we allow subdivisions τ of Γ which consist of rectangular elements. Let $\widetilde{X}$ be a finite dimensional space of piecewise polynomial functions over τ and let $X = \{\phi\circ\gamma^{-1}\,;\,\phi \in \widetilde{X}\} \subset \widetilde{H}^{-1/2}(G)$. Note that no interelement continuity of the functions in $\widetilde{X}$ is required. For simplicity we restrict ourselves at this point to regular subdivisions τ, i.e. we claim that the intersection of two different elements is either a common node or a common edge or is empty. For generalizations to non-regular meshes see Section 3.

The Galerkin equations for solving approximately (1.1) are given by: Find $\psi \in X$ such that

$$V(\psi,\phi) = \langle f,\phi\rangle \qquad \forall\phi \in X\,. \tag{1.2}$$

The bilinear form $V(.,.)$ is defined as $V(\psi,\phi) = \langle V\psi,\phi\rangle$ and $\langle .,.\rangle$ denotes the duality pairing of $H^{1/2}(G)$ and $\widetilde{H}^{-1/2}(G)$.

Let ψ_i, ψ_j be two basis functions in $\widetilde{X}$ and let Γ_i, Γ_j be elements in τ with $\mathrm{supp}(\psi_i) \subset \overline{\Gamma}_i$ and $\mathrm{supp}(\psi_j) \subset \overline{\Gamma}_j$. Now, the entries of the Galerkin matrix and the right hand side vector are

$$V(\psi_j \circ \gamma^{-1}, \psi_i \circ \gamma^{-1}) = \frac{1}{4\pi} \int_{\Gamma_i} \psi_i(x) \left(\int_{\Gamma_j} \frac{\psi_j(x)}{\|\gamma(x) - \gamma(y)\|} J(y)\, dy \right) J(x)\, dx \quad (1.3)$$

$$\langle f, \psi_i \circ \gamma^{-1} \rangle = \int_{\Gamma_i} \psi_i(x)\, f(\gamma(x))\, J(x)\, dx \quad (1.4)$$

where $J(x) = \|\frac{\partial \gamma(x)}{\partial x_1} \times \frac{\partial \gamma(x)}{\partial x_2}\|$ and $\|.\|$ denotes the Euklidean norm in $\mathbb{R}^3$.

It could be shown in [4] for plane surfaces G that the *h-p* version of the boundary element method converges exponentially fast. The *h-p* meshes are geometrically graded towards ∂G and the polynomial degrees of the test and trial functions in $x \in \Gamma$ are small if x is close to $\partial\Gamma$ and are increased perpendicular to $\partial\Gamma$ (for details see [3]. For plane surfaces, however, the integrals (1.3) can be evaluated analytically and the computational cost for the assembly of the Galerkin matrix grows only algebraically with the number N of *h-p* refinement steps, i.e. like $O(N^\alpha)$. Our main goal is now to show that there is a quadrature rule which approximates the singular integrals (1.3) exponentially fast (with N) and which needs $O(N^\alpha)$ kernel evaluations ($\alpha \in \mathbb{N}$ fixed). Furthermore, we give (numerical) evidence that the *h-p* Galerkin method applied to (1.1) in combination with this quadrature rule leads to exponential convergence of the approximate solutions.

Our quadrature rule is basically an application of the variable order composite quadrature rule for singular integrals suggested by Schwab [8] to the inner *and* outer integral in (1.3). To approximate both integrals we have to deal with point and edge singularites. The kernel $\|\gamma(x) - \gamma(y)\|^{-1}$ of the inner integral has obviously a point singularity at $y = x$ whereas the single layer potential $V(\psi \circ \gamma^{-1})$ has singular behaviour at $\gamma(\partial\,\mathrm{supp}(\psi))$.

For point singularities Schwab suggested the following rule for the approximation of the integral $\int_{\Gamma_0} \psi_0(x)\, dx$ where $\Gamma_0 = (0,1)^2$ and ψ_0 is singular at the origin: Given a fixed parameter $\sigma_1 \in (0,1)$ and an integer n one considers geometric subdivisions of Γ_0 into smaller rectangles $R_{l,k}$. We define $z_0 = 0$, $z_k = \sigma_1^{n-k}$, $1 \le k \le n$, and

$$\begin{aligned} R_{1,k} &= (z_{k-1}, z_k) \times (0, z_{k-1}) && \text{for} \quad 2 \le k \le n\,, \\ R_{2,k} &= (z_{k-1}, z_k) \times (z_{k-1}, z_k) && \text{for} \quad 1 \le k \le n\,, \\ R_{3,k} &= (0, z_{k-1}) \times (z_{k-1}, z_k) && \text{for} \quad 2 \le k \le n\,. \end{aligned}$$

For fixed $\varrho_1, \varrho_2 \in \mathbb{N}_0$ let $Q_{l,k}$ denote the tensor product of the $(k+\varrho_1)$–point Gaussian quadrature rule in x_1-direction and the $(k+\varrho_2)$–point Gaussian quadrature rule in x_2-direction, scaled to $R_{l,k}$. Hence

$$Q_{l,k}\psi_0 \approx \int_{R_{l,k}} \psi_0(x)\, dx\,.$$

The composite quadrature rule $Q_n^{(1)}$ is now defined as

$$Q_n^{(1)}\psi_0 = Q_{2,1}\psi_0 + \sum_{k=2}^{n} \sum_{l=1}^{3} Q_{l,k}\psi_0 \approx \int_{\Gamma_0} \psi_0(x)\, dx\,.$$

The quadrature points and the subdivision of Γ_0 for $\sigma_1 = 0.4$, $n = 4$ and $\varrho_1 = \varrho_2 = 0$ are shown in Figure 1.1(a).

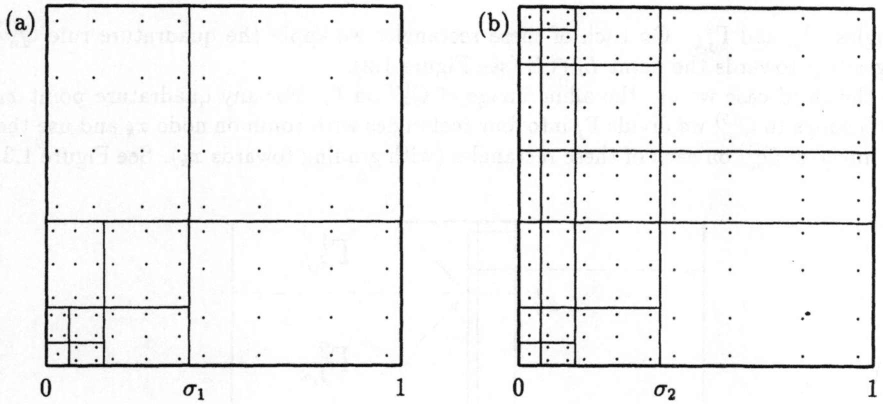

Figure 1.1: Subdivision of $[0,1]^2$ and quadrature points for (a) $Q_n^{(1)}$ and (b) $Q_n^{(2)}$ where $n = 4$, $\sigma_1 = \sigma_2 = 0.4$ and $\varrho_1 = \varrho_2 = 0$.

Remark: *When approximating the Galerkin entries (1.3) we will choose $\varrho_k = p_k$ where p_k is the polynomial degree of ψ_j (or ψ_i) in x_k-direction ($k = 1, 2$).*

For corner-edge singularities we consider again the reference element Γ_0. Let ϕ_0 have corner singularities at the origin $(0,0)$ and at the point $(x_1, x_2) = (0,1)$ and an edge singularity at $x_1 \equiv 0$. We use a geometric subdivision of Γ_0 towards the corners $(0,0)$ and $(0,1)$ and towards the corresponding edge with grading parameter $\sigma_2 \in (0, 1/2)$. This defines a quadrature rule $Q_n^{(2)}$. Hence, we have

$$Q_n^{(2)}\phi_0 \approx \int_{\Gamma_0} \phi_0(x)\,dx\,.$$

For an example see Figure 1.1(b).

If the function ϕ_0 is singular at all four edges of Γ_0 we use a quadrature rule $Q_n^{(3)}$ with geometrical grading towards all the edges of Γ_0 and with grading parameter $\sigma_3 \in (0, 1/2)$ (details are omitted).

To approximate the Galerkin entries (1.3) we have to deal with three critical cases where the kernel becomes singular:

i. Γ_i and Γ_j have a common node
ii. Γ_i and Γ_j have a common edge
iii. $\Gamma_i = \Gamma_j$

In the first case we use affine images of the quadrature rule $Q_n^{(1)}$ on Γ_i and Γ_j with grading towards the common node.

We define the integer $m = \lfloor n^{4/3} \rfloor$.

In the second case we use the affine image of $Q_m^{(2)}$ on Γ_i with grading towards the common edge and the common nodes. Let x_k denote the quadrature points of $Q_m^{(2)}$ on Γ_i. For any of these points x_k we consider the straight line L_k which contains x_k and which is orthogonal to the common edge E of Γ_i and Γ_j. The line L_k divides Γ_j into two rectangles $\Gamma^1_{j,k}$ and $\Gamma^2_{j,k}$. On each of these rectangles we apply the quadrature rule $Q^{(1)}_n$ with grading towards the point $L_k \cap E$ (see Figure 1.2).

In the third case we use the affine image of $Q^{(3)}_m$ on Γ_i. For any quadrature point x_k which belongs to $Q^{(3)}_m$ we divide Γ_j into four rectangles with common node x_k and use the affine image of $Q^{(1)}_n$ on each of these rectangles (with grading towards x_k). See Figure 1.3.

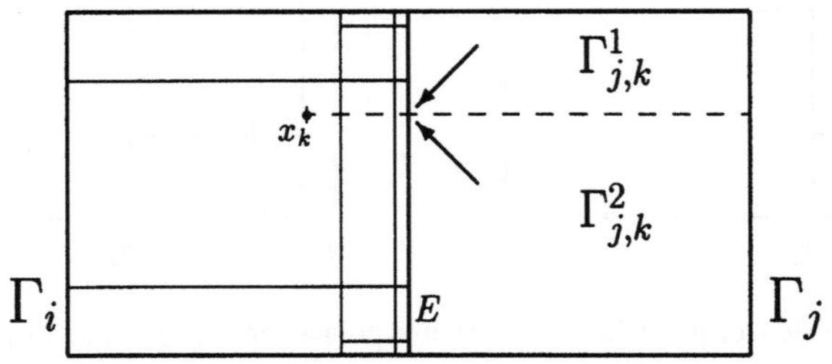

Figure 1.2: Composite quadrature for two elements Γ_i and Γ_j with common edge E. The arrows indicate direction of grading on Γ_j. The grading on Γ_j varies with the location of the quadrature points x_k in Γ_i.

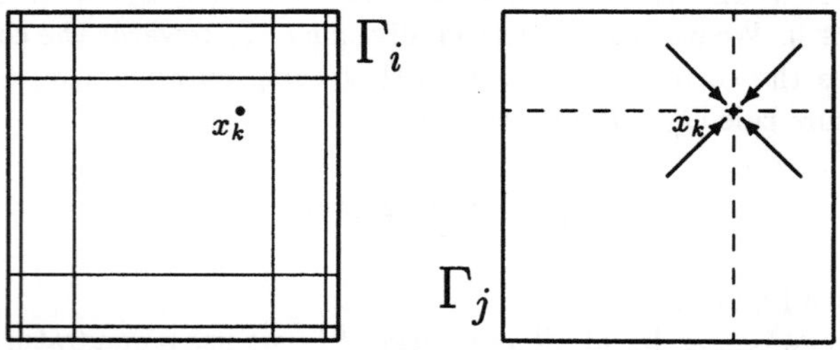

Figure 1.3: Composite quadrature for $\Gamma_i = \Gamma_j$. The grading for the inner quadrature (over Γ_j) varies with the location of the outer quadrature points $x_k \in \Gamma_i$.

2 CONVERGENCE RESULTS

In this section we prove exponential convergence of the quadrature rule introduced in the previous section for a simple example of two square elements in the (x_1, x_2)-plane with a common node. In the proof we restrict ourselves to piecewise constant test and trial functions, i.e. we have $\varrho_1 = \varrho_2 = 0$. Numerical results for higher polynomial degrees, for the case of two elements Γ_i and Γ_j with common edge and for the case $\Gamma_i = \Gamma_j$ are included at the end of the section. The experimental results indicate exponential convergence in any case.

2.1 A SIMPLE EXAMPLE

We consider the elements $\Gamma_1 = (0,1)^2$ and $\Gamma_2 = (-1,0)^2$ and the parameter function $\gamma : [-1,1]^2 \to G$ defined as $\gamma(x_1,x_2) = (x_1,x_2,0)$. Let $\psi_1, \psi_2 \in \widetilde{H}^{-1/2}(G)$ be defined as

$$\psi_j(x) = \begin{cases} 1 & \text{if} \quad x \in \Gamma_j \times \{0\} \\ 0 & \text{if} \quad x \in G \setminus (\Gamma_j \times \{0\}) \end{cases} \qquad (j = 1,2).$$

Hence,

$$V(\psi_1,\psi_2) = \frac{1}{4\pi} \int_{\Gamma_2} \int_{\Gamma_1} \frac{1}{\|x-y\|} \, dy \, dx \,. \tag{2.1}$$

For $0 < \sigma < 1$ and $n \in \mathbb{N}$ define $Q_1 = Q_n^{(1)}$ and let Q_2 be the affine image of $Q_n^{(1)}$ on Γ_2 with grading towards the origin. For $i \in \{1,2\}$ let $x_k^{(j)}$ and $w_k^{(j)}$ be the knots and weights of the rule Q_j, i.e.

$$Q_j g = \sum_{k=1}^{M} w_k^{(j)} g(x_k^{(j)})$$

where $M = 1 + 3\sum_{i=2}^{n} i^2 = n^3 + \frac{3}{2}n^2 + \frac{1}{2}n - 2$.

The following result states exponential convergence of the composite quadrature rule $Q_2 Q_1$ applied to $\|x-y\|^{-1}$.

Theorem 2.1 *For any $0 < \sigma < 1$ there are constants $c_1, c_2 > 0$ such that*

$$\left| V(\psi_1,\psi_2) - \sum_{k=1}^{M} \sum_{j=1}^{M} w_k^{(1)} w_j^{(2)} \|x_k^{(1)} - x_j^{(2)}\|^{-1} \right| \leq c_1 \, e^{-c_2 n} \tag{2.2}$$

for all $n \in \mathbb{N}$.

To prove the theorem we need the theory of countably normed spaces and the result from [8] on the exponential convergence of Q_j:

For a domain $A \subset \mathbb{R}^2 \setminus \{0\}$ and a parameter $0 < \beta < 1$ let $H_\beta^k(A)$ be the closure of $C^\infty(\overline{A})$ with respect to the weighted Sobolev norm

$$\|g\|^2_{H_\beta^k(A)} = \sum_{j=0}^{k} \sum_{|\alpha|=j} \int_A |D^\alpha g(x)|^2 \phi^2_{\beta+j}(x) \, dx$$

where the weight function ϕ_s is defined as $\phi_s(x) = |x|^s$ $(s \in \mathbb{R})$.
The countably normed space $\mathcal{B}_\beta(A)$ is defined as the subspace of all functions g in $L^1(A) \cap \bigcap_{k=0}^{\infty} H_\beta^k(A)$ whose derivatives satisfy the growth condition

$$\left(\int_A |D^\alpha g(x)|^2 \phi^2_{\beta+j}(x) \, dx \right)^{1/2} \leq D_g \, (d_g)^j \, j! \tag{2.3}$$

for all $\alpha \in \mathbb{N}_0^2$ with $|\alpha| = j$. The constants $d_g \geq 1$ and $D_g > 0$ depend on A and g but not on j.

As an example consider the family of functions

$$k_x(y) = \frac{1}{\|x-y\|} \qquad (x \in \Gamma_2).$$

From [8, Prop. 2.2] we know that $k_0 \in \mathcal{B}_\beta(\Gamma_1)$ for $0 < \beta < 1$. Since $k_0(y)$ is analytic for $y \neq 0$ we have $k_0 \in \mathcal{B}_\beta((0,2)^2)$ and, hence, $k_0 \in \mathcal{B}_\beta(A)$ for all domains $A \subset (0,2)^2$ uniformly, i.e. the constants d_{k_0} and D_{k_0} in (2.3) do not depend on A. By a simple shift argument we obtain:

Lemma 2.2 *Is is $k_x \in \mathcal{B}_\beta(\Gamma_1)$ uniformly in x, i.e. for any $0 < \beta < 1$ there are constants $d_0 \geq 1$ and $D_0 > 0$ such that*

$$\left(\int_A |D^\alpha k_x(y)|^2 \phi^2_{\beta+j}(y)\, dy\right)^{1/2} \leq D_0\, (d_0)^j\, j! \tag{2.4}$$

for all $|\alpha| = j \in \mathbb{N}_0$ and for all $x \in \Gamma_2$.

The following result was proved in [8]:

Lemma 2.3 (8, Thm. 4.1) *Let $g \in \mathcal{B}_\beta(\Gamma_1)$ with $\beta > 0$ sufficiently small. Then, for any $0 < \sigma_1 < 1$ there exist constants $b_1, b_2 > 0$ independent of n such that*

$$\left|\int_{\Gamma_1} g(y)\, dy - Q_n^{(1)} g\right| \leq b_1\, e^{-b_2 n} \tag{2.5}$$

where the constants b_1 and b_2 depend only on $\sigma_1, \beta, d_g, D_g$ and Γ_1.

Remark. *In [8] another parameter $\mu > 0$ was introduced which controls the number of quadrature points in $R_{l,k}$ and a condition for the optimal choice of μ was given. For the sake of simplicity we do not elaborate on it in this context, although a proper choice of μ may lead to an improved convergence rate b_2 in (2.5).*

We are now in the position to prove Theorem 2.1.
Proof of Theorem 2.1: The triangle inequality yields

$$4\pi \left|V(\psi_1,\psi_2) - \sum_{k=1}^{M}\sum_{j=1}^{M} w_k^{(1)} w_j^{(2)} \|x_k^{(1)} - x_j^{(2)}\|^{-1}\right| \leq$$

$$\leq \underbrace{\left|\int_{\Gamma_1}\int_{\Gamma_2} \frac{1}{\|x-y\|}\, dy\, dx - Q_2 \int_{\Gamma_1} \frac{1}{\|.-y\|}\, dy\right|}_{=\epsilon_1} + \underbrace{\left|Q_2 \int_{\Gamma_1} \frac{1}{\|.-y\|}\, dy - Q_2^{(x)} Q_1^{(y)} \frac{1}{\|x-y\|}\right|}_{=\epsilon_2}.$$

We will estimate ϵ_2 and ϵ_1 seperately. By definition of Q_i and k_x we have

$$\epsilon_2 = \sum_{j=1}^{M} w_j^{(2)} \left|\int_{\Gamma_1} k_{x_j^{(2)}}(y)\, dy - Q_n^{(1)} k_{x_j^{(2)}}\right|.$$

Form Lemma 2.3 it follows that

$$\epsilon_2 \leq \sum_{j=1}^{M} w_j^{(2)} b_1\, e^{-b_2 n}$$

where b_1 and b_2 are independent of j (Lemma 2.2). Hence,

$$\epsilon_2 \leq |\Gamma_2|\, b_1\, e^{-b_2 n} = b_1\, e^{-b_2 n}. \tag{2.6}$$

To estimate ϵ_1 we have to show that $V\psi_1(x) \in \mathcal{B}_\beta(\Gamma_2)$ for all $0 < \beta < 1$. For plane rectangular elements the single layer potential can be calculated analytically [3]. For $x \in \Gamma_2$ we have:

$$\begin{aligned} V\psi_1(x) &= \left.\left.(y_1 - x_1)\,\text{arsinh}\frac{y_2 - x_2}{|y_1 - x_1|} + (y_2 - x_2)\,\text{arsinh}\frac{y_1 - x_1}{|y_2 - x_2|}\right|_{y_1=0}^{1}\right|_{y_2=0}^{1} \\ &= -x_1 \ln(-x_2 + \|x\|) - x_2 \ln(-x_1 + \|x\|) + g_1(x) \\ &= -(\cos\theta + \sin\theta)\, r \ln r + g_2(r, \theta) \end{aligned}$$

where (r, θ) are the usual polar co-ordinates and g_1, g_2 are analytic in Γ_2.

With an alternative formulation of the growth condition in polar co-ordinates [1] it can be shown easily that $V\psi_1 \in \mathcal{B}_\beta(\Gamma_2)$ for all $0 < \beta < 1$. Hence, from Lemma 2.3 it follows that

$$\epsilon_1 \leq b_3\, e^{-b_4 n}\,. \tag{2.7}$$

From (2.6) and (2.7) we conclude (2.2) with $c_1 = (b_1 + b_3)/(4\pi)$ and $c_2 = \min\{b_2, b_4\}$. ■

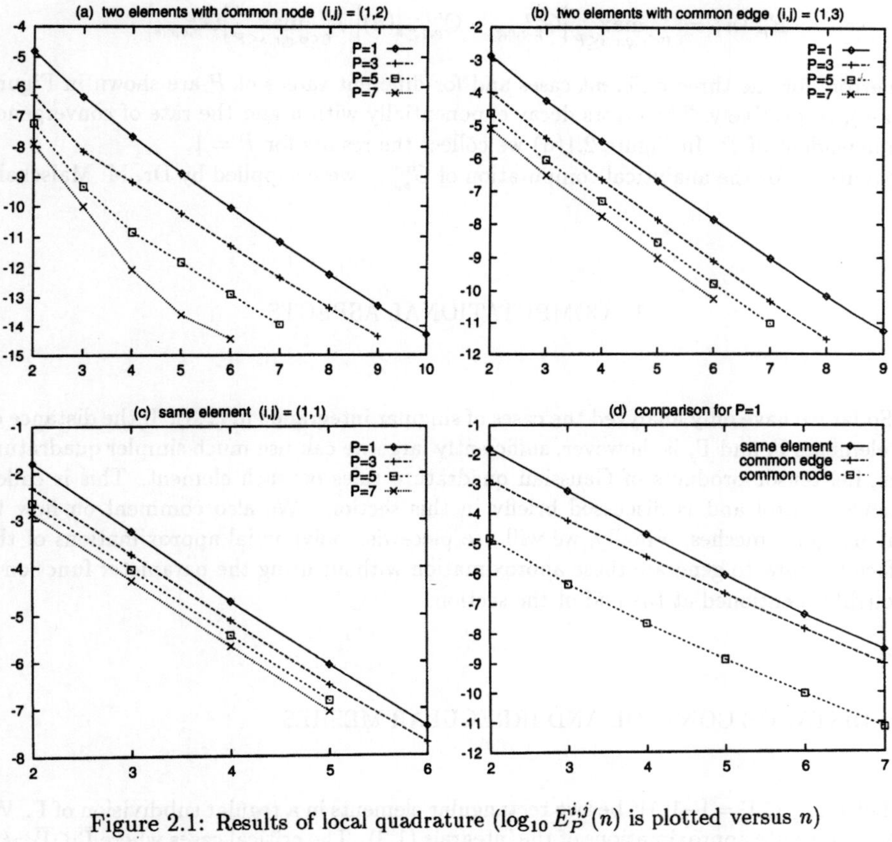

Figure 2.1: Results of local quadrature ($\log_{10} E_P^{i,j}(n)$ is plotted versus n)

2.2 EXPERIMENTAL RESULTS

We consider the elements $\Gamma_1 = (0,1)^2$, $\Gamma_2 = (-1,0)^2$ and $\Gamma_3 = (-1,0) \times (0,1)$. We seek to compute approximations of the Galerkin entries

$$G^{i,j}_{p,q,r,s} = \frac{1}{4\pi} \int_{\Gamma_i} \int_{\Gamma_j} \frac{\phi_{p,q}(x)\, \psi_{r,s}(y)}{\|x-y\|}\, dy\, dx\,. \tag{2.8}$$

The functions $\phi_{p,q}(x)$ and $\psi_{r,s}(y)$ are affine images of $l_p(\eta)\, l_q(\xi)$ and $l_r(\eta)\, l_s(\xi)$, respectively, where l_k denotes the k-the Legendre polynomial and (η, ξ) are the local co-ordinates on the reference element $(-1,1)^2$.

For the approximation of (2.8) we use the quadrature rules introduced in Section 1. We consider subsequently the cases $(i,j) = (1,2)$ (common node), $(i,j) = (1,3)$ (common edge) and $(i,j) = (1,1)$ (same element). We have chosen $\sigma_1 = 0.20$, $\sigma_2 = 0.24$ and $\sigma_3 = 0.28$ which leads to (nearly) the same convergence rates for all three cases (see Figure 2.1). Given $n \in \mathbb{N}$ let $Q^{i,j}_{p,q,r,s}(n)$ be the approximation to $G^{i,j}_{p,q,r,s}$. The error $E^{i,j}_P(n)$ is defined as

$$E^{i,j}_P(n) = \max_{0 \le p,q,r,s \le P} \left| G^{i,j}_{p,q,r,s} - Q^{i,j}_{p,q,r,s}(n) \right| \Big/ \max_{0 \le p,q,r,s \le P} \left| G^{i,j}_{p,q,r,s} \right| .$$

The errors for the three different cases and for different values of P are shown in Figure 2.1(a-c), respectively. The errors decay exponentially with n and the rate of convergence is independent of P. In Figure 2.1(d) we collect the results for $P = 1$.
The routines for the analytical computation of $G^{i,j}_{p,q,r,s}$ were supplied by Dr. M. Maischak.

3 COMPUTATIONAL ASPECTS

So far we have only analyzed the cases of singular integrands in (1.3). If the distance of the elements Γ_i and Γ_j is, however, sufficiently large we can use much simpler quadrature rules, i.e. tensor products of Gaussian quadrature rules on each element. This is called distance control and is discussed briefly in this section. We also comment on how to treat irregular meshes. Finally, we will use piecewise polynomial approximations of the surface G. How to generate these approximation without using the parameter function γ is roughly mentioned at the end of the section.

3.1 DISTANCE CONTROL AND IRREGULAR MESHES

Let $\Gamma_i, \Gamma_j \subset \Gamma = [-1,1]^2$ be two rectangular elements in a regular subdivision of Γ. We seek to compute approximations of the integrals (1.3). The critical cases where $\overline{\Gamma}_i \cap \overline{\Gamma}_j \neq \emptyset$ have already been discussed in Section 1. Nevertheless, the graded quadrature rules may also be applied if $\overline{\Gamma}_i \cap \overline{\Gamma}_j = \emptyset$ but Γ_i and Γ_j are close to each other. For regular meshes

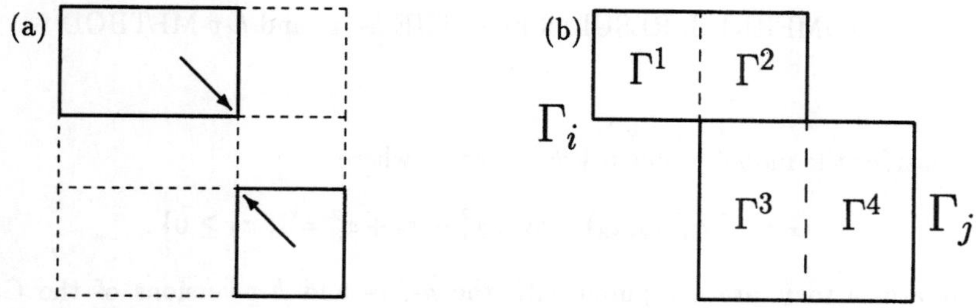

Figure 3.1: (a) Two close elements in a regular mesh. The arrows indicate direction of grading. (b) The irregular case: Instead of computing an approximation to the integral over $\Gamma_i \times \Gamma_j$ we consider subsequently the integrals over $\Gamma^1 \times (\Gamma^3 \cup \Gamma^4)$, $\Gamma^2 \times \Gamma^3$ and $\Gamma^2 \times \Gamma^4$.

we can directly apply the quadrature rules with grading towards the node or edge which is closest to the other element, respectively. For an example see Figure 3.1(a).

To decide whether two elements are close to each other we introduce enlarged elements Λ_i, Λ_j as follows: For $k \in \{1,2\}$ let $h_{i,k}$ be the side length of Γ_i in x_k-direction and let $h_i = \max\{h_{i,1}, h_{i,2}\}$. With m_i we denote the mid point of Γ_i. Now, Λ_i is defined as the rectangle with mid point m_i and side length $h_{i,k} + 2h_i$ in x_k-direction. As a rule of thumb we will use the graded quadrature from Section 1 if $\overline{\Lambda}_1 \cup \overline{\Lambda}_2 \neq \emptyset$ and otherwise we will simply use tensor products of n-point Gaussian quadrature rules on each element.

In the case of irregular meshes it is always possible to apply the graded quadratures rules from Section 1 by subdividing the elements Γ_i and Γ_j to obtain "artificial regularity". In Figure 3.1(b) it is shown how to achieve this goal for a model case. Note that the bisection of Γ_i and Γ_j affects only the quadrature rule while the test and trial functions remain unchanged.

3.2 SURFACE APPROXIMATION

For the numerical experiments in Section 4 we have used piecewise polynomial interpolations of the open curved surface G. We do not need the parameter function γ. This is very useful since the surface may be given implicitly or the given parametrization γ (which is not unique) may have some properties which are not welcome. We need, however, the parametrization of ∂G (preferably by the arclength) to ensure that in the limit $\partial\Gamma$ is mapped to ∂G. For the example in Section 4 we need also the parametrization of the diagonals of Γ (by the arclength) to avoid small angles in our finite element mesh. Whenever a new interpolation point on the surface G is needed (e.g. in each h-, p- or h-p-refinement step) we compute the normal direction at the appropriate point on our current interpolant Γ_h and the intersection of the normal vector with G. Hence, we need a subroutine which computes intersections of straight lines with G. More details and the full algorithm can be found in [6]. The algorithm has the important property that in general uniform subdivisions of Γ are mapped to quasi-uniform subdivisions of G.

4 NUMERICAL RESULTS FOR THE h-, p- and h-p-METHOD

We consider the model problem $V\psi = 1$ on G where

$$G = \{(x_1, x_2, x_3) \in \mathbb{R}^3 \,;\, x_1^2 + x_2^2 + x_3^2 = 1,\, x_3 \geq 0\}\,.$$

Approximations to ψ are computed with the h-, p- and h-p-versions of the Galerkin boundary element method. The initial mesh consists of four elements. For the p and h-p-versions we use isoparametric elements and for the h-version we use superparametric elements, i.e. piecewise linear test and trial functions and piecewise quadratic approximations of G. The superparametric approximation for curved surfaces was suggested by Nedelec [7]. The test and trial functions are tensor products of Legendre polynomials. The mesh for the p-version consists of four elements.

Since ψ is singular at ∂G we use for the h-p-version a geometric mesh in the parameter region $\Gamma = [-1,1]^2$ with grading towards the edges of Γ. The grading parameter is $\sigma_{hp} = 0.17$. We use piecewise linear approximations on all elements on the boundary of Γ and increase the polynomial degrees perpendicular to $\partial\Gamma$ (cf. [3]). The approximation of G after five h-p refinement steps (with $\sigma_{hp} = 0.4$) is shown in Figure 4.1(a).

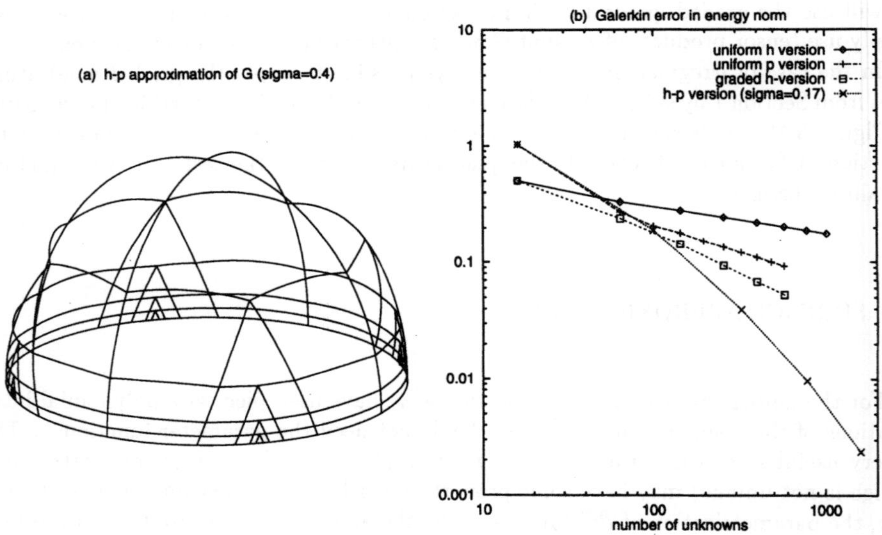

Figure 4.1: Approximation of surface G and Galerkin error in energy norm

We consider two different h-refinements: the uniform h-version and the graded h-version. For the graded h-version we use algebraic grading towards the edges of Γ, i.e. the mesh points at the Nth refinement step are of the form $(\pm 1 \mp (i/N)^3, \pm 1 \mp (j/N)^3)$ for $0 \leq i, j \leq N$.

Let N_h, N_p and N_{hp} be the number of refinement steps for the h-version, the p-version and the h-p-version, respectively. For the approximation of the Galerkin entries we use the quadrature rules from Section 1 and Section 3 with the values of σ_k $(k = 1, 2, 3)$ from Section 2.2 and the following values for n:

$$n_h = \lfloor \log_2 N_h \rfloor + 1, \qquad n_p = \lfloor \log_2 N_p \rfloor + 1, \qquad n_{hp} = N_{hp}.$$

The Galerkin error in energy norm $\|.\|_{-1/2} = \|.\|_{\widetilde{H}^{-1/2}(G)}$ is plotted in Figure 4.1(b). Whereas the error for the h- and p-version decays algebraically, we obtain exponential convergence for the h-p-version. Let $d \in \mathbb{N}$ be the number of unknowns in the Galerkin scheme. The expected convergence rates (cf. [3] for plane surfaces) are

$$\|\psi - \psi_h\|_{-1/2} \sim d^{-1/4+\varepsilon} \qquad \|\psi - \psi_{hg}\|_{-1/2} \sim d^{-3/4+\varepsilon}$$
$$\|\psi - \psi_p\|_{-1/2} \sim d^{-1/2+\varepsilon} \qquad \|\psi - \psi_{hp}\|_{-1/2} \sim \exp(-c\, d^{1/4})$$

where $\psi_h, \psi_{hg}, \psi_p$ and ψ_{hp} are the Galerkin solutions of the uniform h-version, the graded h-version, the uniform p-version and the h-p-version, respectively. The numerical results confirm these expectations.

REFERENCES

[1] BABUŠKA, I. and GUO, B.Q.: "Regularity of the Solution of Elliptic Problems with Piecewise Analytic Data I: Boundary Value Problems for Linear Elliptic Equations of Second Order", SIAM J. Math. Anal. 19, pp. 172-203, 1988.

[2] GUO, B.Q. and BABUŠKA, I.: "The h-p Version of the Finite Element Method, Part 1: The Basic Approximation Results", Comput.Mech. 1, pp. 21-24,1986, "Part 2: General results and applications", *ibd.*, pp. 203-220.

[3] HEUER, N., MAISCHAK, M. and STEPHAN, E.P.: "The h-p-Version of the Boundary Element Method for Screen Problems", Preprint, Universität Hannover (submitted for publication).

[4] HOLM, H., MAISCHAK, M. and STEPHAN, E.P.: "Exponential Convergence of the h-p Version BEM for Mixed BVP's on polyhedrons", Preprint, Universität Hannover, 1996.

[5] LIONS, P.L. and MAGENES, E.: "Non-Homogeneous Boundary Value Problems and Applications I", Springer-Verlag, New York, 1972.

[6] MUND, P.: "An Algorithm for the Interpolation of Curved Surfaces with Rectangular Grids", Preprint, Universität Hannover (in preparation).

[7] NEDELEC, J.C.: "Curved Finite Element Methods for the Solution of Singular Integral Equations in $\mathbb{R}^3$", Comp.Meth.Appl.Mech.Eng. 8, pp. 61-80, 1976.

[8] SCHWAB, C.: "Variable Order Composite Quadrature of Singular and Nearly Singluar Integrals", Computing 53, pp. 173-194, 1994.

Realization of *hp*-Galerkin BEM in $\mathbb{R}^3$

Stefan A. Sauter[1]
Lehrstuhl für Praktische Mathematik
Mathematisches Seminar Bereich II
Christian-Albrechts-Universität zu Kiel
Hermann-Rodewald-Str. 3/1
24098 Kiel, Germany

Christoph Schwab[2]
Seminar für Angewandte Mathematik
ETH Zürich
Rämistrasse 101
CH-8092 Zürich, Switzerland

Summary

It is well known that Galerkin discretizations based on *hp*-Finite Element Spaces are converging exponentially with respect to the degrees of freedom for elliptic problems with piecewise analytic data. However, the question whether these methods can be realized for general situations such that the exponential convergence is preserved also with respect to the computing time is very essential.

In this paper, we will show how the numerical quadrature can be realized in order that the resulting fully discrete *hp-Boundary Element Method* converges exponentially with algebraically growing work. The key point is to approximate the integrals constituting the stiffness matrix by exponentially converging cubature methods.

Introduction

Elliptic boundary value problem in domains $\Omega \subset \mathbb{R}^d$ with piecewise smooth boundaries can be reformulated as Fredholm integral equations on the boundary manifold $\Gamma = \partial\Omega$ provided the fundamental solution is known. In this paper, we will consider Galerkin discretizations of such kinds of integral equations. We will employ *hp*-Finite Element Spaces for the test and trial functions defined on geometrically graded meshes with variable polynomial degree distribution. These finite element spaces exhibit exponential convergence for piecewise analytic functions. Since the convergence of Galerkin discretizations of strongly elliptic problems is quasi-optimal, the Galerkin BEM with exact stiffness matrix converges exponentially, too. For practical realizations of *hp-Boundary Element Methods* one has to approximate the stiffness matrices using numerical cubature techniques (only in very special situations, the system matrix can be computed exactly). Two main questions arise in this context.

[1]partially supported by the NSF under grant DMS 91-20877
[2]partially supported by the AFOSR under grant F49620-J-0100

1. To what precision must the matrix entries be computed in order to preserve the exponential convergence?

2. How can the matrix entries be approximated to the required accuracy with only algebraically growing work?

We show that to preserve the exponential convergence of the Galerkin scheme, the approximate matrix entries $\tilde{A}_{i,j}$ have to satisfy

$$\left|\tilde{A}_{i,j} - A_{i,j}\right| \leq CN^{-1} \exp\left(-bN^{1/4}\right) \tag{1}$$

where N denotes the dimension of the finite element space.

In order to obtain a fully discrete *hp*-BEM satisfying (1) with algebraically growing CPU-time it is therefore necessary to design cubature techniques which converge exponentially with respect to the number of cubature points. The presentation of such cubature methods, designed first in [15], is the purpose of the present paper. For the *h*-version BEM, this problem was already solved in [13], [7], [12] and [14]. For the *hp*-BEM, additional complications arise, namely, the appearance of elements with arbitrarily high aspect ratio coupled with the requirement (1). Finally, we will present the results of numerical experiments which confirm the theoretical investigations.

In this paper, our aim is to present a framework for the analysis of cubature errors in the context of stability and consistency of an underlying Galerkin BEM. The focus is on the practical aspects of the methods, i.e., we present the techniques in an algorithmic way giving insight how *hp*–BEM can be realized efficiently with respect to both, the implementation time as well as the efficiency and versatility of the resulting code. Especially for high-order methods, it is essential to use an adequate data structure and structuring of the algorithm in order to obtain code that runs efficiently for a large class of problems. With proper data structures, the implementation of *hp*-BEM is of comparable complexity as that of the usual *h*-version BEM.

The full theoretical background and proofs of the present results can be found in [15].

Galerkin Boundary Element Methods

Let Ω be a domain in $\mathbb{R}^d$ with orientable, piecewise analytic Lipschitz boundary $\Gamma := \partial\Omega$. By $d\Gamma$ we denote the surface measure defined almost everywhere on Γ. We consider the space $L^2(\Gamma)$ of functions $u : \Gamma \to \mathbb{R}$ which are square integrable with respect to $d\Gamma$. Furthermore, for $0 < s < 3/2$, we define the Sobolev spaces $H^s(\Gamma)$ as the trace spaces of $H^{s+1/2}_{loc}\left(\mathbb{R}^3\right)$ while, for $s < 0$, $H^s(\Gamma)$ is the dual space of $H^{-s}(\Gamma)$. The scalar product in H^s is denoted by $(\cdot,\cdot)_s$ while the corresponding norm is given by $\|\cdot\|_s$.

To apply the Galerkin boundary element method to Fredholm integral equations we write the continuous problem in a variational setting which, in general, takes the form

$$(v, \lambda_1 u)_0 + (v, K_1 u)_0 = (v, \lambda_2 f)_0 + (v, K_2 f)_0, \qquad \forall v \in H^{s_1}(\Gamma) \tag{2}$$

where the scalar product $(\cdot,\cdot)_0 : L^2(\Gamma) \times L^2(\Gamma) \to \mathbb{R}$ is identified with its extension onto $H^{s_1} \times H^{-s_1}$ by means of Riesz' representation theorem. In (2), λ_1, λ_2 are analytic functions (in global variables) and K_i are integral operators given by

$$K_i[w](x) = p.f. \int_\Gamma k_i(x,y)\, w(y)\, d\Gamma_y.$$

These integrals are assumed to exist in the *part-fini* sense [17]. The kernel function and/or the derivatives of k behave singular only for $x = y$. For $i = 1, 2$, the operator $\lambda_i I + K_i$ is assumed to be continuous, i.e., $\|\lambda_i I + K_i\|_{-s_i \leftarrow s_i} \leq C$. The left hand side of (2) defines the bilinear form $a : H^{s_1} \times H^{s_1} \to \mathbf{R}$ while the right hand side defines the functional $F : H^{s_1} \to \mathbf{R}$. The restriction to real-valued (bi-)linear forms is made only to avoid technicalities while generalizations to complex-valued forms are straightforward.

The Galerkin discretization of (2) is based on a dense sequence of finite-dimensional subspaces V_L of $H^{s_1}(\Gamma)$. The Galerkin approximation to (2) is given by seeking $u_L \in V^L$ such that

$$a\left(u^L, v\right) = F(v), \qquad \forall v \in V^L. \tag{3}$$

We assume stability, i.e. that for some $\gamma > 0$ independent of L

$$\inf_{u \in V_L} \sup_{v \in V_L} \frac{|a(u,v)|}{\|v\|_{s_1} \|u\|_{s_1}} \geq \gamma. \tag{4}$$

Under these conditions, for sufficiently large L, (3) is uniquely solvable and the Galerkin solution converges with the optimal rate:

$$\|u - u^L\|_{s_1} \leq C \inf_{v \in V_L} \|u - v\|_{s_1}. \tag{5}$$

In the following we will define subspaces V_L having the property that the infimum above decreases as $\exp\left(-bN_L^{1/4}\right)$ where N_L denotes the dimension of V_L provided the function u lies in a suitable countably normed space B_ϱ. Such spaces were first introduced on two-dimensional domains in [4], [5]. They were generalized to the case of two-dimensional surfaces and operators of order zero in [2], [3], while the case of operators of orders different from zero is considered in [10].

hp-Boundary Elements

In this section we will explain how *hp*-boundary element spaces can be constructed on general surfaces. We impose the following assumption on the surface Γ. There exists a polyhedron $\tilde{\Gamma}$ together with a covering $\tilde{\tau}_0 = \left\{\tilde{K}_1, \tilde{K}_2, \ldots, \tilde{K}_{N_0}\right\}$ consisting of (open) plane quadrangles and triangles which are mutually disjoint and satisfy $\tilde{\Gamma} = \overline{\bigcup_{\tilde{K} \in \tilde{\tau}_0} \tilde{K}}$. Furthermore, we assume that there exists a bi-Lipschitz continuous mapping $\eta : \tilde{\Gamma} \to \Gamma$, i.e.,

$$C_1 \|\tilde{x} - \tilde{y}\| \leq \|\eta(\tilde{x}) - \eta(\tilde{y})\| \leq C_2 \|\tilde{x} - \tilde{y}\|, \qquad \forall \tilde{x}, \tilde{y} \in \tilde{\Gamma},$$

which can be extended analytically to plane extensions $\tilde{K}^{ext}$ of $\tilde{K}$:

$$(\eta\,|_{\tilde{K}})^{ext} \text{ is analytic in } \tilde{K}^{ext} \supset\supset \tilde{K} \text{ for all } \tilde{K} \in \tilde{\Gamma}.$$

It will turn out that, for the cubature methods, the following assumption is very convenient and will be assumed throughout the paper.

Assumption 1 *Let $\mathfrak{E}$ denote the set of (possibly curved) edges of Γ while $\tilde{\mathfrak{E}} = \eta^{-1}(\mathfrak{E})$ is the pull-back on $\tilde{\Gamma}$. We assume that, for every triangle $\tilde{K} \in \tilde{\tau}_0$, the relation $\overline{\tilde{K}} \cap \tilde{\mathfrak{E}} = \emptyset$ holds. Furthermore, we require that, for all elements $\tilde{K} \in \tilde{\tau}_0$, the condition*

if $\tilde{K}$ has an edge $\tilde{e} \subset \tilde{\mathfrak{E}}$ then, the opposite edge is parallel to $\tilde{e}$

is satisfied.

Now, we are ready to define a sequence of meshes $\{\tilde{\tau}_L\}_{L\geq 0}$ for the piecewise plane surface $\tilde{\Gamma}$. These reference meshes, then, induce true surface meshes via $\tau_L := \left\{\eta\left(\tilde{K}\right) : \tilde{K} \in \tilde{\tau}_L\right\}$. The grids $\tilde{\tau}_L$ will be geometrically graded towards the edges of $\tilde{\Gamma}$ and are defined recursively by the following procedure. Let us assume that the mesh $\tilde{\tau}_L$ was already generated. In the following algorithm, the notation $i+1$ stands for $i \bmod 4 + 1$ and $i_2 := (i-1) \bmod 2 + 1$.

Geometric Refinement:

for all $\tilde{K} \in \tilde{\tau}_{L-1}$ **do begin**
  if $\tilde{K}$ is a quadrangle **do begin**
  let $\{X_i\}_{1\leq i\leq 4}$ denote the set of vertices of $\tilde{K}$ (counterclockwise ordering);
  let $e_i := \overline{X_i X_{i+1}}$ denote the set of edges;
  for $i = 1$ **to** 2 **do begin**
    if $e_i \cup e_{i+2} \subset \tilde{\mathfrak{E}}$ **then** connect the midpoints of e_{i+1} and e_{i+3};
  end;
  for $i = 1$ **to** 4 **do begin**
    if $e_i \subset \tilde{\mathfrak{E}}$ **and** $e_{i+2} \cap \tilde{\mathfrak{E}} = \emptyset$ **then** connect $X_{i+1} + \sigma(X_{i+2} - X_{i+1})$
      with $X_i + \sigma(X_{i+3} - X_i)$;
  end;
end;end;

The resulting geometric mesh is denoted by $\tilde{\tau}_L = \left\{\tilde{K}_1, \tilde{K}_2, \ldots, \tilde{K}_{N_L}\right\}$. Note that, for $L > 0$, each element of τ_L has a uniquely determined parent $\mathcal{P}\left(\tilde{K}\right) \in \tau_{L-1}$ characterized by $\tilde{K} \subset \mathcal{P}\left(\tilde{K}\right)$.

Polynomial degree distribution

The subspace V^L consists of piecewise polynomials (in local coordinates) of degrees $p^K = \left(p_1^K, p_2^K\right)$ on $K \in \tau_K$ which we combine in the *linear degree vector* $\delta p = \left\{p^K : K \in \tau_L\right\}$ with initial degree L_0 and slope $\mu > 0$. It is constructed recursively by the following algorithm. The notation $\lfloor m \rfloor$ denotes the largest integer smaller or equal than m.

Polynomial Refinement:

if $L = 0$ **then** define $p^K = (L_0, L_0)$ for all $K \in \tau_0$
else begin
if K is a triangle **then** $p^K = (L, L)$ **else begin**
for all $K \in \tau_L$ **do begin**
  let $\{e_i\}_{1\leq i\leq 4}$ denote the edges of K and $\{E_i\}_{1\leq i\leq 4}$ the edges of $\mathcal{P}(K)$;
  for $i := 1$ **to** 4 **do begin**
    if there is E_j such that $e_i = E_j$ **then**

$$p_{i_2}^K := \min\left(L, \left\lfloor p_{j_2}^{\mathcal{P}(K)} + \mu \right\rfloor\right);$$

else if there is E_j such that $e_i \subsetneq E_j$ **then** $p_{i_2}^K := L_0$

end;

end;end;end;

Remark 2 *Note that the geometric refinement algorithm preserves Assumption 1 for the finer grids* $\tilde{\tau}_L$, $L \geq 0$.

Remark 3 *We assume that the local numbering of the vertices and edges of the panels* K *used in the polynomial refinement algorithm is fixed throughout the paper.*

hp-Finite Element Space

We begin with defining polynomial spaces on reference domains. For $x \in \mathbb{R}^d$, let $|x|_k$ denote the usual l^k-norm. For quadrilaterals, we employ the master element $K_0 = \left\{x \in \mathbb{R}^2 \mid |x|_\infty < 1\right\}$ and, for integers $p = (p_1, p_2) \in \mathbb{N}_0^2$, the polynomial space $P_p(K_0)$ consisting of functions of the form $\sum_{\alpha_1=0}^{p_1} \sum_{\alpha_2=0}^{p_2} c_\alpha x_1^{\alpha_1} x_2^{\alpha_2}$. For triangles, the unit element is given by $K_0 = \left\{x \in \mathbb{R}_+^2 \mid |x|_1 < 1\right\}$, while for $p = (p_1, p_1)$ the space $P_p(K_0)$ consists of functions of the form $\sum_{|\alpha|_1 \leq p_1} c_\alpha x_1^{\alpha_1} x_2^{\alpha_2}$ where $\alpha \in \mathbb{N}_0^2$ is a multi-index.

These spaces are now transported onto the surface panels by the composite mapping

$$\chi_K = \eta \circ \chi_{\tilde{K}}, \tag{6}$$

where $\chi_{\tilde{K}}$ is affine (bi-)linear. For quadrilateral panels K we have to ensure that the edge $\overline{\binom{-1}{-1}, \binom{1}{-1}}$ is mapped onto the first edge of K (cf. Remark 3). The global spaces V_L are composed of the local ones. Let the mesh τ_L and the polynomial distribution δp be constructed by the geometric and polynomial refinement algorithms defined above. We define the intermediate space $V_L^\star$ by

$$V_L^\star := \left\{v : \Gamma \to \mathbb{R} \mid v\mid_K \circ \chi_K \in P_{p^K}(K_0)\right\}.$$

However, some global constraints might be imposed to ensure conformity of the subspaces, i.e., $V_L \subset H^{s_1}(\Gamma)$ and/or for stability reasons in order to apply, e.g., the finite section method. For this reason, it might be necessary to enlarge the space $V_L^\star$ or to impose some global constraints as, e.g., to enforce continuity across the edges. We do not go here further into the details since the main concern of this paper is to present a framework for quadrature techniques and error analysis for *hp*-BEM and not a detailed investigation of general *hp*-spaces. Instead, we will illustrate the situation with the following Examples 4 and 6.

Example 4 *Let us consider the classical double layer potential characterized by* $\lambda_1 = 1$ *and*

$$k_1(x, y) = \frac{\langle n(y), y - x\rangle}{4\pi \|y - x\|^3}.$$

Let the reduced mesh τ_L' *be defined by* $\tau_L' := \left\{K \in \tau_L \mid \overline{K} \cap \mathfrak{E} = \emptyset\right\}$. *Then* V_L *is given by*

$$V_L := \left\{v \in V_L^\star \mid \forall K \in \tau_L \backslash \tau_L' : v\mid_K \equiv 0\right\}.$$

Now we will choose a basis $\{\phi_i\}_{1\leq i\leq N_L}$ of V_L which is stable in $H^{s_1}(\Gamma)$. The details can be found in the following

Assumption 5 *There is a constant C_{stab}, possibly depending on L, such that, for any function $u(x) = \sum_{i=1}^{N_L} \mathbf{u}_i\phi_i(x) \in V_L$, there holds*

$$\sum_{i=1}^{N_L} |\mathbf{u}_i|^2 \leq C_{stab} \|u\|_{H^{s_1}(\Gamma)}^2. \tag{7}$$

In the situation of Example 4, such a basis is given by properly weighted tensor products of Legendre polynomials lifted on the surface patches by using the mapping χ_K. The details can be found in the following

Example 6 *Let τ_L be a quadrilateral surface mesh. On the unit square K_0, a basis of $P_p(K_0)$ is given by tensor product Legendre polynomials $\{\phi_\alpha^0\}_{\substack{0\leq\alpha_1\leq p_1\\0\leq\alpha_2\leq p_2}}$ with $\phi_\alpha^0(\hat{x}) = \sqrt{(\alpha_1+1/2)(\alpha_2+1/2)}L_{\alpha_1}(\hat{x}_1)L_{\alpha_2}(\hat{x}_2)$ and the normalization for the Legendre polynomials $L_\alpha(1) = 1$. The basis of the space V^L defined in Example 4 is given by*

$$V^L = \operatorname{span}\left\{\phi_{K,\alpha}(x) = \left\{\begin{array}{ll}\left(\phi_\alpha^0 \circ \chi_{K'}^{-1}\right)(x)/\sqrt{|K|} & \text{if } K = K' \\ 0 & \text{otherwise}\end{array}\right., \forall K \in \tau_L', \left\{\begin{array}{l}0\leq\alpha_1\leq p_1\\0\leq\alpha_2\leq p_2\end{array}\right\}\right\}$$

where $|K| = \|X_2 - X_1\| \|X_4 - X_1\|$.

In [15, Lemma 9] it was shown that, for this basis, (7) holds in $L^2(\Gamma)$ with C_{stab} independent of p and L.

Note that we have *not* assumed that the C_{stab} is independent of L. The following stability and consistency estimates will be explicit in C_{stab} and a growth of C_{stab} (with respect to L) will lead to exponential convergence reduced by a factor $1/C_{stab}$. This, however, might be satisfactory as well. Note that for some weak formulations of boundary integral equations, the stability (4) trivially holds even though (7) holds only with C_{stab} (weakly) depending on L (as for first kind equations posed in $H^{\pm 1/2}(\Gamma)$).

Using the basis representation, the Galerkin equations (3) are then equivalent to the following system of algebraic equations

$$\mathbf{A}^L\mathbf{u}^L = \mathbf{F}^L$$

where $\mathbf{A}^L$ and $\mathbf{F}^L$ are defined by

$$\mathbf{A}_{i,j}^L := a(\phi_i, \phi_j), \quad \mathbf{F}_i^L := F(\phi_i)$$

and the solution $\mathbf{u}^L$ is linked to the solution of (3) via

$$u^L = \sum_{i=1}^{N_L} \mathbf{u}_i\phi_i. \tag{8}$$

In general, it is *not* possible to compute the matrix entries $\mathbf{A}_{i,j}^L$ and the entries of the vector $\mathbf{F}^L$ exactly. Instead, one has to employ numerical quadrature to approximate those coefficients. In this section, we investigate how the accuracy of this additional approximation has to be chosen such that the convergence order of the Galerkin scheme (without numerical quadrature) is preserved.

The following Lemma concerns how local errors due to numerical quadrature contribute to a global error.

Lemma 7 *Let $\tilde{\mathbf{A}}^L$ denote an approximation to the Galerkin system matrix $\mathbf{A}^L$ and $\tilde{\mathbf{F}}^L$ an approximation to $\mathbf{F}^L$. Assume that the matrix entries satisfy*

$$\left|\tilde{\mathbf{A}}_{i,j}^L - \mathbf{A}_{i,j}^L\right| \leq \Phi_A(L) \tag{9}$$

while for $\tilde{\mathbf{F}}$ we assume

$$\left|\tilde{\mathbf{F}}_i^L - \mathbf{F}_i^L\right| \leq \Phi_F(L). \tag{10}$$

Then, there holds for every $u, v \in V^L$:

$$|a(u,v) - a_L(u,v)| \leq C_{stab} N_L \Phi_A(L) \|u\|_{H^{s_1}(\Gamma)} \|v\|_{H^{s_1}(\Gamma)} \tag{11}$$

where $a_L : V^L \times V^L \to \mathbf{R}$ denotes the bilinear form associated with $\tilde{\mathbf{A}}^L$. For the associated linear form F_L, we get

$$|F_L(u) - F(u)| \leq \sqrt{C_{stab} N_L}\, \Phi_F(L) \|u\|_{s_1}, \qquad \forall u \in V^L. \tag{12}$$

The proof can be found in [15]. We can now estimate the impact of the consistency errors (11) on the asymptotic convergence rate of the solution $\tilde{u}^L$ of the fully discrete Galerkin scheme defined by

$$\tilde{\mathbf{A}}^L \tilde{\mathbf{u}}^L = \tilde{\mathbf{F}}^L, \quad \tilde{u}^L := \sum_{i=1}^{N_L} \tilde{\mathbf{u}}_i^L \phi_i.$$

Theorem 8 *Assume that the exact solution u of (2) lies in the countably normed space B_ϱ. Furthermore, we assume that the Galerkin method is stable, i.e., (4) holds, and that the approximate system matrix $\tilde{\mathbf{A}}^L$ and the right hand side $\tilde{\mathbf{F}}^L$ satisfy (11) and (12) with*

$$\Phi_A(L) = L\sigma^{\varrho L} / (C_{stab} N_L), \qquad \Phi_F(L) = L\sigma^{\varrho L} / \sqrt{C_{stab} N_L} \tag{13}$$

and the grading factor $\sigma \in \left]0, 1/2\right]$ defined above. Then, the fully discrete Galerkin solution $\tilde{u}^L$ satisfies the error estimate

$$\left\|u - \tilde{u}^L\right\|_{s_1} \leq C_u L \sigma^{\varrho L}$$

with $C > 0$ independent of L.

Assume the regularity of the adjoint problem, , i.e., for every $g \in H^{-s_1}(\Gamma) \cap B_\varrho(\Gamma)$, $0 \leq \varrho < 1/2$, the solution φ of the adjoint problem

$$a(v, \varphi) = g(v), \qquad \forall v \in H^{s_1}(\Gamma)$$

exists in $H^{s_1}(\Gamma)$ *and belongs to* $B_\varrho(\Gamma)$ *as well. If the quadrature errors satisfy (13) with* ϱ *replaced by* 2ϱ *then*

$$\left|\langle g,u\rangle-\left\langle g,\tilde{u}^L\right\rangle\right|<CL^2\sigma^{2\varrho L}$$

holds, with a constant C *independent of* L.

Proof. The proof is essentially the same as given in [15, Theorem 12]. ■

Theorem 8 shows that the fully discrete Galerkin method converges with an exponential rate provided the consistency errors (9) and (10) is controlled by conditions (13). This rather tight bound on the cubature error must be achieved with a work W_L of algebraic order in dependence on the problem size N_L since, otherwise, the convergence rate of the fully discrete Galerkin scheme will not be exponential in terms of the work W_L. In the next section, we will present cubature techniques satisfying these requirements.

Cubature for *hp*-Galerkin BEM

The purpose of this section is to develop and analyze a quadrature scheme such that the resulting numerically integrated system matrix $\tilde{\mathbf{A}}^L$ satisfies the consistence estimate

$$\left|\mathbf{A}^L_{i,j}-\tilde{\mathbf{A}}^L_{i,j}\right|\le L^2\sigma^{2\varrho L}/\left(N_LC_{stab}\right)$$

and, for the approximate right hands side, there holds

$$\left|\mathbf{F}^L_i-\tilde{\mathbf{F}}^L_i\right|\le L^2\sigma^{2\varrho L}/\sqrt{N_LC_{stab}}.$$

In the following, we will work out the cubature methods only for quadrilateral meshes. Triangles can be treated exactly in the same fashion, if they are mapped to quadrilaterals and the number of quadrature points is increased by one for each variable.

Before going into the details of the numerical cubature techniques we have to write the (bi-)linear forms a and F in an appropriate way. If the kernel function k_1 is hyper-singular we assume that the concept of partial integration (see [11], [8]) can be applied, i.e.,

$$a_2(u,v)=\sum_{l,m=0}^{1}p.v.\int_{\Gamma\times\Gamma}D_l[u](x)\,D_m[v](y)\,k_1^{l,m}(x,y)\,dydx,\tag{14}$$

where D_j are surface-tangential, differential operators of order j and the kernel functions $k_1^{l,m}$ are Cauchy singular. The abbreviation *p.v.* is short for *principal value*. An analogous representation is assumed if the kernel function k_2 is hyper-singular. Note that other regularization techniques as presented in [6, Section 8.3] and [9] can be applied as well but, in order to avoid too many technicalities, we confine ourselves to (14). In the following, we will consider the approximation of the following integrals. Let K, K_x, and $K_y\in\tau_L$ denote surface panels and define

$$I_K^A:=\textstyle\int_K\lambda_1\phi_i\phi_j dx,\qquad I_K^F:=\int_K\lambda_2\phi_i f\,dx,$$

$$I^A_{K_x\times K_y}:=\lim_{\varepsilon\to 0}\int_{\substack{K_x\times K_y\\ \|x-y\|\ge\varepsilon}}D_l[\phi_i](x)\,D_m[\phi_j](y)\,k_1^{n,m}(x,y)\,dydx,$$

$$I^F_{K_x\times K_y}:=\lim_{\varepsilon\to 0}\int_{\substack{K_x\times K_y\\ \|x-y\|\ge\varepsilon}}D_l[\phi_i](x)\,D_m[f](y)\,k_2^{n,m}(x,y)\,dydx.$$

Since the integrands of the integral $I_K^{A,F}$ are analytic they can be approximated by standard techniques. Let us consider the integrals $I_{K_x \times K_y}^{A,F}$. Since the cubature strategies are the same for both, A and F, we present the details only for $I_{K_x \times K_y}^{A}$ and skip the superscript A in the following. We have to distinguish between the following three basic cases. We assume that there is a constant C_s such that any pair of panels $\tilde{K}_x, \tilde{K}_y \in \tilde{\tau}_L$ satisfies one of the following conditions:

1. $\tilde{K}_x = \tilde{K}_y$, "case of identical panels",

2. Condition **1** is violated and $\operatorname{dist}\left(\tilde{K}_x, \tilde{K}_y\right) \leq C_s \max_{z \in \{x,y\}} \operatorname{diam} \tilde{K}_z$ holds. Furthermore, there exist two plane quadrangles $\tilde{K}_x^\star, \tilde{K}_y^\star \subset \tilde{\Gamma}$ which share exactly one common edge and have the property that, at least, three edges of $\tilde{K}_x$ belong to $\partial \tilde{K}_x^\star$ and, at least, three edges of $\tilde{K}_y$ belong to $\partial \tilde{K}_y^\star$. "edge-parallel case" and

3. Conditions **1** and **2** are violated.

The integral over the surface pieces K_x and K_y has to be pulled back onto a suitable parameter domains in $\hat{K}_x$ and $\hat{K}_y$ in $\mathbb{R}^2$. The transformations have to be chosen such that the geometric situation of the parameter panels $\hat{K}_{x,y}$ is the same as on the surface, e.g., $\hat{K}_x, \hat{K}_y$ share an edge if this is the case for K_x and K_y. Due to the space limitations in these proceedings we will discuss in the following only the case **1**. The cubature techniques and error analysis for the remaining cases can be derived in the same fashion as presented below in combination with the transformations and error estimates of [15].

In the first step, we will apply *relative coordinates* introduced in [13] and [7] which render the integrand tractable for automatic cubature methods. Let $\{X_i\}_{1 \leq i \leq 4}$ denote the set of vertices of K and $\varepsilon_i := \|X_{i+1} - X_i\|$. The mapping of the *reference element* $\hat{K} := (0, \varepsilon_1) \times (0, \varepsilon_2)$ onto the surface panel K is given by $\kappa_K := \chi_K (u_1/\varepsilon_1, u_2/\varepsilon_2)$ with χ_K defined by (6). The kernel in local coordinates on $\hat{K} \times \hat{K}$ is given by

$$k_{loc}(u, v) := k_1^{l,m} (\kappa_K (u), \kappa_K (v)).$$

In these coordinates, the integral $I_{K \times K}$ takes the form

$$I_{K \times K} := \lim_{\varepsilon \to 0} \int_{\substack{\hat{K} \times \hat{K} \\ \|u-v\| \geq \varepsilon}} k_{loc}(u, v)\, B(u, v)\, dv du$$

with

$$B(u, v) := g_K(u)\, g_K(v) \left(D_l [\phi_i] \circ \kappa_K\right)(u) \left(D_m [\phi_j] \circ \kappa_K\right)(v) \tag{15}$$

and g_K denotes the determinant of the Jacobi matrix. In [15, Section 3.3.1], the following Lemma was proved.

Lemma 9 *The integral $I_{K \times K}$ can be rewritten as*

$$I_{K \times K} = \int_0^{\varepsilon_1} \int_0^{\varepsilon_2} \int_0^1 \int_0^1 H(\eta, z)\, d\eta_2 d\eta_1 dz_2 dz_1 \tag{16}$$

with the function $H(\eta, z)$ defined by

$$H(\eta, z) = (\varepsilon_1 - z_1)(\varepsilon_2 - z_2) \sum_{i=1}^{4} k_{loc}\left(u^i, v^i\right) B\left(u^i, v^i\right)$$

and u^i, v^i defined by

$$\begin{pmatrix} u_1^1 \\ u_2^1 \\ v_1^1 \\ v_2^1 \end{pmatrix} = \begin{pmatrix} \eta_1 (\varepsilon_1 - z_1) \\ \eta_2 (\varepsilon_2 - z_2) \\ \eta_1 (\varepsilon_1 - z_1) + z_1 \\ \eta_2 (\varepsilon_2 - z_2) + z_2 \end{pmatrix}, \quad \begin{pmatrix} u_1^2 \\ u_2^2 \\ v_1^2 \\ v_2^2 \end{pmatrix} = \begin{pmatrix} \eta_1 (\varepsilon_1 - z_1) \\ \eta_2 (\varepsilon_2 - z_2) + z_2 \\ \eta_1 (\varepsilon_1 - z_1) + z_1 \\ \eta_2 (\varepsilon_2 - z_2) \end{pmatrix},$$

$$\begin{pmatrix} u_1^3 \\ u_2^3 \\ v_1^3 \\ v_2^3 \end{pmatrix} = \begin{pmatrix} \eta_1 (\varepsilon_1 - z_1) + z_1 \\ \eta_2 (\varepsilon_2 - z_2) \\ \eta_1 (\varepsilon_1 - z_1) \\ \eta_2 (\varepsilon_2 - z_2) + z_2 \end{pmatrix}, \quad \begin{pmatrix} u_1^4 \\ u_2^4 \\ v_1^4 \\ v_2^4 \end{pmatrix} = \begin{pmatrix} \eta_1 (\varepsilon_1 - z_1) + z_1 \\ \eta_2 (\varepsilon_2 - z_2) + z_2 \\ \eta_1 (\varepsilon_1 - z_1) \\ \eta_2 (\varepsilon_2 - z_2) \end{pmatrix}.$$

The integrand of (16) is weakly singular with possibly singular behaviour only at the origin.

Since the ratio $\varepsilon_1/\varepsilon_2$ might be arbitrarily large (or small) the straightforward use of *regularizing* coordinate transforms as, e.g. polar coordinates, will not lead to a robust cubature method. Instead, following [16] we have to subdivide the reference panel $\hat{K}$ into sub-panels $\hat{K} = \bigcup_{j=0}^{q} Q_j$ with the property that, for $j > 0$, $\operatorname{dist}(Q_j, 0) \geq C \operatorname{diam} Q_j$ and, for $j = 0$, $\operatorname{dist}(Q_0, 0) = 0$ and Q_0 is a rectangle with side lengths of the same magnitude. In [15], a sub-division algorithm is explicitly described satisfying $q = \left\lfloor \left| \log \frac{\varepsilon_2}{\varepsilon_1} \right| \right\rfloor$. For the following, we assume w.l.o.g. that $\varepsilon_1 \geq \varepsilon_2$. The integral over $Q_0 = (0, \tilde{\varepsilon}) \times (0, \varepsilon_2)$ with $\tilde{\varepsilon} \sim \varepsilon_2$ has to be transformed furthermore. Since the interval $(0, \varepsilon_2)$ can be mapped onto $(0, \tilde{\varepsilon})$ by a scaling being independent of the discretization parameters we assume w.l.o.g. that $\tilde{\varepsilon} = \varepsilon_2$. The following representation is proved in [15, Section 3.3.1].

$$\int_{Q_0} \int_0^1 \int_0^1 H(\eta, z) \, d\eta dz = \int_0^{\tilde{\varepsilon}} \int_0^1 \int_0^1 \int_0^1 \hat{z}_1 \sum_{i=1}^{2} \hat{H}\left(\eta, z^i(\hat{z})\right) d\eta d\hat{z}$$

with

$$z^1(\hat{z}) = \hat{z}_1 (1, \hat{z}_2)^T, \quad z^2(\hat{z}) = \hat{z}_1 (\hat{z}_2, 1)^T.$$

We summarize that we have derived the following representation of the initial integral $I_{\hat{K} \times \hat{K}}$

$$I_{\hat{K} \times \hat{K}} = \sum_{j=1}^{q} \int_{Q_j} H(\eta, z) \, d\eta dz + \int_0^{\tilde{\varepsilon}} \int_0^1 \int_0^1 \int_0^1 \hat{z}_1 \sum_{i=1}^{2} H\left(\eta, z^i(\hat{z})\right) d\eta d\hat{z}. \tag{17}$$

In [15, Section 3.3.2], we have shown that all integrands of the representation (17) are analytic. The integrals over Q_j will be approximated by 4-dimensional tensor products of properly scaled Gauß-Legendre formulae $G_{\eta_1}^{n_1} G_{\eta_2}^{n_2} G_{z_1}^{n_3} G_{z_2}^{n_4}$. Error estimates will be obtained from classical quadrature error estimates for analytic functions (see [1]). To this end, let $B(u, v)$ be defined by (15). Let further $\mathcal{E}_\rho(a, b)$ denote the ellipse in the complex plane having foci at a and b while ρ denotes the sum of the semi-axes. Let $\vec{\rho} \in \mathbb{R}_+^4$ and the domains $D(\vec{\rho})$ be defined by

$$D(\vec{\rho}) = \mathcal{E}_{\rho_1}(0, \varepsilon_1) \times \mathcal{E}_{\rho_2}(0, \varepsilon_2) \times \mathcal{E}_{\rho_3}(0, \varepsilon_1) \times \mathcal{E}_{\rho_4}(0, \varepsilon_2).$$

Note that, for $\rho = \varepsilon/2$, the ellipse $\mathcal{E}_\rho(0, \varepsilon)$ collapses to the interval $(0, \varepsilon)$. We assume that $B(u, v)$ can be extended analytically into any compact neighbourhood of $\hat{K} \times \hat{K}$. We define the functions $B(\vec{\rho})$, $B'(\vec{\rho})$, and $B^+(\vec{\rho})$ by

$$B(\vec{\rho}) := \max_{(u,v) \in D(\vec{\rho})} |B(u, v)|, \; B'(\vec{\rho}) := \max_{(u,v) \in D(\vec{\rho})} \left| \sum_{|\alpha + \beta| = 1} \frac{\partial B(u, v)}{\partial u^\alpha \partial v^\beta} \right|, \; B^+(\vec{\rho}) := B(\vec{\rho}) + B'(\vec{\rho}).$$

Theorem 10 *Let the cubature error be defined by*

$$E := \left| I_{\hat{K}\times\hat{K}} - \sum_{j=1}^{q} G_{\eta_1}^{n_1} G_{\eta_2}^{n_2} G_{z_1}^{n_3} G_{z_2}^{n_4} [H] - G_{\eta_1}^{\tilde{n}_1} G_{\eta_2}^{\tilde{n}_2} G_{z_1}^{\tilde{n}_3} G_{z_2}^{\tilde{n}_4} \left[\hat{H}\right] \right|$$

where $\hat{H}$ denotes the integrand of the last integral in (17). The error E can be estimated by

$$E \leq C\varepsilon_1\varepsilon_2^2 \left\{ \log\frac{\varepsilon_1}{\varepsilon_2} \left(\sum_{i=1}^{2} B^+(\vec{\rho}_i)(1+\gamma/\varepsilon_i)^{-2n_i} + \sum_{i=3}^{4} B^+(\vec{\rho}_i)(1+\gamma)^{-2n_i} \right) \right.$$
$$\left. + \sum_{i=1}^{2} B^+(\vec{\rho}_{4+i})(1+\gamma/\varepsilon_i)^{-2(\tilde{n}_i-1)} + \sum_{i=3}^{4} B^+(\vec{\rho}_{4+i})(1+\gamma)^{-2n_i} \right\}.$$

The semi-axes $\vec{\rho}_i$ are given by

$$\begin{aligned} 2\vec{\rho}_1 &= (c\operatorname{diam}\Omega, \varepsilon_2, c\operatorname{diam}\Omega, \varepsilon_2), & 2\vec{\rho}_2 &= (\varepsilon_1, c\operatorname{diam}\Omega, \varepsilon_1, c\operatorname{diam}\Omega), \\ 2\vec{\rho}_3 &= (c\varepsilon_1\operatorname{diam}\Omega, \varepsilon_2, c\varepsilon_1\operatorname{diam}\Omega, \varepsilon_2), & 2\vec{\rho}_4 &= (\varepsilon_1, c\varepsilon_2\operatorname{diam}\Omega, \varepsilon_1, c\varepsilon_2\operatorname{diam}\Omega) \\ 2\vec{\rho}_{4+i} &= 2\rho_i, \qquad i = 1,2, & 2\vec{\rho}_7 &= 2\vec{\rho}_8 = c\operatorname{diam}\Omega\,(\varepsilon_1,\varepsilon_2,\varepsilon_1,\varepsilon_2). \end{aligned}$$

The constants $C, c,$ and γ depend only on the mapping η and the initial grid $\tilde{\tau}_0$.

Proof. The proof is the same as in [15, Thms. 17, 20]. ■

Remark 11 *For the basis defined in Example 6 and the classical double layer potential (cf. Example 4), it was shown in [15, Thms. 17, 20] that the quantities $B^+(\vec{\rho}_i)$ can be estimated by*

$$B^+(\vec{\rho}_i) \leq C\pi_K^2\tilde{\pi}_K \frac{(1+\nu_i\gamma)^{\mu_i}}{\varepsilon_1\varepsilon_2}$$

with

$$\mu_i = \begin{cases} 2p_{i_2}^K & \text{for } 1 \leq i \leq 6, \\ 2\left(p_1^K + p_2^K\right) & \text{otherwise.} \end{cases} \qquad \nu_i = \begin{cases} 1/\varepsilon_{i_2} & \text{for } i = 1,2,5,6, \\ 1 & \text{for } i = 3,4,7,8. \end{cases}$$

and $i_2 = (i-1) \bmod 2 + 1$. The functions π_K and $\tilde{\pi}_K$ are given by

$$\pi_K := \sqrt{(p_1^K + 1/2)(p_2^K + 1/2)}, \quad \tilde{\pi}_K := \left\{p_1^K\left(p_1^K+1\right) + p_2^K\left(p_2^K+1\right)\right\}/2.$$

The cubature orders of Theorem (10) have to be adapted to the consistency requirements, in the case of the approximation of the bilinear form as

$$E < L^2\sigma^{2\varrho L}/\left(C_{stab}N_L\right), \tag{18}$$

and for the approximation of the right hand side by

$$E < L^2\sigma^{2\varrho L}/\sqrt{C_{stab}N_L}. \tag{19}$$

In view of the exponential convergence of the cubature methods, it is clear that, under the assumption that C_{stab} grows *algebraically* with respect to the level L, and, in addition, the quantities B, B', and B^+ grow like $\rho^{\mu(p)}$ with a fixed polynomial μ in the polynomial degree p, then, the exponential accuracies (18), (19) can be achieved with algebraically growing numbers of cubature points. Due to the space limitations of these proceedings, we will not work out the edge-parallel and regular cases in detail. This case can be treated in the same setting using the transformations and estimates presented in [15]. Under the above assumptions, the exponential accuracy can be achieved with algebraically increasing number of cubature points also for these cases. For the model problem of Examples 4 and 6, it was shown in [15] that the generation of the approximate system matrix satisfying the global consistency requirements (13) needs $O(N_L^{3.25})$ arithmetic operations. Provided the continuous solution is piecewise analytic so that the Galerkin approximation (without numerical cubature) converges with exponential rate, it follows that the fully discrete Galerkin solution converges with exponential rate with respect to the CPU-time. This was confirmed by our numerical experiments in [15] which also indicated strongly that the above work estimate is sharp.

References

[1] P. Davis and P. Rabinowitz. *Methods of Numerical Integration.* Academic Press, 1984.

[2] J. Elschner. The Double Layer Potential Operator over Polyhedral Domains II: Spline Galerkin Methods. *Math. Meth. Appl. Sci.*, 15:23–37, 1992.

[3] J. Elschner. On the Exponential Convergence of some Boundary Element Methods for Laplace's Equation in Nonsmooth Domains. In M. Costabel, M. Dauge, and S. Nicaise, editors, *Boundary Value Problems and Integral Equations in Nonsmooth Domains*, New York, 1995. Marcel Dekker.

[4] B. Guo and I. Babuška. The hp-Version of the Finite Element Method. Part i: The Basic Approximation Results. *Computational Mechanics*, 1:21–41, 1986.

[5] B. Guo and I. Babuška. The hp-Version of the Finite Element Method. Part II: General Results and Applications. *Computational Mechanics*, 1:203–220, 1986.

[6] W. Hackbusch. *Integral Equations.* ISNM 120, Birkhäuser Publ. Basel, 1995.

[7] W. Hackbusch and S. Sauter. On the Efficient Use of the Galerkin Method to Solve Fredholm Integral Equations. *Applications of Mathematics*, 38(4-5):301–322, 1993.

[8] H. Han. The Boundary Integro-Differential Equations of Three-Dimensional Neuman Problem in Linear Elasticity. *Numer. Math.*, 68(2):269–281, 1994.

[9] C. Lage. *Software Development for Boundary Element Mehtods: Analysis and Design of Efficient Techniques (in German).* PhD thesis, Lehrstuhl Prakt. Math., Universität Kiel, 1995.

[10] M. Maischak and E. Stephan. The hp-Version of the Boundary Element Method in R3 - the Basic Approximation Results. *Math. Meth. Appl. Sci*, to appear, 1996.

[11] J. Nédélec. Integral Equations with Non Integrable Kernels. *Integral Equations Oper. Theory*, 5:562–572, 1982.

[12] S. Sauter. Cubature Techniques for 3-d Galerkin BEM. In W. Hackbusch and G. Wittum, editors, *BEM: Implementation and Analysis of Advanced Algorithms.* Vieweg Verlag, 1996.

[13] S. A. Sauter. *Über die effiziente Verwendung des Galerkinverfahrens zur Lösung Fredholmscher Integralgleichungen.* PhD thesis, Inst. f. Prakt. Math., Universität Kiel, 1992.

[14] S. A. Sauter and A. Krapp. On the Effect of Numerical Integration in the Galerkin Boundary Element Method. Technical Report 95-4, Lehrstuhl Praktische Mathematik, Universität Kiel, Germany, 1995, to appear in Numer. Math.

[15] S. A. Sauter and C. Schwab. Quadrature for hp-Galerkin BEM in 3-d. Technical Report 96-02, Seminar for Appl. Math., ETH Zürich, March 1996.

[16] C. Schwab. Variable order composite quadrature of singular and nearly singular integrals. *Computing* , 53:173-194, 1994.

[17] C. Schwab and W. Wendland. Kernel Properties and Representations of Boundary Integral Operators. *Math. Nachr.*, 156:187–218, 1992.

A SYMMETRICAL BOUNDARY ELEMENT FORMULATION FOR SOUND TRANSMISSION THROUGH CABIN WALLS

H. Schippers and F. P. Grooteman
National Aerospace Laboratory NLR
P.O. Box 153
8300 AD Emmeloord, The Netherlands

SUMMARY

A symmetrical boundary element formulation is described for the transmission of sound through the panels of a cabin wall of an aircraft. The geometrical model consists of a double panel configuration including a cavity partly filled with porous material and with air. The acoustic pressure in the vibrating medium (air and porous material) is modelled by a boundary integral ansatz. The vibrating panels and the vibrating medium are coupled by the boundary condition for flexible walls. Classical boundary integral formulations result in a non-symmetric boundary element matrix, while the discretization of the Helmholtz equation using finite elements would yield a symmetrical stiffness matrix. In the present paper a symmetrical boundary element matrix is derived for the acoustic pressure inside and outside the cabin wall. The boundary elements are put on the panels of the cabin wall and on the interface between the air and the porous material. The computational costs of the present symmetrical boundary element formulation are comparable with the costs of classical boundary element formulations in acoustics, which are based on the direct method.

INTRODUCTION

Investigations on the acoustic field around and inside commercial aircraft are motivated by the urge to reduce cabin noise inside the aircraft. The exterior acoustic field causes vibrations of the cabin wall, which contribute to the interior noise. The cabin noise inside the aircraft can be reduced by putting porous acoustically absorbing material (insulation) just on the inside of the skin panel of the cabin wall.

In this paper a symmetrical boundary element is described for the transmission of sound through the panels of a cabin wall. The geometrical model consists of a double panel configuration including a cavity partly filled with porous material and with air (an illustration is given in figure 1). The lower panel is a stiffened skin panel, while the upper panel is an un-stiffened trim panel. The trim panel is backed by a semi-infinite room modelling the interior of the aircraft. The displacement of the skin panel is assumed to be given by the exterior acoustic field at the outer side of the aircraft. The sound transmission problem requires the calculation of the acoustic pressure in the air and in the porous material as well as the fluid-structure interaction with the un-stiffened trim panel. The acoustic pressure in the air and porous material is modelled by the Helmholtz equation (in the case of glass wool insulation with a complex wavenumber, which models the damping). The porous material is essentially assumed to be described by its porosity, resistivity, effective density and effective speed of sound, all of which are frequency dependent.

The acoustic pressure in the vibrating medium (air and porous material) is modelled by a boundary integral ansatz. The vibrating panels and the vibrating medium are coupled by the boundary condition for flexible walls. Classical boundary integral formulations result in a non-symmetric boundary element matrix, while the discretization of the Helmholtz equation using finite elements would yield a symmetrical stiffness matrix. In the present paper a symmetrical boundary integral formulation is described for the acoustic pressure inside and outside the cabin wall. The formulation maps the normal derivative of the pressure along the boundary on the pressure itself. In the case of a bounded domain (e.g. the volume covered by the porous medium or the volume covered by the air inside the cabin wall) the symmetrical boundary integral formulation has been taken from ref. [1]. The symmetrical boundary element matrix follows from applying Galerkin's method. The present formulation is advantageous from the point of view that the boundary element matrix has the same mapping properties as the finite element matrix of the weak formulation of the Helmholtz equation. These properties are not preserved by the classical boundary element formulation, which yields an asymmetric system matrix. The basis and test functions are given by the classical piecewise linear functions on flat non-overlapping triangular elements.

The boundary elements are put on the panels of the cabin wall and on the interface between the air and the porous material. The computations involve the discretization of a hyper-singular integral operator, a weakly singular integral operator and a regular operator. The hyper-singular integral operator is regularized by integration of parts. Once this operator has been regularized, the computation of the coefficients of the corresponding boundary element matrix requires the same number of kernel evaluations as the computation of the coefficients of the weakly singular operator. Therefore, the computational costs of the present symmetrical boundary element formulation are comparable with the costs of classical boundary element formulations in acoustics, which are based on the direct method.

MATHEMATICAL FORMULATION

In this section a mathematical model is presented for the transmission of sound through panels of a cabin wall. The model is applied to a double wall configuration of a cabin wall panel as shown in figure 1. It contains a closed cavity, partly filled with air and partly with porous material (e.g. glass wool). This cavity is given by the rectangular box Ω_1 in

$\mathbf{R}^3$, with $\Omega_1 = \Omega_{1,a} \cup \Omega_{1,g}$. The boundary of $\overline{\Omega_1}$ is denoted by $\Gamma = \partial\overline{\Omega_1}$. The unit normal vector n on Γ is pointing outwards, i.e. into $\mathbf{R}^3 \setminus \overline{\Omega_1}$. The boundary is split into three parts

$$\Gamma = \Gamma_1 \cup \Gamma_2 \cup \Gamma_3,$$

where Γ_1 corresponds to the stiffened skin panel and Γ_3 corresponds to the un-stiffened trim panel. The double wall configuration is assumed to be baffled in an infinite plate in the plane $z = 0$. The part of this plane that is not occupied by Γ_3 is denoted by Γ_∞.

Let the semi-infinite space given by $z > 0$ be denoted by Ω_i and the space denoted by the complement of $\Omega_i \cup \overline{\Omega_1}$ in $\mathbf{R}^3$ by Ω_e. Here, Ω_i and Ω_e represent the interior and the exterior space of the aircraft cabin.

The transmission problem requires the modelling of the sound pressure in and outside the cabin wall as well as the fluid-structure interaction with the stiffened and un-stiffened panel. The problem is assumed to be time harmonic with angular frequency ω. In the present paper the sound pressure in Ω_e is not modelled. It is assumed that the acoustic field in Ω_e as well as the induced displacement on Γ_1 are known.

The acoustic pressure p in the regions Ω_i and $\Omega_{1,a}$ has to satisfy the Helmholtz equation

$$\Delta p + k^2 p = 0, \qquad k = \frac{\omega}{c}, \tag{1}$$

where c is the speed of sound in air and k is the acoustic wave number. The acoustic pressure p in the region $\Omega_{1,g}$ also has to satisfy the Helmholtz equation, but with a complex wave number γ,

$$\Delta p + \gamma^2 p = 0, \tag{2}$$

where the definition of γ depends on the material properties of the porous medium. The porous material is essentially assumed to be described by its porosity h, resistivity σ, effective density ρ_e and effective speed of sound c_e, all of which are frequency dependent. For wave propagation in limp material (an approximate model for glass wool) the following expression for γ has been derived in Ref. [2]

$$\gamma = (\omega/c_e)\sqrt{h\kappa(\omega)/\rho_e}, \qquad \kappa(\omega) = \rho_e - (\imath\sigma/\omega)/(1 - \frac{\imath h\sigma}{\omega M}), \tag{3}$$

where M is the density of the fibre part. In the limp model the stiffness of the fibres is neglected. The limitation of the model is that the resonances in the porous material can not be described. This would require both stiffness and mass effects to be taken into account. As a consequence, the high frequency range (small wavelength) can not be calculated with sufficient accuracy. When the minimum acoustic wavelength of the computational problem is larger than the thickness of the porous layer, the assumption of the limp material holds.

The appropriate boundary conditions for p are

$$\begin{aligned} \frac{\partial p}{\partial n}(r) &= \kappa(\omega)\omega^2 W(r), \qquad r \in \Gamma_1 && (4)\\ &= 0, \qquad r \in \Gamma_2 \cup \Gamma_\infty && (5)\\ &= \rho_a \omega^2 w(r), \qquad r \in \Gamma_3. && (6) \end{aligned}$$

At boundary Γ_1 the normal displacement W of the skin panel is prescribed. The boundaries Γ_2 and Γ_∞ are assumed to be acoustically hard. At boundary Γ_3 the normal displacement w follows from solving an appropriate elastomechanical model for the trim panel. For reasons of explanation it is assumed in this section that the trim panel is isotropic and that the displacements are small, so that w is governed by the equations for a harmonically vibrating plate

$$\mathcal{L}(w) \equiv D\Delta^2 w - \omega^2 \delta \rho_s w = \mu, \tag{7}$$

where Δ denotes the Laplace operator, D the bending stiffness of the plate, ρ_s its density and δ its thickness. The right-hand side of (7) represents the load acting on the plate caused by the jump μ in the acoustic pressure,

$$\mu = p^+ - p^-, \tag{8}$$

where p^+ (p^-) denotes the pressure on the upper (lower) side of the trim panel. In numerical investigations the elastomechanical operator $\mathcal{L}$ will be modeled by four-noded finite shell elements. Note that the equations (1), (6), (7) and (8) define a coupled fluid-structure interaction problem.

At the fluid surface S between the porous material and the air two boundary conditions have to be satisfied. First, the pressure has to be continuous

$$p_g = p_a \tag{9}$$

and second the mass flow has to be continuous across the interface S. For glass-wool, having a porosity which is almost equal to unity, the mass flow continuity reduces to continuity of the normal displacement,

$$u_g^n = u_a^n = u_S^n. \tag{10}$$

Far away from the double wall configuration it is required that p satisfies the Sommerfeld radiation condition, which requires that

$$\frac{\partial p}{\partial r} - \imath k p = \mathcal{O}(1/r), \qquad r \to \infty. \tag{11}$$

BOUNDARY INTEGRAL FORMULAS FOR THE ACOUSTIC PRESSURE

In the previous section it has been shown that the acoustic pressure in the closed domain Ω_1 (partly filled with porous material and partly with air) and in the half-space Ω_i above the trim panel is governed by the Helmholtz equation (for air (1) and for the porous medium (2)). In this section boundary integral formulas for the acoustic pressure are presented which satisfy the Helmholtz equation. First, a formula is given for the acoustic pressure in the half-space Ω_i and second a symmetric formula is derived for a closed domain Ω (in terms of the previous section Ω is defined either by the domain $\Omega_{1,a}$ or by the domain $\Omega_{1,g}$). In the following sections the Green-function G for the Helmholtz equation in an infinite domain is used,

$$G_\alpha(r, r') = \frac{e^{\imath\alpha|r-r'|}}{4\pi|r-r'|}, \qquad r \neq r', \tag{12}$$

with $\alpha = k$ for waves propagating in air and with $\alpha = \gamma$ (as defined in (3)) for waves propagating in the porous medium.

BOUNDARY INTEGRAL FORMULA FOR THE HALF-SPACE Ω_i

The vibrating trim panel is baffled in a perfectly rigid plane given by Γ_∞. The boundary integral formula for the acoustic pressure in the half-space Ω_i is given by the single layer potential ansatz

$$\begin{aligned} p(r) &= (V_k\phi)(r), \qquad r \in \Omega_i, \\ V_k\phi(r) &= \int_{\Gamma_3} G_k(r,r')\phi(r')\, d\Gamma(r'), \qquad r \in \Omega_i. \end{aligned} \tag{13}$$

with the single layer given by

$$\phi(r) = -2\rho_a\omega^2 w(r), \qquad r \in \Gamma_3. \tag{14}$$

It can be shown that this single layer potential satisfies the Helmholtz equation and the boundary conditions (5), (6) and (11). This follows from classical results of potential theory. The integral operator V_k is weakly singular, so that (13) is also well defined for $r \in \Gamma_3$, i.e.

$$p(r) = (V_k\phi)(r), \qquad r \in \Gamma_3. \tag{15}$$

Note that V_k is a symmetric operator, because G_k is symmetric with respect to r and r'.

BOUNDARY INTEGRAL FORMULA FOR A CLOSED DOMAIN Ω

Let $\Omega \subset \mathbf{R}^3$ be a bounded domain with piecewise smooth boundary $\partial\Omega$. In terms of the previous section Ω is defined either by the sub-domain $\Omega_{1,a}$ or by the sub-domain $\Omega_{1,g}$. Note that the boundary of $\Omega_{1,a}$ is given by $\partial\Omega_{1,a} = \Gamma_{2,a} \cup \Gamma_3 \cup S$, while the boundary of $\Omega_{1,g}$ is given by $\partial\Omega_{1,g} = \Gamma_{2,g} \cup \Gamma_1 \cup S$ ($\Gamma_{2,a}$ and $\Gamma_{2,g}$ are the parts adjoining the air and the porous medium, respectively).

For the acoustic pressure the following representation formula holds

$$p(r) = \int_{\partial\Omega} G_\alpha(r,r')\partial_{n'}p(r')\, d\Gamma(r') - \int_{\partial\Omega} \partial_{n'}G_\alpha(r,r')p(r')\, d\Gamma(r'). \tag{16}$$

Define the following boundary integral operators

$$\begin{aligned} V_\alpha\phi(r) &:= \int_{\partial\Omega} G_\alpha(r,r')\phi(r')\, d\Gamma(r'), \qquad r \in \partial\Omega, \\ K_\alpha\phi(r) &:= \int_{\partial\Omega} \partial_{n'}G_\alpha(r,r')\phi(r')\, d\Gamma(r') \qquad r \in \partial\Omega, \\ K'_\alpha\phi(r) &:= \partial_n \int_{\partial\Omega} G_\alpha(r,r')\phi(r')\, d\Gamma(r') \qquad r \in \partial\Omega, \\ D_\alpha\phi(r) &:= -\partial_n \int_{\partial\Omega} \partial_{n'}G_\alpha(r,r')\phi(r')\, d\Gamma(r'), \qquad r \in \partial\Omega. \end{aligned}$$

The representation formula (16) and the jump relations for single and double layer potentials lead to the following boundary integral equations on $\partial\Omega$

$$
\begin{aligned}
p &= V_\alpha \partial p - K_\alpha p + \frac{1}{2}p = V_\alpha \partial p + (\frac{1}{2}I - K_\alpha)p && (17)\\
\partial p &= D_\alpha p + K_\alpha^{'} \partial p + \frac{1}{2}\partial p \,. && (18)
\end{aligned}
$$

In most boundary element calculations in acoustics the first equation (17) is being used for sound predictions. For some applications also the second one is used. However, the discretization of these boundary integral equations result in general in non-symmetric system matrices, which have fictitious eigenvalues and eigenvectors. The equations (17) and (18) can be rewritten as

$$
\begin{aligned}
p &= (\frac{1}{2}I + K_\alpha)^{-1} V_\alpha \partial p && (19)\\
p &= D_\alpha^{-1}(\frac{1}{2}I - K_\alpha^{'})\partial p \,. && (20)
\end{aligned}
$$

Observe that the operators $(\frac{1}{2}I + K_\alpha)^{-1}V_\alpha$ and $D_\alpha^{-1}(\frac{1}{2}I - K_\alpha^{'})$ define do called Poincaré-Steklov operators mapping

$$\partial p \rightarrow p\,|_{\partial\Omega} \,.$$

The operators in (19) and (20) are not symmetric, while the weak formulation of (1): find $p \in H^1(\Omega)$ such that

$$\int_\Omega (\nabla p \cdot \nabla \psi - \alpha^2 p \psi) dA - \int_{\partial\Omega} \frac{\partial p}{\partial n} \psi dS = 0, \qquad (21)$$

for all test functions $\psi \in H^1(\Omega)$, yields (with restriction of p to $\partial\Omega$) a symmetric Poincaré-Steklov operator. The boundary element discretization of (17) or (18) will result in a non-symmetric boundary element matrix. Therefore, the boundary element discretization of (17) or (18) will not have the same mapping properties as the finite element discretization of (21). It has been shown in Ref. [1] that a symmetric boundary element discretization of a Poincaré-Steklov operator can be obtained by substituting (20) into (17). The result becomes

$$p = T_\alpha \partial p \qquad (22)$$

with

$$T_\alpha = V_\alpha + (\frac{1}{2}I - K_\alpha) D_\alpha^{-1} (\frac{1}{2}I - K_\alpha^{'}). \qquad (23)$$

Note that T_α defines a symmetric Poincaré-Steklov operator, since V_α and D_α are symmetric. The discretization of T_α involves, however, the approximation of the hypersingular operator D_α , the weakly singular operator V_α and the regular operator K_α. The hypersingular operator can be regularized by integration of parts and the integrals reduce to weakly singular integrals. The coefficients of the boundary element matrices of the operators D_α and V_α can be evaluated simultaneously so that the computational costs are comparable with the costs of the discretization of (17) or (18).

BOUNDARY ELEMENT DISCRETIZATION

The integral operators V_k of (15) and T_α of (23) are discretized using a Galerkin boundary element method. The trial spaces are given by the piecewise linear basis functions $\{w_l\}_{l=1}^N$ on a triangular surface mesh with N nodes and M elements $\{S_k\}_{k=1}^M$. Following ref. [1] one obtains the following boundary element discretization for the operator T_α

$$T_{\alpha,h} = M_h^{-1}\mathcal{V}_{\alpha,h}M_h^{-1} + (\frac{1}{2}I_h - M_h^{-1}\mathcal{K}_{\alpha,h})D_{\alpha,h}^{-1}(\frac{1}{2}I_h - \mathcal{K}'_{\alpha,h}M_h^{-1}), \qquad (24)$$

with

$$\begin{aligned}
(M_h)_{l,j} &= (w_l, w_j), \\
(\mathcal{V}_{\alpha,h})_{l,j} &= (V_\alpha w_l, w_j), \\
(\mathcal{K}_{\alpha,h})_{l,j} &= (K_\alpha w_l, w_j), \\
(\mathcal{K}'_{\alpha,h})_{l,j} &= (\mathcal{K}_{\alpha,h})_{j,l} \\
(D_{\alpha,h})_{l,j} &= (D_\alpha w_l, w_j).
\end{aligned}$$

Note that (24) defines a symmetric boundary element matrix, because $\mathcal{V}_{\alpha,h}$ and $D_{\alpha,h}$ are symmetric.

The calculation of the matrix element $(D_{\alpha,h})_{l,j}$ involves the evaluation of a hypersingular integral, since

$$(D_\alpha w_l, w_j) = -\int_\Gamma\int_\Gamma \frac{\partial^2 G_\alpha}{\partial n \partial n'}(r, r')w_l(r)\ w_j(r')\ d\Gamma(r)d\Gamma(r'),$$

where the second integral must be interpreted as a finite part integral for $r \in \Gamma$. The hypersingular integral can be regularized by integration of parts using the fact that G_α is the fundamental solution of the Helmholtz operator (See e.g. ref. [3]). It can be shown that

$$\begin{aligned}
(D_\alpha w_l, w_j) &= \int_\Gamma\int_\Gamma G_\alpha(r, r')\langle n_r \times \nabla_r w_l(r)\ , n'_r \times \nabla_{r'} w_j(r')\ \rangle d\Gamma(r)d\Gamma(r') \\
&\quad -\alpha^2 \int_\Gamma\int_\Gamma w_l(r)\ w_j(r')\ G_\alpha(r, r')\langle n_r, n_{r'}\rangle\ d\Gamma(r)d\Gamma(r'). \qquad (25)
\end{aligned}$$

Observe that (25) only contains weakly singular integrals.

NUMERICAL IMPLEMENTATION

Let the nodes of the finite element mesh be given by $\{r_l\}_{l=1}^N$. On each triangular element S_k three local linear functions $\psi_i^k(r)$, $i = 1, 2, 3$ are introduced, which yield the value 1 at the i-th vertex of S_k and 0 at the other two vertices of S_k. Indirect addresses are used to associate vertices of a triangular element with the node numbers of the mesh, i.e. there exists an array $nn(i,k)$ yielding the node number of the i-th vertex of the k-th triangular element. The basis and test functions are continuous functions of the form

$$\mu(r) = \sum_{k=1}^{M} \sum_{i=1}^{3} \mu_i^k \psi_i^k(r),$$

where μ_i^k is the value of the function μ at the node r_l with $l = nn(i,k)$.

With this set of basis and test functions the form $(D_\alpha \xi, \mu)$ can be rewritten as

$$(D_\alpha \xi, \mu) = \sum_{l=1}^{M} \sum_{j=1}^{3} \sum_{k=1}^{M} \sum_{i=1}^{3} D_{j,i}^{l,k} \xi_j^l \mu_i^k \tag{26}$$

where

$$\begin{aligned} D_{j,i}^{l,k} &= \int_{S_l} \int_{S_k} \alpha^2 \langle n_{S_l}, n_{S_k} \rangle \psi_i^k(r') \psi_j^l(r) G_\alpha(r,r')\; d\Gamma(r') d\Gamma(r), \\ &- \int_{S_l} \int_{S_k} \langle n_{S_l} \times \nabla \psi_j^l, n_{S_k} \times \nabla \psi_i^k \rangle G_\alpha(r,r')\; d\Gamma(r') d\Gamma(r). \end{aligned} \tag{27}$$

The matrix elements $\mathcal{V}_{j,i}^{l,k}$, $\mathcal{K}_{j,i}^{l,k}$ and $M_{j,i}^{l,l}$, which follow from substituting the basis and test functions into $(V_\alpha \xi, \mu)$, $(K_\alpha \xi, \mu)$ and the mass-matrix (ξ, μ), can be obtained in a similar way,

$$\mathcal{V}_{j,i}^{l,k} = \int_{S_l} \int_{S_k} \psi_i^k(r') \psi_j^l(r) G_\alpha(r,r')\; d\Gamma(r') d\Gamma(r), \tag{28}$$

$$\mathcal{K}_{j,i}^{l,k} = \int_{S_l} \int_{S_k} \frac{-1 + \imath\alpha|r-r'|}{4\pi} \frac{e^{\imath\alpha|r-r'|}}{|r-r'|^2} \langle n(r'), \frac{r'-r}{|r'-r|} \rangle \psi_i^k(r') \psi_j^l(r)\; d\Gamma(r') d\Gamma(r), \tag{29}$$

$$M_{j,i}^{l,l} = \int_{S_l} \psi_j^l(r) \psi_i^l(r)\; d\Gamma(r). \tag{30}$$

The coefficients of the global boundary element matrices of the previous section (related to the unknown quantities at the nodes) are obtained by assembling the local boundary element matrices, e.g. the coefficients $(\mathcal{V}_{\alpha,h})_{p,q}$ follow from summing the influence coefficients $V_{j,i}^{l,k}$ for which $nn(j,l) = p$ and $nn(i,k) = q$, i.e.

$$(\mathcal{V}_{\alpha,h})_{p,q} = \sum_{l=1}^{M} \sum_{\substack{j=1 \\ nn(j,l)=p}}^{3} \sum_{k=1}^{M} \sum_{\substack{i=1 \\ nn(i,k)=q}}^{3} V_{j,i}^{l,k}.$$

Analogously $(D_{\alpha,h})_{p,q}$, $(\mathcal{K}_{\alpha,h})_{p,q}$ and $(M_h)_{p,q}$ are obtained. The matrix $\mathcal{K}_{\alpha,h}'$ can be obtained from transposing the matrix $\mathcal{K}_{\alpha,h}$.

Note that the derivatives of $\psi_j^l(r)$ are constant for $r \in S_l$, since the basis functions are piecewise linear. As a consequence the terms between brackets $\langle\ \rangle$ in (27) are constant. Therefore, the evaluation of the influence coefficients $D_{j,i}^{l,k}$ in (27), $\mathcal{V}_{j,i}^{l,k}$ in (28) and $\mathcal{K}_{j,i}^{l,k}$ in (29) requires only the calculation of the following boundary integrals

$$\int_{S_l} \int_{S_k} \psi_i^k(r') \psi_j^l(r) G_\alpha(r,r')\; d\Gamma(r') d\Gamma(r), \tag{31}$$

$$\int_{S_l} \int_{S_k} G_\alpha(r,r')\; d\Gamma(r') d\Gamma(r), \tag{32}$$

and

$$\int_{S_l}\int_{S_k} \frac{-1+\imath\alpha|r-r'|}{4\pi}\frac{e^{\imath\alpha|r-r'|}}{|r-r'|^2}\langle n(r'),\frac{r'-r}{|r'-r|}\rangle\psi_i^k(r')\psi_j^l(r)\,d\Gamma(r')d\Gamma(r). \qquad (33)$$

The calculation of the singular integrals in (27) and (28) has been described in Ref. [4]. They are approximated by use of a Gauss quadrature rule for the outer integral and a regularizing coordinate transformation for the inner integral. This approach is also used for the evaluation of the nearly singular integrals in (27) and (28), i.e. the regular integrals for which the distance of S_k to S_l is small.

It remains to calculate $\mathcal{K}_{j,i}^{l,k}$ for $k \neq l$. Note that $\mathcal{K}_{j,i}^{k,k} = 0$ for flat elements since $\langle n(r'),\frac{r'-r}{|r'-r|}\rangle = 0$ for $r, r' \in S_k$. The coefficients $\mathcal{K}_{j,i}^{l,k}$ are computed by applying a Gauss quadrature rule, i.e. $\mathcal{K}_{j,i}^{l,k}$ is approximated by

$$\frac{4|S_l||S_k|}{4\pi}\sum_{n=1}^{N_G}\sum_{m=1}^{N_G} w_n w_m(-1+\imath\alpha|r_n-r'_m|)\frac{e^{\imath\alpha|r_n-r'_m|}}{|r_n-r'_m|^2}\langle n_{S_k},\frac{r'_m-r_n}{|r'_m-r_n|}\rangle\psi_i^k(r'_m)\psi_j^l(r_n), \quad (34)$$

where $|S_k|$ is the area of the triangular element S_k, n_{S_k} the unit normal on S_k and N_G the number of Gauss points. The weights of the Gauss quadrature rule sum up to $\frac{1}{2}$.

The integral $M_{j,i}^{l,l}$ is easily calculated. For $i = j$ the exact value is $|S_l|/18$ and for $i \neq j$ the exact value is $|S_l|/24$.

NUMERICAL RESULTS

The symmetrical boundary element formulation has been implemented in the modular finite element program B2000 and has been coupled with the existing finite element code for the structural analysis of plates and shells. The boundary element formulation is applied to the acoustic-structural analysis of a double panel configuration. The structure consists of two identical aluminium plates with a length of 1.46 m and a width of 0.76 m (see figure 2), which are clamped at all edges. The distance between the plates is 0.1 m. The gap between the plates is either filled by air or by glass-wool. The eigenfrequencies of this double panel configuration and its environment are computed. The results are compared with a finite element code where the acoustic pressure was obtained by the numerical solution of (21) using 20-noded acoustic elements.

The elastomechanics of the plates is modelled by putting 100 linear quadrilateral finite shell elements on each plate. The basic properties of the aluminium plates are: modulus of elasticity = 7.0E+10 N/m^2, Poisson's ratio = 0.3, density = 2800 kg/m^3 and thickness = 0.0025 m.

In the first example the cavity between the plates is filled with air. The air surrounding the plates can also be modelled, but is left out here, to be able to compare the results with the finite element acoustic calculations. The properties of air are: $\rho_a = 1.2$ kg/m^3 and $c = 340$ m/s.

The boundary element surface discretization consists of 200 triangular elements on each plate (in such a way that the nodes correspond with the nodes of the structural quadrilateral elements) and 320 triangular elements on the four vertical faces to close the cavity between the plates. The first five eigenfrequencies computed are given in the second column of table 1. In the third column of table 1 the results are shown of the finite element acoustic calculations. A good agreement of the results is observed between both methods. The small differences are due to differences in the numerical methods, e.g. integration rules.

In the second example the cavity between the plates is filled with glass-wool, having the following properties: $\rho_e = 1.2$ kg/m^3, $c_e = 290$ m/s, $M = 2267$ kg/m^3, $h = 0.9955$ and $\sigma = 2.3E4$ Ns/m^3. Note that the complex wavenumber (as defined in (3)) depends on the frequency ω, so that the coupled eigenvalue problem becomes non-linear. The same number of elements has been used as in the first example. The first five eigenfrequencies as computed by the symmetrical boundary element approach and the acoustic finite limp element model are given in the fourth end fifth column of table 1. The eigenfrequencies are complex valued due to energy dissipation. Again small differences are observed between the two numerical methods. For this example the differences are also due to the non-linearity of the eigenvalue problem. In order to solve the eigenvalue problem an initial value of ω has to be specified to calculate $\kappa(\omega)$ in (3). The results of table 1 have been obtained taking a value close to the real part of the eigenfrequency. When, however, a value of 30 Hz ($\omega = 60\pi$ rad /s) would have been chosen for the calculation of $\kappa(\omega)$, the first eigenfrequency would have been 7.84+i*0.190, showing the dependency of the problem on the initial value of $\kappa(\omega)$.

Table 1: Eigenfrequencies [Hz] of double panel configuration

medium:	air		glass-wool	
MODE	BEM	FEM	BEM	FEM
1	11.95	12.27	7.46+i*0.161	7.84+i*0.081
2	13.16	13.16	12.73+i*0.019	12.62+i*0.018
3	20.83	21.02	14.97+i*0.188	15.05+i*0.153
4	21.93	21.93	21.21+i*0.054	21.93+i*0.049
5	36.98	37.14	29.61+i*0.657	29.29+i*0.466

CONCLUSIONS

A symmetrical boundary element formulation has been described for the sound transmission through a cabin wall of an aircraft. The computational costs of the presented boundary element formulation are comparable with the costs of classical boundary element formulations in acoustics. This could be achieved by regularizing the hypersingular integral operator and by computing the coefficients of the boundary element matrices of the weakly singular and hypersingular integral operators simultaneously.

The symmetrical boundary element formulation has been applied to compute the eigenfrequencies of the coupled acoustic-structural problem of a double panel configuration where the cavity between the panels was filled by either air or by glass-wool. The computed eigenfrequencies of this coupled problem show a fair agreement with results of a code where both the elastomechanics of the panels and the acoustic pressure was modelled by finite elements. In a forthcoming paper the described boundary element formulation will be applied to a double panel configuration where the cavity is partly filled by air and partly by glass-wool.

REFERENCES

[1] SCHMIDT, G.: "Boundary element discretization of Poincaré-Steklov operators", Numer. Math., Vol. 69, 1994, pp. 83-101.

[2] GORANSSON, P.: "Acoustic Finite Element Formulation of a Flexible Porous Material - A Correction for Inertial Effects", Journal of Sound and Vibration, Vol. 185, 1995, pp. 559-580.

[3] MAUE, A.W.: "Toward formulation of a general diffraction problem via an integral equation", Zeitschrift für Physik, Vol. 126, 1949, pp. 601-618.

[4] KOELINK, H.T., SCHIPPERS, H., HEIJSTEK, J.J., DERKSEN, J.J.: "Modal analysis of solar arrays using boundary integral equations", NLR TP 92281 L (1992), in Boundary Elements XIV, Vol 2: Stress Analysis and Computational Aspects, C.A. Brebbia, J. Dominguez, F. Paris, eds., Computational Mechanics Publ. & Elsevier Science Publ., 1992, pp. 515-525.

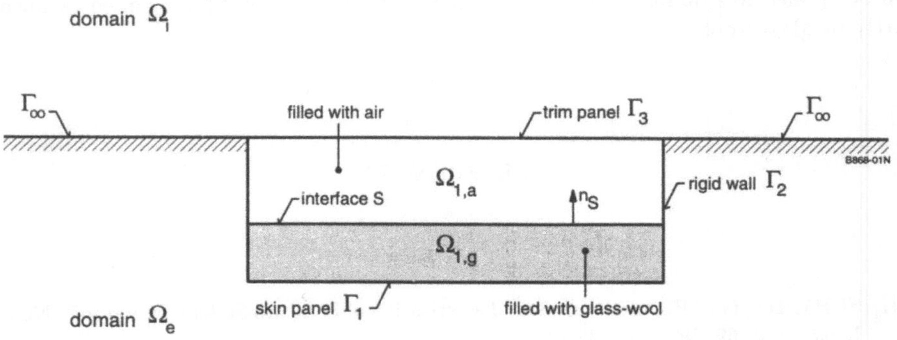

Fig. 1 Cross section of baffled double wall configuration

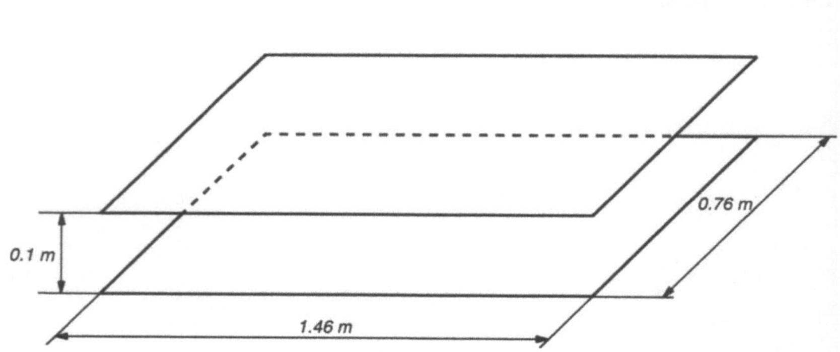

Fig. 2 Two panels clamped at the edges at a distance from each other

An Extraction Technique for Boundary Element Methods

H. Schulz
Universität Stuttgart
D–70569 Stuttgart
Germany

C. Schwab
ETH Zürich
CH–8092 Zürich
Switzerland

W. L. Wendland
Universität Stuttgart
D–70569 Stuttgart
Germany

SUMMARY

Here we present the twodimensional version of a bootstrapping algorithm for the extraction of potentials and arbitrary derivatives of the Cauchy-data for the Laplacian in connection with corresponding boundary integral equations. The method uses the derivatives of the Green's representation formula in terms of Cauchy singular or weakly singular integrals and their compositions with derivatives. We find a method, which allows the recursive numerical extraction of potentials and their derivatives near and up to the boundary curve.

INTRODUCTION

The computation of potentials and their derivatives near the supporting boundary curve is one of the difficult tasks in boundary integral equation methods. If the observation point x is sufficiently far away from the boundary curve Γ then one even gets superconvergent results due to the smoothness of the kernels. If x is near the boundary then severe numerical difficulties arise due to the oscillatory nature of the kernels. The numerical techniques developed so far for this problem usually have limited success, since the analytical jump relations of the potentials are ignored. In [11] one finds a numerical example, which shows rather clearly this behaviour. In order to give a simple illustration of our ideas let us consider a bounded two dimensional domain $\Omega \subset \mathbb{R}^2$ with a simple closed Lipschitz boundary $\Gamma = \partial\Omega$, which is given by a finite number of $C^\infty-$ or real analytic pieces. In the paper [13] the three-dimensional case is treated; and the present technique is also related to other approaches found in the engineering literature (see [13] for further references).
As a simple model equation let us consider the Dirichlet problem for the Laplacian

$$PU = -\Delta U = 0 \text{ in } \Omega, \quad U_{|\Gamma} = \phi. \tag{1}$$

Then it is well known ([6][3]), that the solution U can be represented in a direct way by

$$U(x) = (V\partial_n U)(x) - (WU)(x) \quad \text{for } x \in \Omega \tag{2}$$

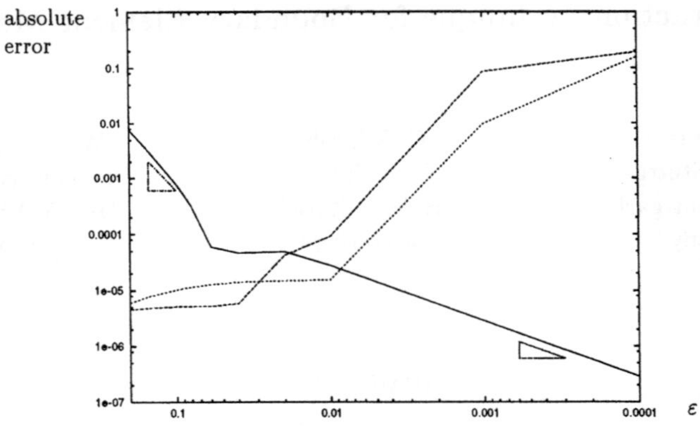

Figure 1:
Potential error near on the boundary on a line $x = x_0 - \varepsilon n(x_0)$
based on piecewise constant Galerkin BEM and $N = 64$ boundary elements
Broken lines: Potential representation evaluated using various numbers of Gauss points (5, 10).
Solid line: Second order Taylor formula with respect to ε and extraction.

the sum of a single layer and a double layer boundary potential. Let $\psi := \partial_n U$ denote the boundary flux or normal derivative. Then, for (1), the representation formula (2) yields the boundary integral equation

$$(V\psi)(x) = \frac{1}{2}\phi(x) + (K\phi)(x) =: F(x) \quad \text{for } x \in \Gamma. \tag{3}$$

The single layer potential operator is defined by

$$(V\psi)(x) := -\frac{1}{2\pi} \int\limits_{\Gamma} \ln|x - \chi(s)| \psi(s)\, ds \tag{4}$$

and the operator corresponding to the double layer potential is given by

$$K\phi(x) := \frac{1}{2\pi} \int\limits_{\Gamma} \frac{(x - \chi(s)) \cdot n(s)}{|x - \chi(s)|^2} \phi(s)\, ds. \tag{5}$$

Here, let $y = \chi(s)$ be the parametric representation with respect to the arc length. We now consider the approximate solution of equation (3) by boundary element methods; for instance Galerkin's method or the traditional point collocation and let us suppose, that we have an efficient numerical method available for resolving the approximate boundary integral equation (3) with respect to ψ. Then the traditional evaluation of $U(x)$ is done via (2) and Fig. 1 shows clearly, that the convergence near to the boundary deteriorates significantly. An alternative is the Taylor formula

$$U(x) = \phi(x_0) + v\psi(x_0) + \frac{1}{2}v^2 \partial_n^2 U(x_0) + \frac{1}{3!}v^3 \partial_n^3 U(x_0) + \ldots \tag{6}$$

about a boundary point $x_0 \in \Gamma$. If this formula is to be exploited, one needs the higher order normal derivatives $\partial_n^k U(x_0)$. In order to find these derivatives let us introduce the diffeomorphism $(s, v) \to y = \chi(s) + vn(s)$, which maps the strip $|v| < \varepsilon_0$, $\varepsilon_0 > 0$

sufficiently small, bijectively onto a tubular neighbourhood of the boundary curve Γ. Here $n(s)$ denotes the exterior normal to Γ. Then we have with the Frenet formula [9]

$$\dot{n} = -\kappa\dot{\chi}, \qquad \ddot{\chi} = \kappa n, \tag{7}$$

the derivatives

$$\frac{\partial y}{\partial s} = \dot{\chi} + v\dot{n} = (1 - v\kappa(s))\dot{\chi} \quad \text{and} \quad \frac{\partial y}{\partial v} = n(s) \tag{8}$$

which define the Riemannian tensor

$$g_{11} = (1 - v\kappa(s))^2, \quad g_{12} = g_{21} = 0, \quad g_{22} = 1 \tag{9}$$

with the determinant $g = (1 - v\kappa(s))^2$. Note that here the orientation of $\dot{\chi}$ and n is chosen a priori; hence, the curvature in (7) is negative for a convex boundary curve ([9]). The Laplacian in the boundary coordinates becomes

$$\Delta U = \frac{1}{1 - v\kappa(s)} \cdot \left\{ \frac{\partial}{\partial s}\left(\frac{1}{(1 - v\kappa(s))} \frac{\partial U}{\partial s} \right) - \kappa(s)\frac{\partial U}{\partial v} + (1 - v\kappa(s))\frac{\partial^2 U}{\partial v^2} \right\} = 0. \tag{10}$$

It can be resolved in the tubular neighbourhood of Γ with respect to

$$\partial_v^2 U = -\frac{1}{(1 - v\kappa)} \left\{ \partial_s \left(\frac{1}{(1 - v\kappa)} \partial_s U \right) - \kappa(s)\partial_v U \right\}. \tag{11}$$

This relation shows that on the right hand side we find tangential derivatives of the second order but only a first order normal derivative. If these are known, then already the second normal derivative can be computed. Further differentiation with respect to v yields

$$\begin{aligned} \partial_v^3 U = {} & \frac{2\kappa}{(1 - v\kappa)^2}\partial_v^2 U - \frac{1}{(1 - v\kappa)} \left\{ \left(\partial_s \left(\frac{\kappa}{(1 - v\kappa)^2} \right) \right) \partial_s U \right. \\ & \left. + \frac{v\dot{\kappa}}{(1 - v\kappa)^2}\, \partial_{sv}^2 U - \frac{\kappa}{(1 - v\kappa)^2}\partial_s^2 U + \frac{1}{(1 - v\kappa)}\partial_{ssv}^3 U \right\} \end{aligned} \tag{12}$$

where on the right hand side we find at most second order derivatives with respect to v and further tangential derivatives. In the three-dimensional case, see [8][13].

Lemma 1. *Taking $v = 0$ we find on Γ recursion formulae with tangential derivatives of the Cauchy data ϕ and ψ in the form*

$$\partial_n^2 U_{|\Gamma} = -\ddot{\phi} + \kappa\psi \tag{13}$$

$$\partial_n^3 U_{|\Gamma} = 2\kappa\partial_n^2 U_{|\Gamma} - \dot{\kappa}\dot{\phi} - \kappa\ddot{\phi} + \ddot{\psi}. \tag{14}$$

Note that further differentiation yields Cauchy's algorithm as in the Cauchy-Kowalewski procedure for solving the Cauchy-problem. Hence all we need for performing this algorithm is the tangential derivatives of the Cauchy data.

Lemma 2. *Let A be a pseudodifferential operator of order a on the curve Γ. Then the tangential derivatives can be computed with the recursion formula*

$$\partial_s^k A\psi = \sum_{j=0}^{k} \binom{k}{j} A_{(j)} \partial_s^{(k-j)} \psi \tag{15}$$

and the recursively defined commutators

$$A_{(0)} := A, \quad A_{(j+1)} := [\partial, A_{(j)}] = \partial_s A_{(j)} - A_{(j)}\partial_s. \tag{16}$$

Here for every $j = 1, \ldots, k$, $A_{(j)}$ is a pseudodifferential operator of order a.

Note that this procedure defines the triangular system of equations

$$A\psi = F \tag{17}$$

$$A\partial_s^k \psi = \partial^k F - \sum_{j=1}^{k} \binom{k}{j} A_{(j)} \partial_s^{k-j} \psi \quad \text{for } k = 1, 2, \ldots, M. \tag{18}$$

M will be chosen later on according to the accuracy of the desired result in Taylor's formula.

ERROR ESTIMATES

In practice, the system (17), (18) is solved numerically by some boundary element method. Note that in all these equations only A is to be inverted. For simplicity let us consider here a Galerkin method although corresponding results with collocation are also valid, correspondingly (see [1], [10]).

Lemma 3. *Let A satisfy a Gårding inequality*

$$\operatorname{Re}\langle \psi, A\psi \rangle_\Gamma \geq \Lambda_1 \|\psi\|_{\mathcal{H}}^2 - \operatorname{Re} C(\psi, \psi) \tag{19}$$

with $\Lambda_1 > 0$ and $C : \mathcal{H} \times \mathcal{H} \to \mathbb{C}$ a compact sesquilinearform. Then the triangular system (17),(18) with the matrix $\mathcal{A}$ of operators also satisfies a Gårding inequality:

$$\operatorname{Re}\langle \vec{\psi}, \mathcal{A}\vec{\psi} \rangle \geq \Lambda_0 \|\vec{\psi}\|_{\mathcal{H}^M}^2 - \operatorname{Re} \mathcal{C}(\vec{\psi}, \vec{\psi}) \tag{20}$$

with $\Lambda_0 > 0$ and some compact matrix $\mathcal{C}$ of operators.

This lemma concerns a special case of strong ellipticity, see [14]. For Gårdings inequality see also [5]. Now let us introduce a family of grid decompositions of the boundary with maximum mesh width h defined by a $S_h^{d,r}(\Gamma)$-system in the sense of Babuška and Aziz [2] and let us consider the Galerkin boundary element approximation of the triangular system (17), (18), it is

$$\begin{aligned}
\langle \mu_h, A\psi_h \rangle_\Gamma &= \langle \mu_h, F \rangle, \\
\langle \mu_h, A\psi_h^{(1)} \rangle_\Gamma &= \langle \mu_h, \dot{F} \rangle - \langle \mu_h, A_{(1)}\psi_h \rangle, \\
&\vdots \\
\langle \mu_h, A\psi_h^{(M)} \rangle_\Gamma &= \langle \mu_h, \overset{(M)}{F} \rangle - \sum_{j=1}^{M} \binom{M}{j} \langle \mu_h, A_{(j)} \psi_h^{(M-j)} \rangle \qquad \forall \mu_h \in \mathcal{H}_h := S_h^{d,r}(\Gamma).
\end{aligned} \tag{21}$$

In our example (3) we have $A = V$ and $a = -1$. Then the Galerkin method for the triangular system (21) converges quasioptimally, and the approximations $\psi_h^{(j)}$ of the tangential derivatives $\partial_s^j \psi$ satisfy the asymptotic error estimate

$$\|\partial_s^j \psi - \psi_h^{(j)}\|_{H^t} \leq c(j) h^{\tau - t} \|\psi\|_{H^{\tau + j}} \quad \text{where} \ -d-1 \leq t \leq -1/2 \leq \tau \leq d. \tag{22}$$

The proof follows immediately from [7], since we have Gårding's inequality (20) for the whole triangular system (see also [14], [15]).

It is clear that the efficiency of the method (21) strongly depends on the numerical realization of the commutators. In the next section we will show how to calculate the commutators.

CALCULATION OF THE COMMUTATORS

The calculation of the commutators can be done in a very straightforward and simple way by using the chain rule in the ascertion of the following theorem.

Theorem 4. *Let B_a be a pseudodifferential operator of order a, given in the form*

$$B_a \psi = \mathcal{P}\psi + \text{p.f.} \int_\Gamma \tilde{k}_a(u, s-u) \psi(s)\, ds, \tag{23}$$

where

$$\mathcal{P}\psi = \sum_{0 \leq l \leq a} b_l(u) \partial_u^l \psi(u) \tag{24}$$

denotes a differential operator and p.f. stands for the Hadamard finite part integral. Then the commutator can be expressed in the simple form

$$[\partial, B_a] = [\partial, \mathcal{P}]\psi + \text{p.f.} \int_\Gamma (\partial_u \tilde{k}_a(u,t))_{|t=s-u} \psi(s)\, ds. \tag{25}$$

Proof. The proof rests on the pseudohomogeneous asymptotic representation of the kernel $\tilde{k}_a$ of a classical pseudodifferential operator in the form

$$\tilde{k}_a(u, s-u) = \sum_{l=0}^{a} c_l(u)(s-u)^{l-a-1} + c_{-1}(u) \log|s-u| + \tilde{k}_R(u, s-u) \tag{26}$$

with smooth functions $c_l(u)$ and with a remainder having integrable first derivatives (see [12]). Then, for $c_l = 0$, $l = -1, ..., a$, (25) follows by interchanging differentiation and integration, from the form of the kernel $\tilde{k}_R(u, s-u)$ and integration by parts. Hence, in view of the product rule, it remains to show (25) for the special case $k_a = (s-u)^{l-a-1}$, i.e. for the convolution of the corresponding distribution with ψ. In this case, however, (25) corresponds to the definition of the derivative of a distribution. □

For a surface Γ, the proof is little more involved [13].

Example: With (25), the first commutator for the single layer potential can be expressed in the form

$$(V_{(1)}\psi)(x) = \frac{1}{4\pi} \int_\Gamma \frac{(\chi(s) - x) \cdot (\dot{\chi}(s) - \dot{\chi}(x))}{|\chi(s) - x|^2} \psi(s)\, ds. \tag{27}$$

POINTVALUE EXTRACTION

The Galerkin solutions $\psi_h^{(j)}$ of (21) are possibly discontinuous and not suitable for the evaluation of $(\partial_n^k U)(x_0)$ as required in (6). Superconvergent point values $\hat{\psi}_h^{(j)}(x_0)$ can, however, be obtained by postprocessing as follows.

Let $x_0 \in \Gamma$ correspond to arc length $s = 0$ and let $\Xi_{x_0}(s) \in C^\infty([0, L))$ be such that

$$\dot{\Xi}_{x_0}(s) \in C^\infty(\Gamma), \quad \Xi_{x_0}(0^+) - \Xi_{x_0}(L^-) = -1, \qquad \text{where } L = \int_\Gamma ds. \tag{28}$$

Then for any distribution ψ which is smooth at x_0, one has

$$\psi(x_0) = \int_0^L \psi(s) \dot{\Xi}_{x_0}(s)\, ds + \int_0^L \dot{\psi}(s) \Xi_{x_0}(s)\, ds. \tag{29}$$

Repeated integration by parts gives

$$\psi(x_0) = \sum_{j=0}^{l} \int_\Gamma G_j^{(l)}(s) \psi^{(j)}(s)\, ds, \quad l \geq 1, \tag{30}$$

where the extraction kernels satisfy

$$G_j^{(l)} \in H_{\text{loc}}^{l-1/2-\varepsilon}(\Gamma). \tag{31}$$

Then approximate $\hat{\psi}_h(x_0)$ are obtained by inserting $\psi_h^{(j)}$ into (30). Exploiting the Galerkin orthogonality as in [7], we find (in the general case of operators of order a)

$$\text{1. For } l < 1/2 + \varepsilon - a + d: \quad |\psi(x) - \hat{\psi}_h(x)| \leq c h^{l+d-1/2-\varepsilon} ||\psi||_{d+l} \tag{32}$$

and

$$\text{2. for } l \geq 1/2 + \varepsilon - a + d: \quad |\psi(x) - \hat{\psi}_h(x)| \leq c h^{2d-a} ||\psi||_{d+l}. \tag{33}$$

Consider the potential $U(x)$, $x \in \Omega$ with distance $\varepsilon > 0$ to the boundary. We can represent x by $x = x_0 - \varepsilon n(x_0)$, $x_0 \in \Gamma$. As in (6) we have the Taylor expansion (with $v = -\varepsilon$)

$$U(x) = \sum_{k=0}^{M'} (-1)^k \frac{1}{k!} \varepsilon^k w^{(k)}(x_0) + \mathcal{O}(\varepsilon^{M'+1}), \quad M' \in \mathbb{N} \tag{34}$$

with $w^{(k)}(x_0) := (\partial_n^{(k)} U)(x_0)$. We define an approximation for $U(x)$ by

$$\hat{U}_h(x) := \sum_{k=0}^{M'} (-1)^k \frac{1}{k!} \varepsilon^k \hat{w}_h^{(k)}(x_0). \tag{35}$$

From (34) and (35) it follows that

$$U(x) - \hat{U}_h(x) = \sum_{k=0}^{M'} (-1)^k \frac{1}{k!} \varepsilon^k \left(w^{(k)}(x_0) - \hat{w}_h^{(k)}(x_0) \right) + \mathcal{O}(\varepsilon^{M'+1}). \tag{36}$$

If the tangential derivatives of ϕ up to the order k and of ψ up to the order $k-1$ at the point x_0 are known it is possible to compute the normal derivatives for (35) ([13], Theorem 3.1). In practice we compute (possibly discontinuous) approximate tangential derivatives by BEM-discretization of (17), (18) and obtain than approximate pointvalues as in Section 4. This gives the error estimates for $|U(x_0) - \hat{U}_h(x_0)|$ by postprocessing.
We illustrate this with the second order Taylor formula

$$U(x) = \phi(x_0) - \varepsilon\psi(x_0) + \varepsilon^2 \left(\kappa\psi - \ddot{\phi} \right)(x_0) + \mathcal{O}(\varepsilon^3). \tag{37}$$

For the *Dirichlet problem* ϕ is known, so the exact derivative $\ddot{\phi}$ can be computed by hand. For ψ we use the extracted and postprocessed solution, resulting in the approximation

$$\hat{U}_h(x) = \phi(x_0) - \varepsilon\hat{\psi}_h(x_0) + \varepsilon^2 \left(\kappa\hat{\psi}_h - \ddot{\phi} \right)(x_0). \tag{38}$$

For the error at the point $x = x_0 - \varepsilon n(x_0)$ we have from (32), (33)

$$\begin{aligned} |U(x) - \hat{U}_h(x)| &\leq c(\varepsilon + \kappa\varepsilon^2)|\psi(x_0) - \hat{\psi}_h(x_0)| + \mathcal{O}(\varepsilon^3) \\ &\leq c(|\kappa|)\varepsilon h^\alpha ||\psi||_{d+l} + \mathcal{O}(\varepsilon^3) \end{aligned} \tag{39}$$

with $\alpha = l + d - 1/2 - \varepsilon$ for $l < 3/2 + \varepsilon + d$ and $\alpha = 2d + 1$ for $l \geq 3/2 + \varepsilon + d$.
For the case of the *Neumann Problem* ϕ is unknown, so for a second order Taylor formula we have to compute in addition to ϕ an approximation for the second order tangential derivative $\ddot{\phi}$ by the steps

1. compute ϕ_h and $\phi_h^{(2)}$ by the extraction technique similar to (21),

2. compute $\hat{\phi}_h(x_0)$ and $\hat{\phi}_h^{(2)}(x_0)$ by pointvalue extraction,

otherwise we obtain a first order Taylor formula only.

NUMERICAL EXAMPLES

For the numerical experiments we consider the elliptical domain

$$\Omega := \left\{x \in \mathbb{R}^2 : \ x_1^2/a^2 + x_2^2/b^2 < 1\right\} \tag{40}$$

with $a = 0.6$, $b = 0.3$ and the model problem (1) with the given Dirichlet datum ϕ. For the boundary Γ we have the parametric representation

$$\Gamma = \left\{x \in \mathbb{R}^2 : \ x_1 = a\,\cos(t),\, x_2 = b\,\sin(t),\, t = 0, ..., 2\pi\right\}. \tag{41}$$

The following algorithm was realized:

1. Compute a first approximation ψ_h for the unknown Neumann datum by solving equation (3) approximately with a piecewise constant ansatz and the Galerkin Method.
2. Find an approximation $\psi_h^{(1)}$ for $\partial_s\psi$ by solving the second equation of (21) approximately with the same piecewise constant trial functions and the Galerkin Method.
3. Obtain superconvergent pointvalues $\hat{\psi}_h(x_0)$ via the postprocessing formula (29).
4. Compute the approximation $\hat{U}_h(x)$ via (38) by using the postprocessed $\hat{\psi}_h(x_0)$. For comparison we have also computed the potential representation for $U(x)$ (with $\hat{\psi}_h$ instead of $\partial_n U$ on the right-hand side and 5 Gauss points per boundary element).

For the following examples we choose the Dirichlet datum in form of some concrete function ϕ, so the unknown Neumann datum ψ and the derivative $\partial_s\psi$ is computable by hand and it is possible to control the numerical errors of the steps 1.-4. directly.

First example: We choose

$$\phi(x) = x_1. \tag{42}$$

Fig. 2 (Tab. 1) and Fig. 3 (Tab. 2) show the L^2-error resp. the L^∞-error of the Galerkin solution, of the tangential derivative and the postprocessed solution with respect to the number N of degrees of freedom. We clearly see the superconvergence expected from (32) and (33).

Second example: We choose now

$$\phi(x) = \log|x-y|, \qquad y \notin \bar{\Omega} \tag{43}$$

with $y = y_0 + \delta n(y_0)$, $y_0 = \left(0.6\cdot\cos\left(\frac{3}{4}\pi\right), 0.3\cdot\sin\left(\frac{3}{4}\pi\right)\right)$, $\delta = 0.4$. The unknown Neumann datum ψ and the derivative $\partial_s\psi$ is explicitly known again. Fig. 4 (Tab. 3) and Fig. 5 (Tab. 4) show the L^2-error resp. the L^∞-error of the Galerkin solution, of the tangential derivative and the postprocessed solution with respect to the number N of degrees of freedom.

Fig. 1 shows the convergence behaviour of the potential representation (2) and of the Taylor expansion (38) with fixed $N = 64$ and $\varepsilon \to 0$. Fig. 6, Fig. 7 and Fig. 8 illustrate the behaviour of the Taylor expansion and of the potential representation with fixed distance 0.2 (Fig. 6), 0.01 (Fig. 7) and 0.001 (Fig. 8) to the boundary in dependence on the number N of degrees of freedom in the Galerkin approximation.

For Fig. 6 the distance to the boundary is equal to 0.2, so the term $\mathcal{O}(\varepsilon^3)$ in (39) dominates the first part of (39) $c(\kappa)\varepsilon h^\alpha||\psi||_{d+l}$, $\alpha = 1.5 - \epsilon$, $l = 1$. The convergence behavior of the potential representation agrees with the expected superconvergence $\mathcal{O}(h^3)$. For $\varepsilon = 0.01$ (Fig. 7) the situation is changed already and for $\varepsilon = 0.001$ (Fig. 8) the distance to the boundary is much too small for the Gauss quadrature rule (5 points) in the numerical potential representation. The term $\mathcal{O}(\varepsilon^3)$ in (39) is very small and so the convergence of the Taylor formula is determined by the first part of (39).

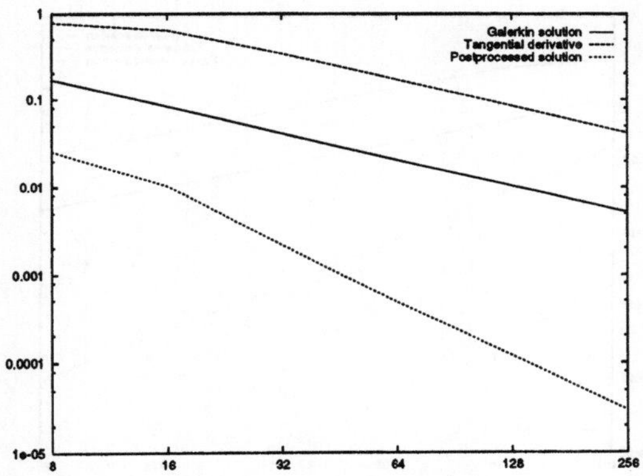

Figure 2: L^2-error versus N (Example 1)

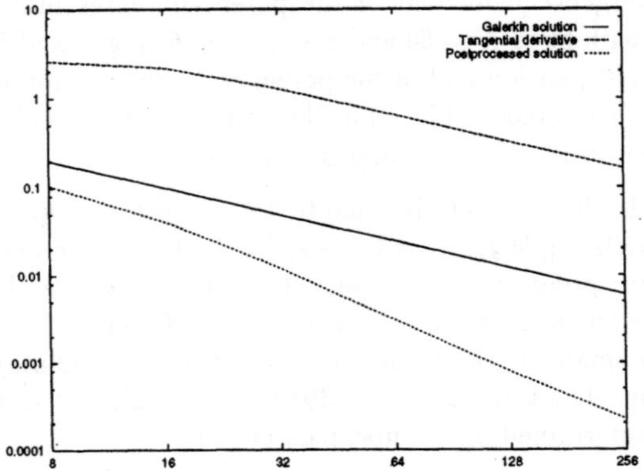

Figure 3: L^∞-error versus N (Example 1)

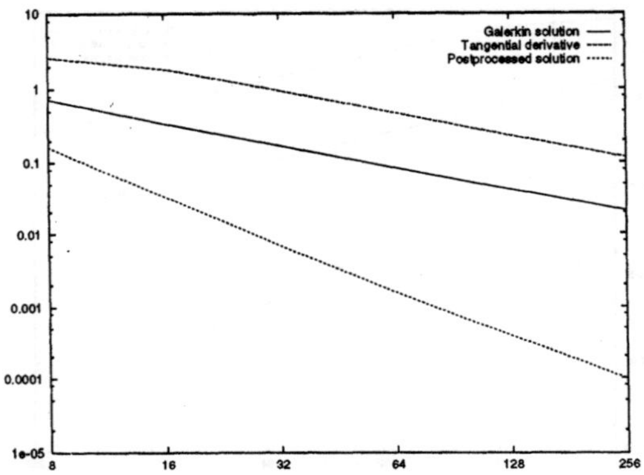

Figure 4: L^2-error versus N (Example 2)

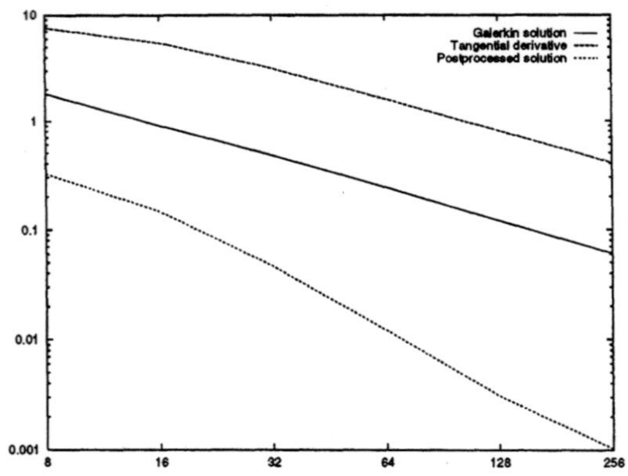

Figure 5: L^∞-error versus N (Example 2)

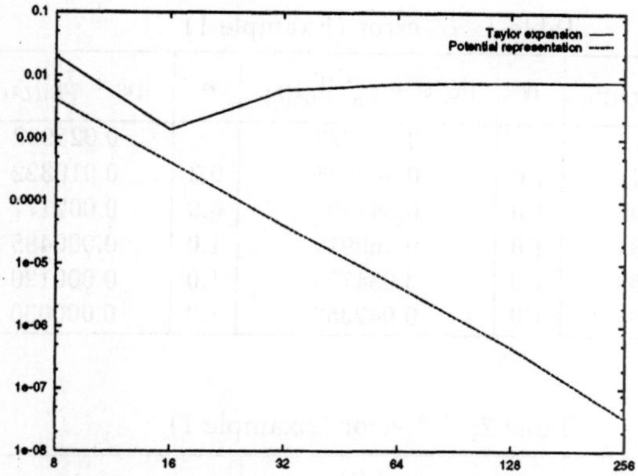

Figure 6: Error at the point $(0, 0.1)$ versus N (Example 2)

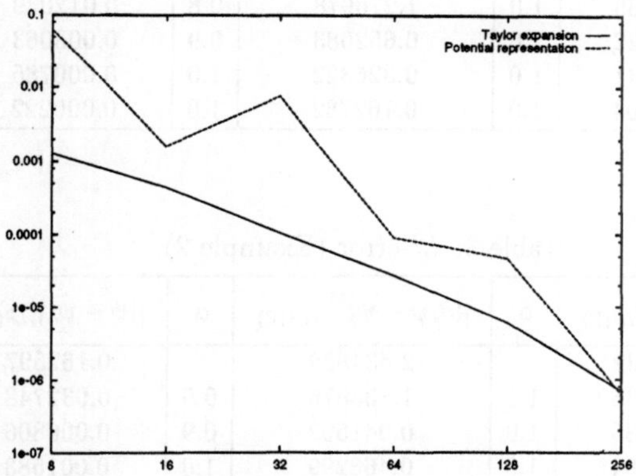

Figure 7: Error at the point $(0, 0.29)$ versus N (Example 2)

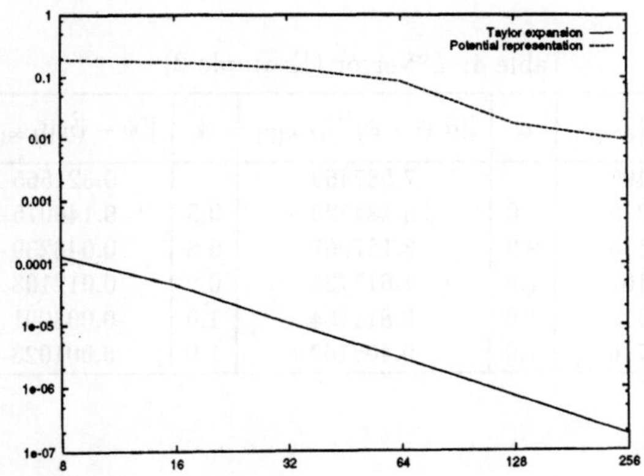

Figure 8: Error at the point $(0, 0.299)$ versus N (Example 2)

Table 1: L^2-error (Example 1)

N	$\|\|\psi - \psi_h\|\|_{L^2(\Gamma)}$	α	$\|\|\partial_s\psi - \psi_h^{(1)}\|\|_{L^2(\Gamma)}$	α	$\|\|\psi - \hat{\psi}_h\|\|_{L^2(\Gamma)}$	α
8	0.168272		0.785733		0.025992	
16	0.083441	1.0	0.643296	0.3	0.010322	1.3
32	0.041600	1.0	0.341493	0.9	0.002177	2.2
64	0.020784	1.0	0.169919	1.0	0.000485	2.2
128	0.010389	1.0	0.084773	1.0	0.000120	2.0
256	0.005194	1.0	0.042359	1.0	0.000030	2.0

Table 2: L^∞-error (Example 1)

N	$\|\|\psi - \psi_h\|\|_{L^\infty(\Gamma)}$	α	$\|\|\partial_s\psi - \psi_h^{(1)}\|\|_{L^\infty(\Gamma)}$	α	$\|\|\psi - \hat{\psi}_h\|\|_{L^\infty(\Gamma)}$	α
8	0.199758		2.610295		0.104826	
16	0.098761	1.0	2.181953	0.3	0.041052	1.4
32	0.049135	1.0	1.276978	0.8	0.012039	1.8
64	0.024522	1.0	0.652083	0.9	0.003063	2.0
128	0.012241	1.0	0.326322	1.0	0.000785	2.0
256	0.006104	1.0	0.162782	1.0	0.000222	1.8

Table 3: L^2-error (Example 2)

N	$\|\|\psi - \psi_h\|\|_{L^2(\Gamma)}$	α	$\|\|\partial_s\psi - \psi_h^{(1)}\|\|_{L^2(\Gamma)}$	α	$\|\|\psi - \hat{\psi}_h\|\|_{L^2(\Gamma)}$	α
8	0.724432		2.634889		0.162597	
16	0.338135	1.1	1.833876	0.5	0.032748	2.3
32	0.167685	1.0	0.941592	0.9	0.006806	2.2
64	0.083584	1.0	0.468299	1.0	0.001583	2.1
128	0.041751	1.0	0.233676	1.0	0.000389	2.0
256	0.020869	1.0	0.116770	1.0	0.000097	2.0

Table 4: L^∞-error (Example 2)

N	$\|\|\psi - \psi_h\|\|_{L^\infty(\Gamma)}$	α	$\|\|\partial_s\psi - \psi_h^{(1)}\|\|_{L^\infty(\Gamma)}$	α	$\|\|\psi - \hat{\psi}_h\|\|_{L^\infty(\Gamma)}$	α
8	1.829495		7.587469		0.321565	
16	0.908275	1.0	5.484529	0.5	0.146075	1.1
32	0.474115	0.9	3.157067	0.8	0.046239	1.7
64	0.241107	1.0	1.617724	0.9	0.012108	1.9
128	0.121313	1.0	0.811184	1.0	0.003091	1.9
256	0.060710	1.0	0.405102	1.0	0.001023	1.6

α: Numerical order of convergence with respect to $h = \frac{1}{N}$.

REFERENCES

[1] D. N. Arnold and W. L. Wendland: *On the asymptotic convergence of collocation methods.* Math. Comp. **41** (1983) 349-381.

[2] I. Babuška and A.K. Aziz: *Survey lectures on the mathematical foundations of the finite element method.* In: The Mathematical Foundation of the Finite Element Method with Applications to Partial Differential Equations (ed. A. K. Aziz), Academic Press, New York (1972) 3–359.

[3] M. Costabel: *Boundary integral operators on Lipschitz domains – elementary results*, SIAM J. Math. Anal. **19** (1987) 613–626.

[4] M. Costabel, E. P. Stephan: *Boundary integral equations for mixed boundary value problems in polygonal domains and Galerkin Approximation.* Banach center Publications **15** PN-Polish Scientific Publ., Warsaw, 1985, pp. 175-251.

[5] M. Costabel and W. L. Wendland: *Strong ellipticity of boundary integral operators*, Crelles Journ. Reine Angew. Math. **372** (1986) 34-63.

[6] G. C. Hsiao and W. L. Wendland: *A finite element method for some integral equations of the first kind*, J. Math. Anal. Appl. **58** (1977) 449–481.

[7] G. C. Hsiao and W. L. Wendland: *The Aubin–Nitsche lemma for integral equations*, J. Integral Equations **3** (1981) 299–315.

[8] G. C. Hsiao and W. L. Wendland: Variational Methods for Boundary Integral Equations. In preparation.

[9] W. Klingenberg: Eine Vorlesung über Differentialgeometrie. Springer-Verlag Berlin-Heidelberg-New York, 1973.

[10] S. Prößdorf and B. Silbermann: *Numerical Analysis for Integral and Related Operator Equations.* Birkhäuser–Verlag, Basel, 1991.

[11] C. Schwab: *Variable order composite quadrature of singular and nearly singular integrals.* Computing **53** (1994) 173–194.

[12] C. Schwab and W. L. Wendland: *Kernel properties and representations of boundary integral operators.* Mathematische Nachrichten **156** (1992) 187–218.

[13] C. Schwab and W. L. Wendland: *On the Extraction technique in Boundary Integral Equations* Preprint No. 96-3, Mathematics Institute A, Stuttgart University (1996).

[14] E. P. Stephan and W. L. Wendland: *Remarks to Galerkin and least squares methods with finite elements for general elliptic problems.* In: Manuscripta Geodaetica **1** (1976) 93-123.

[15] W. L. Wendland: *Strongly elliptic boundary integral equations.* In: The State of the Art in Numerical Analysis (eds. A. Iserles, M. Powell), Clarendon Press, Oxford (1987) 511–561.

FAST SOLVERS FOR THE SYMMETRIC BOUNDARY ELEMENT METHOD

Olaf Steinbach
Mathematisches Institut A
Universität Stuttgart
Pfaffenwaldring 57
70569 Stuttgart, Germany

SUMMARY

The symmetric formulation of boundary integral equations is well suited to solve mixed boundary value problems by boundary element methods based on a Galerkin approximation. This leads to a skew–symmetric and positive definite system of linear equations, which may be solved by different solution strategies via iterative schemes. We give optimal preconditioners based on an operator approach, which are in general usable for boundary element methods based on the symmetric formulation. Alternatively, one may use Toeplitz type preconditioners, too. In the end, a numerical example for a mixed boundary value problem in plane linear elasticity is given.

MIXED BOUNDARY VALUE PROBLEMS

For a selfadjoint and elliptic differential operator of 2nd order,

$$L(x)u(x) = -\sum_{k,\ell=1}^{d} \frac{\partial}{\partial x_k}\left(a_{k\ell}(x)\cdot\frac{\partial}{\partial x_\ell}u(x)\right) + c(x)u(x), \tag{1}$$

and a bounded domain $\Omega \subset I\!R^d$ with $d = 2,3$, we consider a mixed boundary value problem, where the Lipschitz continuous boundary $\Gamma = \partial\Omega$ is decomposed into two distinct parts Γ_D and Γ_N, where boundary conditions of Dirichlet or Neumann type, respectively, are given:

$$\left.\begin{array}{rcll} L(x)u(x) &=& f(x) & \text{for } x\in\Omega, \\ \gamma_0 u(x) &=& g(x) & \text{for } x\in\Gamma_D, \\ \gamma_1 u(x) &=& h(x) & \text{for } x\in\Gamma_N. \end{array}\right\}. \tag{2}$$

Here, the trace operators are of the form

$$\gamma_j u(x) = \partial_n^j u(x) \;\; \text{for } x\in\Gamma, j=0,1, \tag{3}$$

where the conormal derivative is given by

$$\partial_n u(x) = \sum_{k,\ell=1}^{d} n_\ell a_{k\ell} \frac{\partial}{\partial x_k} u(x) \tag{4}$$

and $n(x) = (n_k)_{k=1}^d$ denotes the outer normal vector in $x \in \Gamma$.

Let us suppose, that we have given a fundamental solution $U^*(x,y)$ for the differential operator (1), so that we can write the boundary integral equations of the direct formulation in the usual way using the Calderon projector:

$$\begin{pmatrix} \gamma_0 u \\ \gamma_1 u \end{pmatrix} = \begin{pmatrix} \frac{1}{2}I - K & V \\ D & \frac{1}{2}I + K' \end{pmatrix} \begin{pmatrix} \gamma_0 u \\ \gamma_1 u \end{pmatrix} + \begin{pmatrix} N_0 f \\ N_1 f \end{pmatrix}. \tag{5}$$

Here, V, K, D denote the standard boundary integral operators for the single layer potential, the double layer potential and the hypersingular integral operator, which are pseudodifferential operators of orders $+1, 0, -1$, respectively. In addition, we introduce the volume or Newton potentials for $x \in \Gamma$ by

$$(N_i f)(x) = \gamma_i \int_\Omega f(y) \cdot U^*(x,y)\, ds_y\,. \tag{6}$$

From (5) one can derive various boundary integral formulations to solve the mixed boundary value problem (2) by a boundary element method. Here we will restrict ourself to the so–called symmetric formulation only, which will be discretized by a Galerkin scheme. The symmetric formulation of boundary integral equations for a mixed boundary value problem originally goes back to Sirtori [13]. Today, this formulation is frequently used in boundary element domain decomposition methods [14] or in the symmetric coupling with the finite element method [2,7], too.

However, there are also formulations based on the first boundary integral equation in (5) only, which allows us, to use collocation schemes for discretization, which may be handled simpler in the numerical realization.

We rewrite the first boundary integral equation in (5) for $x \in \Gamma_D$, while the second one is used for $x \in \Gamma_N$. With $u := \gamma_0 u, t := \gamma_1 u$ we have

$$\begin{aligned} (Vt)(x) &= \frac{1}{2} g(x) + (Ku)(x) - (N_0 f)(x) \quad \text{for } x \in \Gamma_D, \\ (Du)(x) &= \frac{1}{2} h(x) - (K't)(x) - (N_1 f)(x) \quad \text{for } x \in \Gamma_N. \end{aligned}$$

If we split an arbitrary boundary integral operator by

$$\begin{aligned} (Au)(x) &= \int_\Gamma k(x,y)u(y)ds_y \\ &= \int_{\Gamma_D} k(x,y)u(y)ds_y + \int_{\Gamma_N} k(x,y)u(y)ds_y = (A_D u)(x) + (A_N u)(x), \end{aligned}$$

we can write these equations as a coupled system

$$\begin{pmatrix} V_D & -K_N \\ K'_D & D_N \end{pmatrix} \begin{pmatrix} t \\ u \end{pmatrix} = \begin{pmatrix} -V_N & \frac{1}{2}I + K_D \\ -\frac{1}{2}I + K'_N & D_D \end{pmatrix} \begin{pmatrix} h \\ g \end{pmatrix} - \begin{pmatrix} N_0 f \\ N_1 f \end{pmatrix} =: \begin{pmatrix} f_1 \\ f_2 \end{pmatrix}$$

with a given right hand side.

Defining the bilinear form

$$a(t,u;\tau,v) := \langle V_D t, \tau\rangle_{L^2(\Gamma_D)} - \langle K_N u, \tau\rangle_{L^2(\Gamma_D)} + \langle K_N' t, v\rangle_{L^2(\Gamma_D)} + \langle D_D u, v\rangle_{L^2(\Gamma_D)}, \quad (7)$$

then it follows, that the bilinear form $a(\cdot,\cdot)$ is $\tilde{H}^{-1/2}(\Gamma_D) \times \tilde{H}^{1/2}(\Gamma_N)$–elliptic, i.e.

$$a(t,u;t,u) \geq \gamma_1 \cdot \left\{ ||t||^2_{\tilde{H}^{-1/2}(\Gamma_D)} + ||u||^2_{\tilde{H}^{1/2}(\Gamma_N)} \right\} \quad (8)$$

holds for all $(t,u) \in \tilde{H}^{-1/2}(\Gamma_D) \times \tilde{H}^{1/2}(\Gamma_N)$, where the Sobolev spaces $\tilde{H}^s(\Gamma)$ are defined in the usual way [8]. Hence, the variational formulation

$$a(t,u;\tau,v) = f(\tau,v) \quad \forall(\tau,v) \in \tilde{H}^{-1/2}(\Gamma_D) \times \tilde{H}^{1/2}(\Gamma_N) \quad (9)$$

with the linear form

$$f(\tau,v) := \langle f_1, \tau\rangle_{L^2(\Gamma_D)} + \langle f_2, v\rangle_{L^2(\Gamma_N)}$$

is uniquely solvable due to the Lax–Milgram theorem.

Let us introduce finite–dimensional subspaces

$$V_h = \operatorname{span}\{\varphi_k^\nu\}_{k=1}^N \subset \tilde{H}^{-1/2}(\Gamma_D), \quad W_h = \operatorname{span}\{\varphi_k^\mu\}_{k=1}^M \subset \tilde{H}^{1/2}(\Gamma_N)$$

with piecewise polynomial trial functions of degree ν and μ, respectively. The Galerkin discretization of (9) leads to the skew–symmetric and positive definite system of linear equations

$$\begin{pmatrix} V_h & -K_h \\ K_h^\top & D_h \end{pmatrix} \begin{pmatrix} \underline{t} \\ \underline{u} \end{pmatrix} = \begin{pmatrix} \underline{f}_1 \\ \underline{f}_2 \end{pmatrix} \quad (10)$$

with the block matrices defined by

$$\begin{aligned} V_h[\ell,k] &= \langle V_D \varphi_k^\nu, \varphi_\ell^\nu\rangle_{L^2(\Gamma_D)}, \\ K_h[\ell,i] &= \langle K_N \varphi_i^\mu, \varphi_\ell^\nu\rangle_{L^2(\Gamma_N)}, \\ D_h[j,i] &= \langle D_N \varphi_i^\mu, \varphi_j^\mu\rangle_{L^2(\Gamma_N)} \end{aligned}$$

for $k,\ell = 1,\ldots,N, i,j = 1,\ldots,M$ and we have the quasi–optimal error estimates

$$||u-u_h||_{\tilde{H}^{1/2}(\Gamma_N)} + ||t-t_h||_{\tilde{H}^{-1/2}(\Gamma_D)} \leq c \cdot \left\{ \inf_{v_h \in W_h} ||u - v_h||_{\tilde{H}^{1/2}(\Gamma_N)} + \inf_{\tau_h \in V_h} ||t - \tau_h||_{\tilde{H}^{-1/2}(\Gamma_D)} \right\}$$

due to (8). In what follows we want to discuss efficient preconditioned iterative schemes to solve the linear system (10).

Note, that in the case of pure Dirichlet of Neumann boundary conditions, the system (10) turns out to be the Galerkin discretization for the classical first kind boundary integral equations, where the resulting stiffness matrices are symmetric and positive definite for the Dirichlet problem and positive semi–definite for a Neumann problem.

Let us suppose, that we have given a preconditioner C_V for the discrete single layer potential V_h and a second one, C_S, for the Schur complement $S_h = D_h + K_h^\top V_h^{-1} K_h$ satisfying the spectral equivalence inequalities

$$c_1^V \cdot (C_V \underline{t}, \underline{t}) \leq (V_h \underline{t}, \underline{t}) \leq c_2^V \cdot (C_V \underline{t}, \underline{t}) \tag{11}$$

for all $\underline{t} \in \mathbb{R}^N$ and

$$c_1^S \cdot (C_S \underline{u}, \underline{u}) \leq (S_h \underline{u}, \underline{u}) \leq c_2^S \cdot (C_S \underline{u}, \underline{u}) \tag{12}$$

for all $\underline{u} \in \mathbb{R}^M$. In chapter 3 we will construct such preconditioners C_V and C_S satisfying (11) and (12), where the constants does not depend on the discretization parameters. Later we will see, that we can replace the Schur complement S_h in (12) by the discrete hypersingular integral operator D_h only.

A first choice to solve (10) by an iterative method is to use a method of generalized conjugate gradients, i.e. of biorthogonal directions like the BiCGStab algorithm [15,20]. As a preconditioner, one can use the block diagonal matrix

$$C_M = \begin{pmatrix} C_V & 0 \\ 0 & C_S \end{pmatrix}, \tag{13}$$

where a scaling of the used preconditioners is necessary to establish the inequalities (11) and (12) such that $c_1^V = c_1^S = 1$ is satisfied.

Applying a block Gauss–Seidel step to (10), i.e.

$$\underline{t} = V_h^{-1} \underline{f}_1 + V_h^{-1} K_h \underline{u},$$

we have to solve the Schur complement system

$$\left[D_h + K_h^\top V_h^{-1} K_h \right] \underline{u} = \underline{f}_2 - K_h^\top V_h^{-1} \underline{f}_1$$

with a symmetric and positive definite matrix S_h. Due to (12) one can use the preconditioned conjugate gradient method to solve this system in a very efficient way. To realize the matrix multiplication of an arbitrary vector $\underline{v}$ with the matrix S_h, one has to solve a linear system $V_h \underline{t} = \underline{w}$ to realize the matrix times vector multiplication $\underline{t} = V_h^{-1} \underline{w}$ once per global iteration step. Due to (11), one may use any preconditioned iterative scheme to realize this action, preferably by the conjugate gradient method again. Note, that this method does not require any scaling of the used preconditioners, however, due to the inner iteration scheme to realize one global iteration step, this method may be disadvantegeous. Therefore we are looking for a method to keep the nice convergence behaviour of the Schur complement method without the realization of V_h^{-1} in any iteration step. Following [1], the transformed matrix

$$\tilde{M}_h = \begin{pmatrix} C_V^{-1} & 0 \\ -K_h^\top C_V^{-1} & I \end{pmatrix} \begin{pmatrix} V_h & -K_h \\ K_h^\top & D_h \end{pmatrix}$$

is selfadjoint and positive definite with respect to the special inner product

$$[(\underline{t}, \underline{u}); (\underline{\tau}, \underline{v})] := ((V_h - C_V)\underline{t}, \underline{\tau}) + (\underline{u}, \underline{v}),$$

if we assume $c_1^V > 1$ in (11). Moreover, $\tilde{M}_h$ is spectrally equivalent to the block diagonal matrix

$$C_{\tilde{M}} = \begin{pmatrix} I & 0 \\ 0 & C_S \end{pmatrix}.$$

Obviously, this approach depends strongly on the symmetry of the block matrix V_h to define the modified inner product. Therefore, this method can not be applied directly to the non–symmetric formulation with collocation discretization, where V_h is no longer symmetric. To overcome this, one can use the transformed matrix

$$\hat{M}_h = \begin{pmatrix} V_h^\top C_V^{-1} - I & 0 \\ -K_h^\top C_V^{-1} & I \end{pmatrix} \begin{pmatrix} V_h & -K_h \\ K_h^\top & D_h \end{pmatrix},$$

which is symmetric with respect to the usual ℓ_2–scalar product, if the block matrices V_h and D_h are symmetric. Moreover, the matrix $\hat{M}_h$ is spectrally equivalent to the block diagonal preconditioner (13), if $c_1^V > 1$ in (11) is satisfied. From this one can construct efficient iterative solvers in the nonsymmetric formulation, too [18].

PRECONDITIONERS

As mentioned in the previous section, one has to construct an efficient preconditioner C_V for the discrete single layer potential V_h as well as a preconditioner C_S for the Schur complement S_h. At the beginning we start with a general statement to construct a preconditioner for the matrix

$$\tilde{T}_h = B_h^\top A_h^{-1} B_h\,, \tag{14}$$

which may be considered as discrete representation of an operator

$$T = B^* A^{-1} B \tag{15}$$

in suitable Sobolev spaces, whose Galerkin discretization is given by the matrix

$$T_h[\ell, k] = \langle T\varphi_k^\mu, \varphi_\ell^\mu \rangle$$

for $k, \ell = 1, \ldots, M$. If

$$\begin{aligned} A_h[j,i] &= \langle A\varphi_i^\nu, \varphi_j^\nu \rangle, \\ B_h[j,k] &= \langle B\varphi_k^\mu, \varphi_j^\nu \rangle \end{aligned}$$

are given for $i, j = 1, \ldots, N, k = 1, \ldots, M$ and if A is positive definite, then we have the upper spectral equivalence inequality [19]

$$(\tilde{T}_h \underline{u}, \underline{u}) \leq (T_h \underline{u}, \underline{u}) \tag{16}$$

for all $\underline{u} \in I\!R^M$. From this it follows directly, that the Schur complement S_h is spectrally equivalent to the Galerkin discretization of the hypersingular integral operator. Hence we can construct a preconditioner for the Schur complement as a preconditioner for D_h only.

For the lower spectral equivalence inequality

$$\gamma_1 \cdot (T_h \underline{u}, \underline{u}) \leq (\tilde{T}_h \underline{u}, \underline{u}) \tag{17}$$

we have to assume the BBL condition for the trial spaces V_h, W_h and for the operator B, i.e.

$$c_1^B \cdot ||u_h||_{W_h} \leq \sup_{\tau_h \in V_h} \frac{|\langle B u_h, \tau_h \rangle|}{||\tau_h||_{V_h}} \tag{18}$$

for all $u_h \in W_h$, where the constant c_1^B is independent on h. Note, that the lower spectral equivalence inequality (17) and therefore the condition (18) is not necessary to establish the spectral equivalence of S_h and D_h.

Following [17,19] we now want to give a general technique to construct preconditioners for the Galerkin discretization of a selfadjoint, positive definite and bounded pseudodifferential operator

$$A : H^s(\Gamma) \to H^{s-2\alpha}(\Gamma) . \tag{19}$$

For example, let A be the single layer potential operator V with $s = -\frac{1}{2}$ and $2\alpha = -1$. Then one has to suppose, that there is given a pseudodifferential operator of opposite order,

$$B : H^{s-2\alpha}(\Gamma) \to H^s(\Gamma),$$

which is again selfadjoint, positive definite and bounded. From the properties of the pseudodifferential operators then follow immediately the spectral equivalence inequalities

$$c_1 \cdot \langle B^{-1}u, u \rangle_{H^{s-\alpha}(\Gamma)} \leq \langle Au, u \rangle_{H^{s-\alpha}(\Gamma)} \leq c_2 \cdot \langle B^{-1}u, u \rangle_{H^{s-\alpha}(\Gamma)}$$

for all $u \in H^s(\Gamma)$. Furtheron, the Galerkin discretization

$$C_h[k, \ell] = \langle B^{-1}\varphi_k^\nu, \varphi_\ell^\nu \rangle_{H^{s-\alpha}(\Gamma)} \tag{20}$$

is spectrally equivalent to the Galerkin discretization A_h of the operator A with the same constants c_1 and c_2, unfortunately, the operator B^{-1} admits no explicit representation in general.

Since the operator to be discretized is of the form (15) we can define the approximation

$$\tilde{C}_h = M_h^\top B_h^{-1} M_h \tag{21}$$

with

$$M_h[j, k] = \langle \varphi_k^\nu, \varphi_j^\mu \rangle_{H^{s-\alpha}(\Gamma)}$$
$$B_h[j, i] = \langle \varphi_i^\mu, \varphi_j^\mu \rangle_{H^{s-\alpha}(\Gamma)}$$

for $k = 1, \ldots, N, i, j = 1, \ldots, M$.

To ensure the spectral equivalence of C_h and $\tilde{C}_h$, one has to assume condition (18), i.e. the invertibility of the mass matrix M_h. Hence one has to set $N = M$ and to choose the trial functions used in an appropriate way to ensure the existence of M_h^{-1}. If we take $\nu = \mu$, i.e. piecewise linear trial functions for the discretization of both operators, the mass matrix M_h will be diagonal dominant and therefore invertible. Moreover, M_h^{-1} can be approximated very well using some steps of a classical iterative method like the Jacobi iteration, since M_h is also sparse. The action of the proposed preconditioner is now given by

$$\tilde{C}_h^{-1} = M_h^{-1} B_h M_h^{-\top} .$$

The remaining question here is the right choice of the operator B, since the hypersingular integral operator is positive definite with respect to the semi-norm $|\cdot|_{H^{1/2}(\Gamma)}$, the operator

$$B = \alpha \cdot I + D$$

will be positive definite for any positive α. Now, the preconditioner for the discrete single layer potential is given by

$$C_h^{-1} = M_h^{-1}(M_h + D_h)M_h^{-\top} .$$

Note, that in the symmetric formulation both boundary integral operators have to be discretized and therefore, they are available in the preconditioning process without any additional work. Of course, we can use the same technique to construct a preconditioner for the discrete hypersingular operator D_h by

$$C_S^{-1} = M_h^{-1} V_h M_h^{-\top} .$$

In Table 3.1 there are given numerical results for both proposed preconditioners in the case of the two-dimensional Laplace equation, if Γ is the boundary of the L-shape domain $\Omega = [0, 0.5]^2 \backslash [0.25]^2$. All boundary integral operators are discretized by a Galerkin scheme with piecewise linear trial and test functions.

Table 1: Numerical results for the proposed preconditioners

	V_h			D_h		
N	$\kappa(C_V^{-1}V_h)$	Iter	sec	$\kappa(C_S^{-1}D_h)$	Iter	sec
64	1.90	8	0.01	1.77	8	0.02
128	1.94	8	0.04	1.80	8	0.05
256	1.99	9	0.20	1.82	8	0.18
512	2.03	9	0.96	1.82	8	0.86
1024	2.07	9	3.70	1.83	8	3.31

Here, $\kappa(C_V^{-1}V_h)$ denotes an approximate bound of the spectral condition number of the preconditioned system, while Iter and sec are the number of iterations and the time needed to get a relative error reduction of $\varepsilon = 10^{-8}$ in the cg iteration scheme.

The approach described is based on a concept of pseudodifferential operators of opposite orders. Hence, this technique is generally usable for preconditioning discrete boundary integral operators, how they appear in the symmetric formulation in a natural way. However, one has to fulfill some assumptions to the degrees of the used trial functions to ensure the invertibility of the mass matrix.

Note, that the proposed preconditioner depends not on the mesh grading, therefore, our method is stable for adaptive meshes. Moreover, there is no restriction to the space dimension and to the structure of Γ, i.e. this method can be applied also, if Γ is only a part of a closed surface, as it appears for mixed boundary value problems.

A second approach to construct efficient preconditioners for discrete boundary integral operators on a curve is based on circulant matrices arising by the corresponding discretization on a circle [10]. These matrices may be handled by the fast fourier transformation with optimal order of work. However, for mixed boundary value problems, the corresponding block matrices are no longer circulant, but they are of Toeplitz structure, hence we can use the Levinson algorithm [4] to solve a preconditioned system of dimension m with $O(m^2)$ multiplications, which is in the same order as a matrix multiplication.

A NUMERICAL EXAMPLE

As a simple model problem we consider the mixed boundary value problem in plane linear elasticity,

$$\left.\begin{aligned} -\Delta u(x) - (\lambda+\mu)\operatorname{grad}\operatorname{div} u(x) &= 0 && \text{for } x \in \Omega, \\ u(x) &= 0 && \text{for } x \in \Gamma_D, \\ t(x) &= h(x) && \text{for } x \in \Gamma_N^1 \\ t(x) &= 0 && \text{for } x \in \Gamma_N^2 = \Gamma \backslash (\Gamma_D \cup \Gamma_N^1) \end{aligned}\right\} \tag{22}$$

where the domain and the boundary conditions are given as in Figure 1:

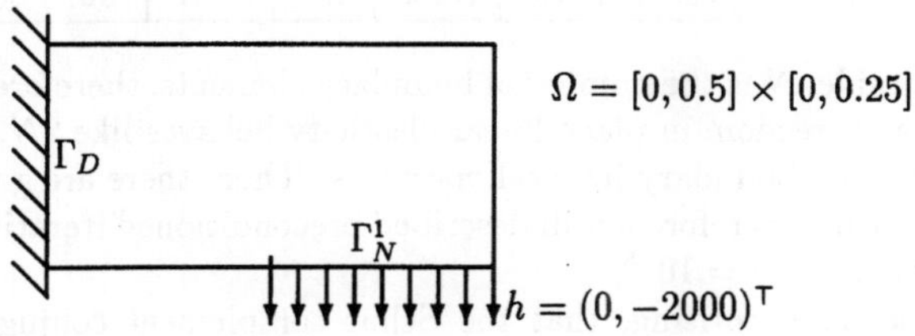

Figure 1: *Domain and boundary conditions*

λ and μ are the Lamé constants given by

$$\lambda = \frac{E\nu}{(1+\nu)(1-2\nu)}, \quad \mu = \frac{E}{2(1+\nu)}$$

from Youngs modulus $E = 200\,000$ and Poissons ratio $\nu = 0.3$.

The fundamental solution to the differential operator in (22) is given by the Kelvin solution

$$U^*(x,y) = \frac{\lambda+\mu}{4\pi\mu(\lambda+2\mu)}\left[-\frac{\lambda+3\mu}{\lambda+\mu}\delta_{k\ell}\log|x-y| + \frac{(x_k-y_k)(x_\ell-y_\ell)}{|x-y|^2}\right]$$

for $k, \ell = 1, 2$. For the bilinear form of the corresponding hypersingular integral operator we have the representation [9]

$$\langle Du, v\rangle_{L^2(\Gamma)} = -\sum_{k,\ell=1}^{2}\int_\Gamma\int_\Gamma G_{ij}(x-y)\frac{d}{ds_x}u_k(x)\,\frac{d}{ds_y}v_\ell(y)\,ds_x ds_y$$

with

$$G_{k\ell}(x-y) = \frac{\mu(\lambda+\mu)}{\pi(\lambda+2\mu)}\left[\delta_{k\ell}\log|x-y| - \frac{(x_k-y_k)(x_\ell-y_\ell)}{|x-y|^2}\right].$$

Hence, the Galerkin discretization of D with piecewise linear trial functions can be done using the Galerkin discretization of the single layer potential V with piecewise constant trial functions, only the constants in the assembling process have to be changed. A description of a full analytic Galerkin approximation scheme for polygonal bounded domains is given in [16]. Here we used piecewise constant trials for the boundary stress approximation and piecewise linear trial functions for the boundary displacements. As preconditioners we use the block diagonal circulant matrices generated by corresponding integral operators on a circle [10]. To scale the preconditioners as required we use a gradient type method to get bounds for the minimal and maximal eigenvalues of the preconditioned system [11].

Table 2: Iterative solvers for the symmetric formulation

	Matrix	Schur CG		BiCGStab		BPCG		CGT	
N	sec	Iter	sec	Iter	sec	Iter	sec	Iter	sec
24	0.08	14	0.08	17	0.09	22	0.09	24	0.09
48	0.28	16	0.24	19	0.30	24	0.25	26	0.28
96	1.13	17	0.87	22	1.33	26	1.04	27	1.13
192	4.57	18	3.77	22	5.24	27	4.48	29	4.82
384	18.99	19	19.42	24	25.53	29	21.41	30	22.19
768	76.74	19	75.38	24	98.78	30	86.91	31	89.83

In this table, N is the number of boundary elements, therefore, the number of all unknown degrees of freedom in plane linear elasticity behaves like $2N$, Matrix denotes the time to discretize all boundary integral operators. Then, there are given the number of iterations and the time therefore for all described preconditioned iterative solvers, where the relative error bound is $\varepsilon = 10^{-8}$.

It seems surprising, that the Schur complement conjugate gradient method is the best one, however, this may depend strongly on the example and the preconditioners available. Moreover, there appear some questions how to scale the preconditioners to use the transformation methods. Numerical examples for the nonsymmetric formulation of a mixed boundary value problem for the Laplace equation show, that the BiCGStab algorithm applied to the transformed matrix $\hat{M}_h$ would be the best method. In general, all methods described here, are comparable.

Figure 2 shows the deformed body resulting from the mixed boundary value problem (22).

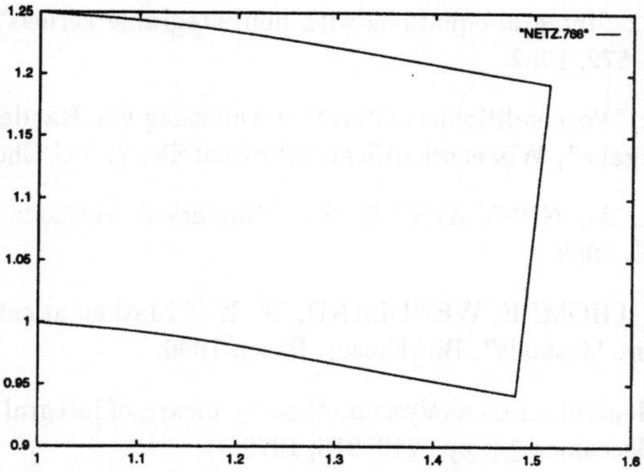

Figure 2: *Displacement field for 768 boundary elements*

All numerical examples were computed on a workstation with an experimentally computed Mflops rate of 1.925.

REFERENCES

[1] BRAMBLE, J. H., PASCIAK, J. E.: "A preconditioning technique for indefinite systems resulting from mixed approximations of elliptic problems", Math. Comput. 50, pp. 1–17, 1988.

[2] COSTABEL, M.: "Symmetric methods for the coupling of finite elements and boundary elements", In: Boundary Elements IX (C. A. Brebbia, G. Kuhn, W. L. Wendland eds.), Springer–Verlag, Berlin, pp. 411–420, 1987.

[3] COSTABEL, M.: "Boundary integral operators on Lipschitz domains: Elementary results", SIAM J. Math. Anal. 19, no. 3, pp. 613–626, 1988.

[4] GOLUB, G. H., VAN LOAN, C. F.: "Matrix Computations", John Hopkins University Press, Baltimore, 1989.

[5] HSIAO, G. C., STEPHAN, E. P., WENDLAND, W. L.: "On the integral equation method for the plane mixed boundary value problem of the Laplacian", Math. Meth. Appl. Sc. 1, pp. 265–321, 1979.

[6] HSIAO, G. C., WENDLAND, W. L.: "A finite element method for some integral equations of the first kind", J. Math. Anal. Appl. 58, pp. 449–481, 1977.

[7] LANGER, U.: "Parallel iterative solution of symmetric coupled fe/be–equations via domain decomposition", Contemp. Math. 157, pp. 335–344, 1994.

[8] LIONS, J. L., MAGENES, E.: "Non-homogeneous boundary value problems and applications I", Springer-Verlag, Berlin, 1972.

[9] NEDELEC, J. C.: "Integral equations with non integrable kernels", J. Int. Eqns. Oper. Ther. 5, pp. 562–572, 1982.

[10] RJASANOW, S.: "Vorkonditionierte iterative Auflösung von Randelementgleichungen für die Dirichlet-Aufgabe", Wissenschaftliche Schriftenreihe 7, TU Chemnitz, 1990.

[11] SAMARSKIJ, A. A., NIKOLAEV, E. S.: "Numerical Methods for Grid Equations", Birkhäuser, Basel, 1989.

[12] SCHATZ, A. H., THOMÉE, WENDLAND, W. E.: "Mathematical Theory of Finite and Boundary Element Methods", Birkhäuser, Basel, 1990.

[13] SIRTORI, S.: "General stress analysis method by means of integral equations and boundary elements", Meccanica 14, pp. 210–218, 1979.

[14] STEINBACH, O.: "Parallel iterative solvers for symmetric boundary element domain decomposition methods", In: Fast Solvers for Flow Problems (W. Hackbusch, G. Wittum eds.), Vieweg-Verlag, Braunschweig, pp. 263–272, 1995.

[15] STEINBACH, O.: "Theorie und Algorithmen für Iterationsverfahren zur Lösung linearer Gleichungssysteme", Preprint 96–4, Universität Stuttgart, 1996.

[16] STEINBACH, O.: "Galerkin- und Kollokationsdiskretisierungen für Randintegralgleichungen in 2D. Dokumentation", Preprint 96-5, Universität Stuttgart, 1996.

[17] STEINBACH, O., WENDLAND, W. L.: "Efficient preconditioners for boundary element methods and their use in domain decomposition methods", To appear in: The Proceedings of the Eighth International Conference on Domain Decomposition Methods in Science and Engineering, Peking, 1995.

[18] STEINBACH, O.: "Fast solvers for boundary integral equations related to mixed boundary value problems", In preperation.

[19] STEINBACH, O.: "The construction of optimal preconditioners in the boundary element method", In preperation.

[20] VAN DER VORST, H. A.: "Bi-CGStab: A fast and smoothly converging variant of Bi-CG for the solution of nonsymmetric linear systems", SIAM J. Sci. Stat. Comput. 13, no. 2, pp. 631–644, 1992.

A TWO GRID METHOD FOR COUPLING FEM AND BEM IN ELASTICITY

Karsten Türke and Eckart Schnack
Institute of Solid Mechanics
University of Karlsruhe (TH)
Kaiserstraße 12
76128 Karlsruhe, Germany

SUMMARY

This paper presents a coupling algorithm of FEM and BEM for solving mixed boundary value problems in elastostatics. A domain decomposition of the bounded domain Ω leads to a basic substructure Ω_1 with a skeleton H, where the FEM is applied, and several macroelements $\Omega_i \quad (i = 2, \ldots, p)$, where the BEM is used on a fine grid h.
As fundamental equations the well known energy bilinear form is used for the whole domain Ω and, additionally, a second bilinear form on the macroelements Ω_i as a coupling condition. This way a symmetric and nonconforming (i.e. different, independent grids H and h) coupling algorithm can be obtained. This can be realized by applying the *Poincaré–Steklov* operator on the macroelement surfaces in the strong singular form and by avoiding the use of the hypersingular boundary integral equation.
The construction of a robust and reliable numerical algorithm depends on the adaptive control of symmetry and definiteness of the coupling matrix. Therefore we use an iterative method for solving the boundary integral equation by expanding the *Calderon* projector in a *Neumann* series. Finally the convergence of this expansion is proved, on the basis of the fundamental work of *Kupradze* et.al. [13].
Numerical results in 2D and 3D will show the preciseness and efficiency of the method.

INTRODUCTION

In this paper a special strategy for coupling the Finite Element Method (FEM) and the Boundary Element Method (BEM) is presented. The aim is to find precise and efficient numerical solutions for elliptic boundary value problems. In particular, we want to resolve local zones with high gradients for problems in linear elasticity. In this field the BEM offers the possibility to approximate very accurately high stress concentrations in compact domains with a low number of degrees of freedom. Of special interest is the so called macro–element technique where the BEM substructures are interpreted as special finite elements. This strategy allows the implementation of the coupling method as an extension module in existing FEM software.

CONTINUOUS FORMULATION

We consider an isotropic interior domain Ω in $\mathbf{R}^m$, $(m = 2, 3)$, with a *Lipschitz* boundary $\Gamma = \partial\Omega$. As a boundary value problem, we analyze the *Lamé* system of linear elasticity with mixed boundary conditions. This leads to the *Navier* equation

$$\Delta^* \boldsymbol{u} = 0 \qquad \text{for } \boldsymbol{\xi} \in \Omega \tag{1}$$

as the describing differential equation for the displacement field $\boldsymbol{u} \in [H^1(\Omega)]^m$. Δ^* denotes an elliptic partial differential operator. As boundary conditions we have given displacements on the *Dirichlet* boundary Γ_u as well as tractions on the *Neumann* boundary Γ_t:

$$\boldsymbol{u}(\boldsymbol{\xi}) = \bar{\boldsymbol{u}}(\boldsymbol{\xi}) \qquad \text{for } \boldsymbol{\xi} \in \Gamma_u = \Gamma \setminus \Gamma_t \tag{2}$$

$$\boldsymbol{t}(\xi) = \Pi\, \boldsymbol{u}(\xi) = \bar{\boldsymbol{t}}(\xi) \qquad \text{for } \boldsymbol{\xi} \in \Gamma_t \quad . \tag{3}$$

Π denotes the traction operator.
According to an initial guess we decompose the structure Ω in $i = 1, \ldots, p$ nonoverlapping subdomains Ω_i

$$\Omega \subset \bigcup_{i=1}^{p} (\Omega_i \cup \Gamma_i) \quad , \quad \Omega_i \cap \Omega_j = \emptyset \quad \text{for } i \neq j \quad , \quad \Gamma_i = \partial\Omega_i \tag{4}$$

For the sake of simplicity we restrict ourselves to the case of $p = 2$ and define

$$\Gamma_c = \partial\Omega_1 \cap \partial\Omega_2 \tag{5}$$

as the coupling interface (see Fig. 1).
The values on Γ_c coming from Ω_2 will be indicated with a "+" sign and from Ω_1 with a "-" sign.

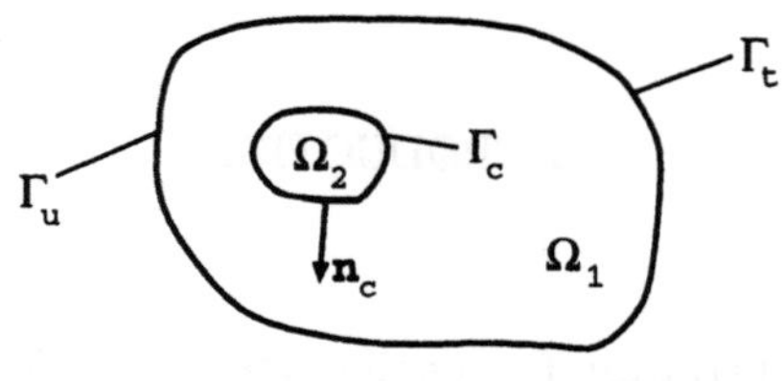

Figure 1: Decomposed domain Ω

The basis of the following coupling strategy is given by a hybrid stress method, formulated 1973 by *Schnack* [1] and *Tong, Pian* and *Lasry* [2]. It realizes a mixed FE method by

the transformation of domain integrals in boundary integrals. 1984 *Schnack* [3] carried out a generalization of this method by incorporating the boundary integral equation in the procedure. In the following years further developments and applications on two and three dimensional problems in elasticity were published by *Schnack* [4], [5], *Carmine* [6] and *Karaosmanoglu* [7].
As basic equations for the coupling strategy we have at first the normal variational formulation of the boundary value problem for Ω_1 and Ω_2:

$$a_1(\boldsymbol{u}_1, \boldsymbol{v}_1) + a_2(\boldsymbol{u}_2, \boldsymbol{v}_1) = f(\boldsymbol{v}_1) \qquad \forall \quad \boldsymbol{v}_1 \in [H_0^1(\Omega)]^m \tag{6}$$

with the semi-definite energy bilinear form

$$a_i(\boldsymbol{u}, \boldsymbol{v}) = \int\limits_{\Omega_i} \sigma_{ij}(\boldsymbol{u})\, e_{ij}(\boldsymbol{v})\, d\Omega \tag{7}$$

and the boundary integral f over the *Neumann* boundary Γ_t

$$f(\boldsymbol{v}_1) = \int\limits_{\Gamma_t} \tilde{\boldsymbol{t}}^T \boldsymbol{v}_1\, d\Gamma \quad . \tag{8}$$

The test functions $\boldsymbol{v}_1$ in Eq. (6) are defined in the whole domain Ω. Additionally we use a second variational principle as a coupling equation:

$$a_2(\boldsymbol{v}_2, \boldsymbol{u}_2) = g(\boldsymbol{v}_2) \qquad \forall \quad \boldsymbol{v}_2 \in [H_0^1(\Omega_2)]^m \tag{9}$$

with test functions $\boldsymbol{v}_2$ defined only in the subdomain Ω_2. The boundary integral

$$g(\boldsymbol{v}_2) = \int\limits_{\Gamma_c} \tilde{\boldsymbol{t}}_+^T \boldsymbol{u}_-\, d\Gamma \qquad \text{with } \tilde{\boldsymbol{t}}_+ = \Pi\, \boldsymbol{v}_2 \tag{10}$$

combines the tractions coming from Ω_2 and the displacements coming from Ω_1. Eq. (10) leads to a variational compatibility conditions and offers the possibility to use nonconforming grids on the discrete level. With the side condition that Ω_2 exactly represents an equilibrium system:

$$\sigma_{ij,i} = 0 \qquad \forall \quad \xi \in \Omega_2 \ , \tag{11}$$

we can transform the bilinear forms a_2 in (6) and (9) into boundary integrals over Γ_c by using *Green*'s theorem. Finally the kinematic and the static fields $\boldsymbol{u}_+$ and $\boldsymbol{t}_+$ on Γ_c coming from Ω_2 are coupled by the *Poincaré–Steklov* operator (cf. *Costabel* and *Wendland* [8])

$$\boldsymbol{u}_+ = \Pi^{-1} \boldsymbol{t}_+ = \left[\left(\frac{1}{2}\boldsymbol{I} + \boldsymbol{K} \right)^{-1} \boldsymbol{V} \right] \boldsymbol{t}_+ \quad . \tag{12}$$

In Eq. (12) $\boldsymbol{V}$ denotes the weak singular integral operator and $\boldsymbol{K}$ the strong singular integral operator derived from *Kelvin*'s fundamental solution.

DISCRETE FORMULATION

For the discrete formulation of the coupling method we start with the coarse grid H. As an illustration, in Fig. 2 a plate with a circular notch is depicted. The substructure

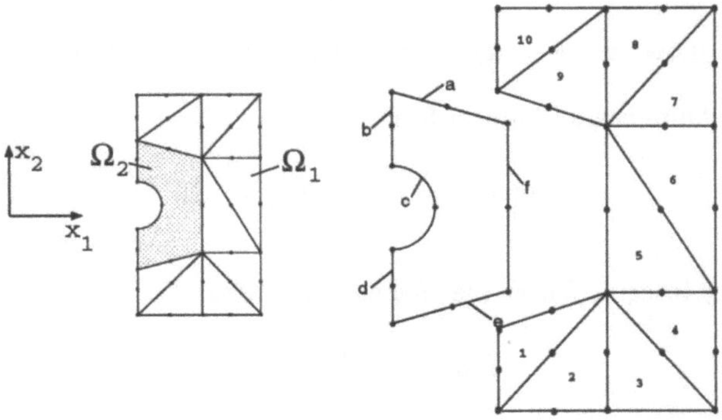

Figure 2: 2D problem with coarse grid H

Ω_2 is defined in the near field of the notch. The remaining part Ω_1 is discretized with triangular finite elements using a displacement ansatz

$$\boldsymbol{u}_H = \boldsymbol{N}_H\,\boldsymbol{\alpha} \qquad \text{in } \Omega_1 \tag{13}$$

with polynomial shape functions $\boldsymbol{N}_H$ and the discrete parameters α. As a subset of (13) we obtain an equivalent displacement ansatz

$$\boldsymbol{u}_H = \boldsymbol{N}_H\,\boldsymbol{\alpha}_c \qquad \text{on } \Gamma_c \tag{14}$$

on the interface (in Fig. 2: $\Gamma_c = a \cup b \cup c \cup d \cup e \cup f$) where $\boldsymbol{\alpha}_c$ denotes the displacements of the coarse grid nodes on the coupling interface.
Independent of (14) we define an equilibrium ansatz for the stress tensor $\boldsymbol{\sigma}_2$ in Ω_2 with

$$\left(\sigma_2\right)_{ij,i} = 0 \tag{15}$$

and obtain a traction ansatz

$$\boldsymbol{t}_H = \boldsymbol{R}_H\,\boldsymbol{\beta} \qquad \text{on } \Gamma_c \tag{16}$$

on the coupling interface with discrete parameters $\boldsymbol{\beta}$ that are independent of $\boldsymbol{\alpha}_c$. For details concerning the construction of the static equilibrium system (16) we refer to *Türke* [9].
On the interface Γ_c a second grid h is defined, which is independent of H (cf. Fig. 3).
We can transform the traction ansatz $\boldsymbol{t}_H$ on the coarse grid into $\boldsymbol{t}_h$ on the fine grid with the same parameters $\boldsymbol{\beta}$:

$$\boldsymbol{t}_H = \boldsymbol{R}_H\,\boldsymbol{\beta} \quad \Longrightarrow \quad \boldsymbol{t}_h = \boldsymbol{R}_h\,\boldsymbol{\beta} \quad . \tag{17}$$

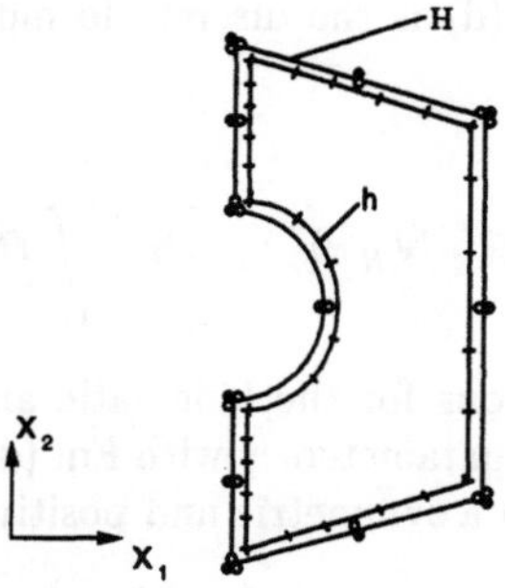

Figure 3: Nonconforming grids h and H

In Eq. (17) $\boldsymbol{t}_h$ is a discretization of $\boldsymbol{t}_H$ according to the fine grid h. The corresponding *Dirichlet* data $\boldsymbol{u}_h$ can be computed with the *Poincaré–Steklov* operator (12). With the initial formulation

$$\boldsymbol{u}_h = \boldsymbol{N}_h \boldsymbol{\gamma} \quad , \tag{18}$$

the discretization of the *Poincaré–Steklov* operator leads to

$$\boldsymbol{A}\boldsymbol{\gamma} = \boldsymbol{B}\boldsymbol{\beta} \quad . \tag{19}$$

In Eq. (18) $\boldsymbol{N}_h$ denotes the matrix of shape functions for the kinematic field on the fine grid h, and $\boldsymbol{\gamma}$ is the vector of the discrete nodal displacements on h. Summarized we obtain the following algorithm: In the first step a mixed ansatz $(\boldsymbol{u}_H, \boldsymbol{t}_H)$ on the coarse grid H is defined. The second step realizes the transformation of the *Neumann* data from the coarse to the fine grid $(\boldsymbol{t}_H \longrightarrow \boldsymbol{t}_h)$. In the third step, the fine grid problem is solved by the boundary integral equation $(\boldsymbol{t}_h \longrightarrow \boldsymbol{u}_h)$. We obtain static and kinematic fields on the fine and the coarse grid that can be inserted as test and trial functions in the coupling equation (9). This leads in the fourth step to a weak compatibility condition on the interface Γ_c

$$\tilde{\boldsymbol{\beta}}^T \int_{\Gamma_c} \boldsymbol{R}_H^T \boldsymbol{N}_H \, d\Gamma \, \boldsymbol{\alpha}_c = \tilde{\boldsymbol{\beta}}^T \int_{\Gamma_c} \boldsymbol{R}_h^T \boldsymbol{N}_h \, d\Gamma \, \boldsymbol{\gamma} \quad . \tag{20}$$

Finally we obtain from the variational principle (6) the global system

$$\int_{\Omega_1} \boldsymbol{\sigma}^T(\boldsymbol{u}_H)\, \boldsymbol{e}(\boldsymbol{v}_H)\, d\Omega + \int_{\Gamma_c} \boldsymbol{t}_H^T \boldsymbol{v}_H \, d\Gamma = \int_{\Gamma_t} \bar{\boldsymbol{t}}^T \boldsymbol{v}_H \, d\Gamma \quad , \tag{21}$$

defined only on the coarse grid H with a domain integral over Ω_1 and an interface integral belonging to Ω_2.
These equations can be assembled to the following block structure

$$\begin{bmatrix} \boldsymbol{K}_{ii} & \boldsymbol{K}_{ic} & \boldsymbol{0} & \boldsymbol{0} \\ \boldsymbol{K}_{ci} & \boldsymbol{K}_{cc} & \boldsymbol{T}^T & \boldsymbol{0} \\ \boldsymbol{0} & \boldsymbol{T} & \boldsymbol{0} & -\boldsymbol{S} \\ \boldsymbol{0} & \boldsymbol{0} & \boldsymbol{B} & -\boldsymbol{A} \end{bmatrix} \begin{bmatrix} \boldsymbol{\alpha}_i \\ \boldsymbol{\alpha}_c \\ \boldsymbol{\beta} \\ \boldsymbol{\gamma} \end{bmatrix} = \begin{bmatrix} \boldsymbol{F}_i \\ \boldsymbol{F}_c \\ \boldsymbol{0} \\ \boldsymbol{0} \end{bmatrix} \begin{matrix} (a) \\ (b) \\ (c) \\ (d) \end{matrix} \tag{22}$$

where the rows (a) and (b) denote the global system, (c) is the coupling equation connecting the displacement parameters $\boldsymbol{\alpha}_c$ on the coarse grid with the displacement parameters

γ on the fine grid, and finally (d) is the discrete formulation of the boundary integral equation.
The operators $\boldsymbol{T}$ and $\boldsymbol{S}$ with

$$\boldsymbol{T} = \int_{\Gamma_c} \boldsymbol{R}_H^T \boldsymbol{N}_H \, d\Gamma \quad ; \quad \boldsymbol{S} = \int_{\Gamma_c} \boldsymbol{R}_h^T \boldsymbol{N}_h \, d\Gamma \tag{23}$$

are integrals of the trial functions for the kinematic and static fields. In a sequential procedure we can eliminate the parameters $\boldsymbol{\gamma}$ with Eq. (d), then eliminate the parameters $\boldsymbol{\beta}$ with Eq. (c) and end up with a symmetric and positive definite global system

$$\begin{bmatrix} \boldsymbol{K}_{ii} & \boldsymbol{K}_{ic} \\ \boldsymbol{K}_{ci} & \overline{\boldsymbol{K}}_{cc} \end{bmatrix} \begin{bmatrix} \boldsymbol{\alpha}_i \\ \boldsymbol{\alpha}_c \end{bmatrix} = \begin{bmatrix} \boldsymbol{F}_i \\ \boldsymbol{F}_c \end{bmatrix} \tag{24}$$

in dependence on the displacement parameters $\boldsymbol{\alpha}$ on the coarse grid H. The equivalent stiffness operator for the subdomain Ω_2 is defined by

$$\overline{\boldsymbol{K}}_{cc} = \boldsymbol{K}_{cc} + \boldsymbol{T}^T \boldsymbol{H}^{-1} \boldsymbol{T} \quad . \tag{25}$$

From the elimination process we can obtain symmetry and positive definiteness of the coupling operator

$$\boldsymbol{H} = \boldsymbol{S} \boldsymbol{A}^{-1} \boldsymbol{B} \tag{26}$$

with a previously defined accuracy. This is realized by using the expansion of the fine grid problem in a *Neumann* series (cf. next chapter).
A proof of the existence and the uniqueness for the solution of the global system (24) is given by *Hsiao, Schnack* and *Wendland* [10] under the condition that the solution of the local *Neumann* problem on the fine grid (cf. Eq. (d) in the block structure (22)) is stable. This is guaranteed for a sufficiently small grid parameter h and using a *Galerkin* approximation or a *Neumann* series for the numerical treatment of the *Poincaré–Steklov* operator.

A NEUMANN SERIES FOR THE POINCARÉ–STEKLOV OPERATOR

We reformulate the *Neumann* problem (12) for a general interior domain Ω_+ with the *Calderon* projector

$$\boldsymbol{u}(\boldsymbol{\xi}) = (\frac{1}{2}\boldsymbol{I} - \boldsymbol{K})\, \boldsymbol{u}(\boldsymbol{\xi}) + \boldsymbol{V}\, \boldsymbol{t}_+(\boldsymbol{\xi}) \quad \forall \quad \boldsymbol{\xi} \in \Gamma_c = \partial\Omega_+ \tag{27}$$

with given tractions $\boldsymbol{t}_+(\boldsymbol{\xi})$ and unknown displacements $\boldsymbol{u}(\boldsymbol{\xi})$. If the solution $\boldsymbol{u}(\boldsymbol{\xi})$ in Ω_+ is found, it can be continuously extended to the exterior domain $\Omega_- = \mathbf{R}^m \backslash \overline{\Omega_+}$ with boundaries at infinity. For the construction of an algorithm to calculate the solution of (27) by iteration we start with some definitions:
The linear function space X is defined as the space of elastic states that includes all stress and deformation states in the interior domain Ω_+ and in the exterior domain Ω_- caused by a loading of Ω_+ on $\partial\Omega_+$ with an admissible traction field $\boldsymbol{t}_+$. Admissible loadings are

defined by the side condition (11).
We call the static field $\boldsymbol{\sigma}(\boldsymbol{\xi})$ and the kinematic field $\boldsymbol{u}(\boldsymbol{\xi})$ with $\boldsymbol{\xi} \in \overline{\Omega}_+ \cup \Omega_-$ the representatives of the elastic states: $\boldsymbol{g} \in X$.
A linear manifold $[\boldsymbol{g}]$ of these elements $\boldsymbol{g}$ is defined by adding an arbitrary constant $\boldsymbol{c}$, in our problem given by rigid body motions: $[\boldsymbol{g}] = \{\boldsymbol{g} + \boldsymbol{c} \,:\, \boldsymbol{c} \in X'\}$. As a consequence the solution is defined in the factor space $\mathring{X} = X \,/\, X'$, if the space of elastic states is represented by the displacement field $\boldsymbol{u}(\boldsymbol{\xi})$. *Rieder* [11] has shown that the linear spaces X and $\mathring{X}$ are real *Banach* spaces. For the explicit algorithmic construction of the factor space $\mathring{X}$ we refer to *Türke* [9].
A general formulation of the *Calderon* projector (27) is given by

$$\boldsymbol{g} - \lambda \mathcal{T} \boldsymbol{g} = \boldsymbol{V} \boldsymbol{t}_+ \tag{28}$$

with $\lambda = +1$ and $\mathcal{T} = \left(\frac{1}{2}\boldsymbol{I} - \boldsymbol{K}\right)$. If we divide Eq. (28) by λ, we obtain

$$\boldsymbol{g}^* = (\mu - \mathcal{T})\, \boldsymbol{g} \tag{29}$$

with the scalar value $\mu = 1/\lambda$ and the inhomogenious part $\boldsymbol{g}^* = \mu \, \boldsymbol{V} \boldsymbol{t}_+$. For solving Eq. (29) we use a fundamental theorem from functional analysis (cf. *Taylor* [12]) for the *Banach* space $\mathring{X}$:

Theorem 1: If $\mathcal{T} \in [\mathring{X}]$, the resolvent $R_\mu = (\mu - \mathcal{T})^{-1}$ is given by

$$R_\mu = \sum_{k=1}^{\infty} \mu^{-k} \mathcal{T}^{k-1} \quad , \tag{30}$$

if $|\mu|$ is larger than the spectral radius $r_\sigma(\mathcal{T})$.

In our case with $\mu = +1$ we obtain convergence for the *Neumann* series

$$\boldsymbol{g} = (1 - \mathcal{T})^{-1} \boldsymbol{g}^* = \sum_{k=1}^{\infty} \mathcal{T}^{k-1} \boldsymbol{g}^* \quad , \tag{31}$$

if the spectral radius r_σ is smaller than "1" or if all eigenvalues λ_i of the homogenious part of Eq. (28) are larger than "1". The properties of the eigenvalues of the integral operator $\mathcal{T}$ were investigated by *Kupradze* et.al. [13] and can be summarized in

Theorem 2:
- The eigenvalues λ_i are real.
- The eigenvalues λ_i are simple poles of the resolvent.
- The eigenvalues λ_i are not smaller than unity in absolute value:

$$|\lambda_i| = \frac{a_+(\boldsymbol{u}, \boldsymbol{u}) + a_-(\boldsymbol{u}, \boldsymbol{u})}{|a_-(\boldsymbol{u}, \boldsymbol{u}) - a_+(\boldsymbol{u}, \boldsymbol{u})|} \geq 1 \quad . \tag{32}$$

According to the representation formula (32) the value $\lambda = +1$ can only be obtained, if the stress–strain energy a_+ or a_- in Ω_+ or Ω_- is zero, i.e. in the case of a rigid body motion. In the factor space $\mathring{X}$ the rigid body motion is defined as a zero element, so that $\lambda = +1$ is a regular value of our problem. Because the eigenvalues are isolated, the inequality

$$\lambda_i \geq \lambda_{i-1} \geq \ldots \geq \lambda_1 > 1 \tag{33}$$

is valid and the convergence for the *Neumann* series (31) is guaranteed. This is a similar argumentation to *Haack* and *Wendland* [14] for the convergence of the *Neumann* series for the *Laplace* equation.
As a consequence we can calculate the displacement vector $\boldsymbol{u}$ on $\partial\Omega_+$ as the limit of the approximations

$$\boldsymbol{u}(\boldsymbol{\xi}) = \lim_{m\to\infty} \boldsymbol{u}^{(m)}(\boldsymbol{\xi}) = \sum_{k=1}^{\infty} (\frac{1}{2}\boldsymbol{I} - \boldsymbol{K})^{k-1} \boldsymbol{V}\, \boldsymbol{t}_+(\boldsymbol{\xi}) \quad \forall \quad \boldsymbol{\xi} \in \partial\Omega_+ \quad , \tag{34}$$

where an improved approximation $\boldsymbol{u}^{(m)}(\boldsymbol{\xi})$ can be calculated by the recursive scheme

$$\boldsymbol{u}^{(m)}(\boldsymbol{\xi}) = (\frac{1}{2}\boldsymbol{I} - \boldsymbol{K})\, \boldsymbol{u}^{(m-1)}(\boldsymbol{\xi}) + \boldsymbol{V}\, \boldsymbol{t}_+(\boldsymbol{\xi}) \tag{35}$$

with the trivial start solution

$$\boldsymbol{u}^{(0)} = \boldsymbol{0} \quad . \tag{36}$$

NUMERICAL RESULTS FOR A 3D NOTCH PROBLEM

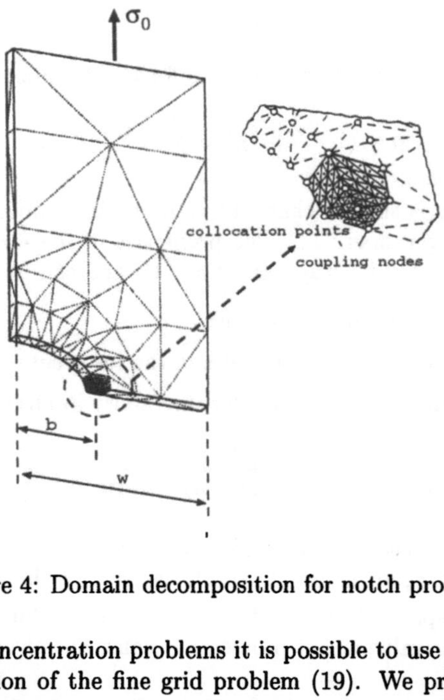

Figure 4: Domain decomposition for notch problem

For regular stress concentration problems it is possible to use the collocation method for the numerical solution of the fine grid problem (19). We present numerical results for the simulation of a 3D plate with an elliptical cutout under uniform tension σ_0. In

Fig. 4 one eighth of the plate and the position of the FE–substructure Ω_1 and the BE–substructure Ω_2 is depicted (cf. *Schnack* [15]). We investigate the stress concentration factor $\alpha = \sigma_{max}/\sigma_0$ and the CPU time in dependence on the geometrical ratio b/w.

Fig. 5 and 6 make evident the advantages of the coupling method compared with the

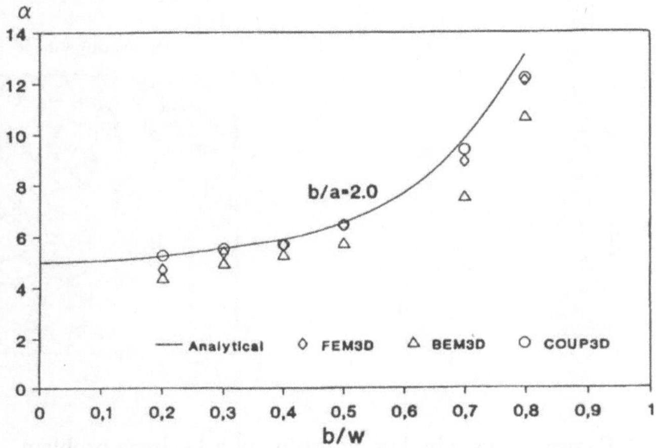

Figure 5: Stress concentration factor α

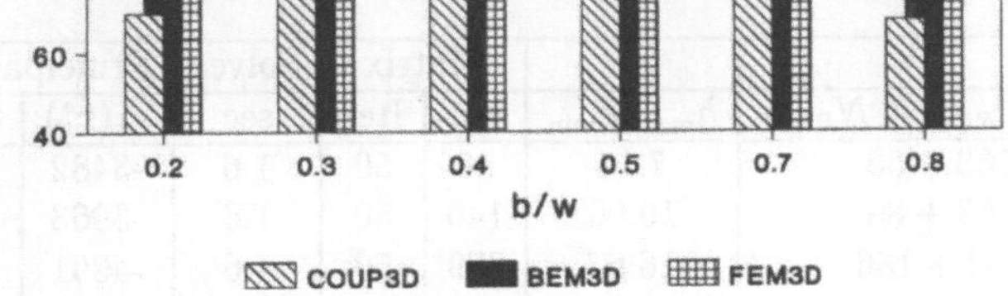

Figure 6: CPU time on an IBM 3090

separate use of FEM and BEM for both accuracy and efficiency.

In order to achieve convergence of the coupling algorithm also for problems with corners we use the *Neumann* series to solve the fine grid problem (19) iteratively. The

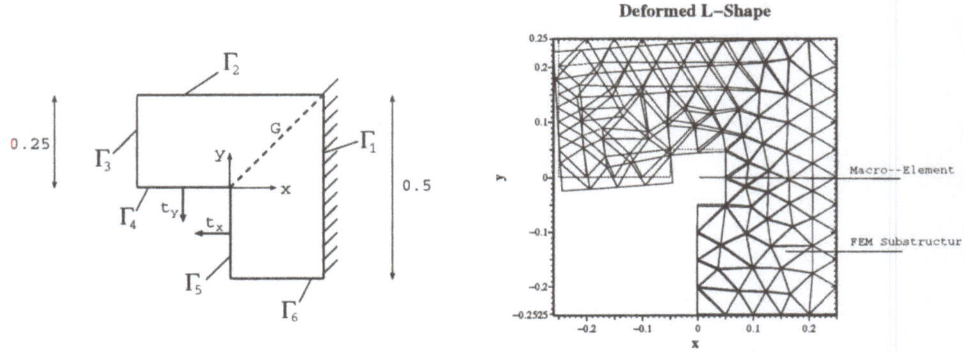

Figure 7: Domain decomposition of a L–shape problem

L–shape in Fig. 7 is loaded with tractions $t_x = t_y = -2000N$, the macro–element is placed around the loaded corner. For the computation a SUN workstation SPARC 20 with 120 MHz and a peak performance of 60 Mflops was used. The first part of Table 1 shows the results of the simulation for a constant coarse grid and an increasing number of degrees of freedoms for the fine grid N_{BEM}. The matrix generation includes the solution of the local *Neumann* problem with 50 iterations in the *Neumann* expansion while the solver is reduced to a very small global system. The principal stresses σ_1 and σ_2 are computed at a point x^* near the corner: $x(x^*) = y(x^*) = 0.01$. The second part of Table 1 shows results with an FEM algorithm for comparison. Fig. 8 shows the principal stress σ_1 and the CPU

Table 1: Results for L–shape simulation

		Matrix		Solver	Principal stresses	
$N = N_{FEM} + N_{BEM}$	$h_{\max}/h_{\min}$	sec	Iter	sec	$\sigma_1(x^*)$	$\sigma_2(x^*)$
$163 + 56$	7.14	66	50	1.6	-3482	-8391
$163 + 80$	10.00	140	50	1.6	-3963	-8512
$163 + 136$	16.67	389	50	1.6	-4091	-8563
$163 + 160$	20.00	609	50	1.6	-4110	-8574
$N = N_{FEM}$	$h_{\max}/h_{\min}$	sec	Iter	sec	$\sigma_1(x^*)$	$\sigma_2(x^*)$
308	5.00	0.6	–	7.4	-4390	-9161
354	10.00	0.8	–	11.8	-4850	-9087
430	16.67	1.0	–	17.7	-5409	-8974
532	33.33	1.3	–	22.0	-5371	-9055

time for the solution of the fine grid problem in dependence on the number of iterations for the *Neumann* expansion. In this case the fine grid density has the value $N_{BEM} = 168$. Finally Fig. 9 is a visualization of the stress concentration obtained with the coupling algorithm using the expansion of the *Calderon* projector in a *Neumann* series.

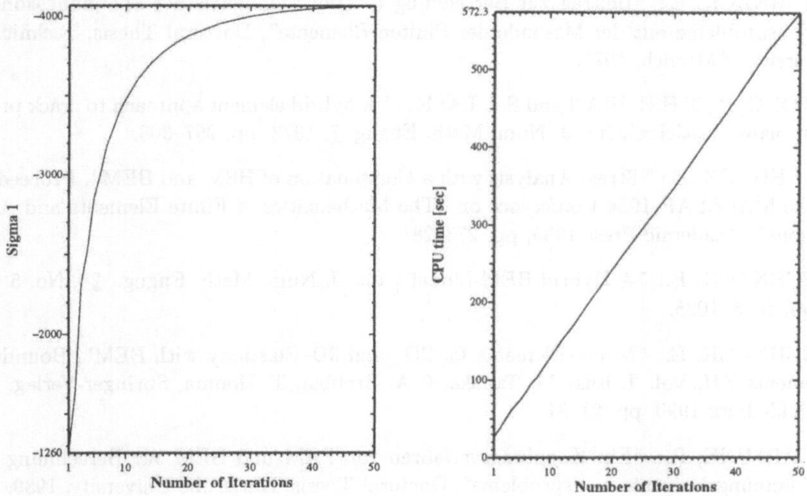

Figure 8: σ_1 and CPU time depending on Neumann iteration

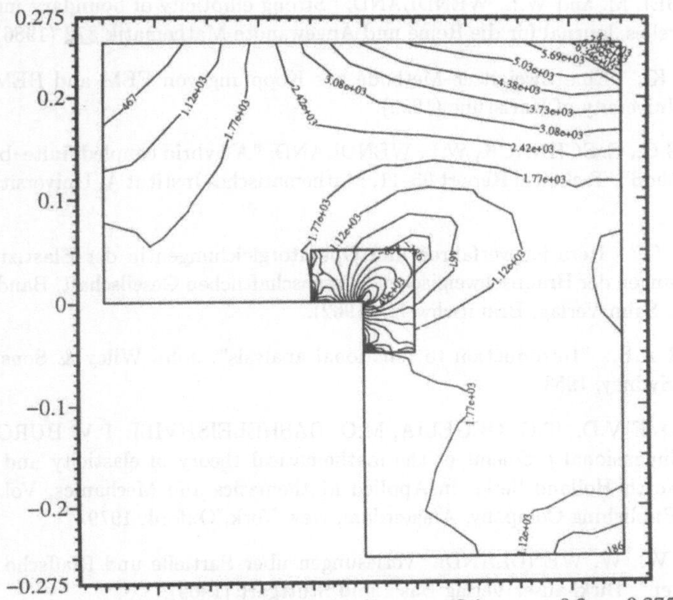

Figure 9: Distribution of the principal stress σ_1

REFERENCES

[1] SCHNACK, E.: "Beitrag zur Berechnung rotationssymmetrischer Spannungskonzentrationsprobleme mit der Methode der Finiten Elemente", Doctoral Thesis, Technical University of Munich, 1973.

[2] TONG, P., T.H.H. PIAN and S.J. LASRY: "A hybrid element approach to crack problems in plane elasticity", Int. J. Num. Meth. Engng. 7, 1973, pp. 297–308.

[3] SCHNACK, E.: "Stress Analysis with a Combination of HSM and BEM", Proceedings of the MAFELAP 1984 Conference on "The Mathematics of Finite Elements and Applications", Academic Press 1985, pp. 273–281.

[4] SCHNACK, E.: "A Hybrid BEM-Model", Int. J. Num. Meth. Engng., 24, No. 5 (1987), pp. 1015–1025.

[5] SCHNACK, E.: "Macro-Elements for 2D- and 3D-Elasticity with BEM", Boundary Elements XII, Vol. I. Eds: M. Tanaka, C.A. Brebbia, T. Honma, Springer-Verlag, Berlin, Heidelberg 1990, pp. 21–31.

[6] CARMINE, R.: "Ein Kopplungsverfahren von FEM und BEM zur Berechnung ebener Spannungskonzentrationsprobleme", Doctoral Thesis, Karlsruhe University, 1989.

[7] KARAOSMANOGLU, N.: "Kopplung von Randelement– und Finite–Element– Verfahren für dreidimensionale elastische Strukturen", Doctoral Thesis, Karlsruhe University, 1989.

[8] COSTABEL M. and W.L. WENDLAND: "Strong ellipticity of boundary integral operators". Krelles Journal für die Reine und Angewandte Mathematik 372 (1986), pp. 34–63.

[9] TÜRKE K.: "Eine Zweigitter–Methode zur Kopplung von FEM and BEM", Doctoral Thesis, University of Karlsruhe (1995).

[10] HSIAO G.C., E. SCHNACK, W.L. WENDLAND: "A hybrid coupled finite–boundary element method", Technical Report 95–11, Mathematisches Institut A, Universität Stuttgart, 1995.

[11] RIEDER G.: "Iterationsverfahren und Operatorgleichungen in der Elastizitätstheorie", Abhandlungen der Braunschweigischen Wissenschaftlichen Gesellschaft, Band XIV, Fried. Vieweg & Sohn Verlag, Braunschweig, (1962).

[12] TAYLOR A.E.: "Introduction to functional analysis", John Wiley & Sons, New York, London, Sydney, 1958.

[13] KUPRADZE V.D., T.G. GEGELIA, M.O. BASHELEISHVILI, T.V. BURCHULADZE: "Three dimensional problems of the mathematical theory of elasticity and thermoelasticity", North–Holland Series in Applied Mathematics and Mechanics, Vol. 25, North–Holland Publishing Company, Amsterdam, New York, Oxford, 1979.

[14] HAACK W., W. WENDLAND: "Vorlesungen über Partielle und Pfaffsche Differentialgleichungen", Birkhäuser Verlag Basel und Stuttgart (1969).

[15] SCHNACK E.: "Kopplung von Finite–Element– und Randelementmethoden für dreidimensionale elastische Strukturen", Report of the German Research Foundation (DFG), Project Schn 245/10–1, Karlsruhe University, 1991.

A MULTIPOLE BOUNDARY ELEMENT METHOD FOR TWO DIMENSIONAL ELASTOSTATICS

Yosihiro Yamada
Software Development Center, Shimadzu Corporation
1, Kuwabara-cho, Nishinokyo, Nakagyo-ku, Kyoto, 604, Japan
Email: yamada@shimadzu.co.jp

Ken Hayami
Department of Mathematical Engineering and Information Physics
Faculty of Engineering, University of Tokyo, 7-3-1, Hongo
Bunkyo-ku, Tokyo 113, Japan
Email: hayami@simplex.t.u-tokyo.ac.jp

SUMMARY

The boundary element method requires discretization only on the boundary. However, $O(N^3)$ work and $O(N^2)$ memory is usually required, where N is the number of boundary elements. To avoid these problems, Rokhlin [6] proposed the multipole method for the potential problem. This paper extends the method to the two dimensional elastostatic problem. Unlike the biharmonic equation formulation by Greenbaum et al. [3], a direct formulation in terms of displacement and traction variables is presented. The method requires $O(N \log N)$ work and memory. Numerical examples demonstrate the efficiency of the proposed method compared to the standard techniques.

INTRODUCTION

The Boundary Element Method (BEM) [1] has become a popular simulation method in science and engineering in recent years. The main advantage of the method is that discretization is required only on the boundary.

However, the method is not necessarily computationally efficient for large scale problems. This is because the method involves solving a system of linear equations for the unknowns on the boundary, where the matrix is dense and nonsymmetric. If the boundary is discretized into N boundary elements, the standard LU decomposition procedure would require $O(N^3)$ computation and $O(N^2)$ memory, which becomes prohibitive as N becomes large, as in three dimensional problems, and even for two dimensional problems if the geometry is complex, as in regions with many holes.

The last decade has seen the development of efficient iterative solvers for nonsymmetric systems of linear equations such as GCR [2], GMRES [7] and Bi-CGSTAB [8], in conjunction with the use of efficient preconditioners. Although these methods have been mainly addressed for sparse systems arising in the finite difference or finite element methods, they may also be applied to BEM. This would reduce the amount of computation, provided the number of iterations is less than $O(N)$. However, the memory required to compute the matrix vector product is still $O(N^2)$, and it is this fact, more than the amount of computation, that hinders the use of BEM for large scale problems, since the actual elapsed time is governed by the memory size.

In order to resolve this problem, Rokhlin proposed the multipole expansion method for the two dimensional potential problem [6]. The method computes the above matrix vector product approximately (to any required accuracy), with $O(N \log N)$ computation and memory, by clustering the effect of the distant boundary elements using multipole expansions. Greengard et al. improved the method to $O(N)$ (the fast multipole method), and generalized it to three dimensional potential problems [5] and multiply connected domains [4].

In this paper, we propose a multipole expansion method for the two dimensional elastostatic problem. The elastostatic problem is computationally more intensive compared to the potential problem, since it requires d-unknowns per boundary element for the d-dimensional problem, instead of one. Moreover, the fundamental solution of the elastostatic problem is more complicated compared to the potential problem, so that a new formulation is required in order to apply the multipole method. Greenbaum et al. [3] have proposed the biharmonic equation formulation, but we will give a direct formulation in terms of displacement and traction variables, which seems more convenient for general boundary conditions and applications.

BOUNDARY ELEMENT FORMULATION OF TWO DIMENSIONAL ELASTOSTATICS [1]

Throughout this paper, the Greek characters α, β and γ will be used to indicate the coordinate components, and the Roman characters i and j will be used to represent the boundary elements, i.e. $\alpha, \beta, \gamma = 1, 2$, and $i, j = 1, 2, \ldots, N$, where N is the total number of boundary elements.

Let $\boldsymbol{x}$ be a point in a domain Ω in $\mathbf{R}^2$, i.e. $\boldsymbol{x} \equiv (x_1, x_2)^{\mathrm{T}} \in \Omega$. The partial differential equation governing the two dimensional elastostatic problem is given by

$$\sum_{\beta} \frac{\partial \sigma_{\alpha\beta}(\boldsymbol{x})}{\partial x_\beta} + b_\alpha(\boldsymbol{x}) = 0, \qquad \alpha, \beta = 1, 2, \tag{1}$$

where $(\sigma_{\alpha\beta})$ represents the stress tensor. To simplify the argument, the body force term b_γ is ignored in the following. The relation between the stress tensor and the displacement

$$\boldsymbol{u}(\boldsymbol{x}) \equiv \begin{pmatrix} u_1(\boldsymbol{x}) \\ u_2(\boldsymbol{x}) \end{pmatrix}$$

is given by

$$\sigma_{\alpha\beta} = \lambda\delta_{\alpha\beta}\sum_{\gamma}\epsilon_{\gamma\gamma} + 2\mu\epsilon_{\alpha\beta}, \qquad \epsilon_{\alpha\beta} \equiv \frac{1}{2}\left(\frac{\partial u_\alpha}{\partial x_\beta} + \frac{\partial u_\beta}{\partial x_\alpha}\right), \tag{2}$$

where $\epsilon_{\alpha\beta}$ is the strain tensor, $\delta_{\alpha\beta}$ is the Kronecker's delta, and λ, μ are Lamé's constants.

The boundary condition for (1) is given by

$$u_\alpha = \bar{u}_\alpha \qquad \boldsymbol{x} \in \Gamma_\alpha^{(\mathrm{D})}, \tag{3}$$

$$p_\alpha \equiv \sum_\beta \sigma_{\alpha\beta} n_\beta = \bar{p}_\alpha \qquad \boldsymbol{x} \in \Gamma_\alpha^{(\mathrm{N})}, \tag{4}$$

where

$$\Gamma_\alpha^{(\mathrm{D})} + \Gamma_\alpha^{(\mathrm{N})} = \Gamma \equiv \partial\Omega,$$

$\boldsymbol{n}$ is the unit outward normal vector at $\boldsymbol{x} \in \Gamma$, and $\boldsymbol{p} \equiv p_\alpha$ is the traction at $\boldsymbol{x} \in \Gamma$.

Let $u^*_{\alpha\beta}(\boldsymbol{x};\boldsymbol{x}_0)$ be the fundamental solution of (1) in terms of the displacement, and let $p^*_{\alpha\beta}(\boldsymbol{x};\boldsymbol{x}_0)$ be the traction caused by $u^*_{\alpha\beta}(\boldsymbol{x};\boldsymbol{x}_0)$. The closed forms of $u^*_{\alpha\beta}(\boldsymbol{x};\boldsymbol{x}_0)$ and $p^*_{\alpha\beta}(\boldsymbol{x};\boldsymbol{x}_0)$ are given by

$$u^*_{\alpha\beta}(\boldsymbol{x};\boldsymbol{x}_0) = \frac{1}{8\pi\mu(1-\nu)}\left\{(3-4\nu)\log\left(\frac{1}{r}\right)\delta_{\alpha\beta} + \frac{r_\alpha r_\beta}{r^2}\right\},$$

$$p^*_{\alpha\beta}(\boldsymbol{x};\boldsymbol{x}_0) = \frac{-1}{4\pi(1-\nu)r}\left[\frac{\partial r}{\partial n}\left\{(1-2\nu)\delta_{\alpha\beta} + \frac{2r_\alpha r_\beta}{r^2}\right\} - (1-2\nu)\frac{n_\alpha r_\beta - n_\beta r_\alpha}{r}\right],$$

where $\boldsymbol{r} \equiv \boldsymbol{x} - \boldsymbol{x}_0$, $r \equiv |\boldsymbol{r}|$ and $\partial/\partial n$ stands for the normal derivative at $\boldsymbol{x} \in \Gamma$. Besides, ν is the Poisson's ratio given by $2\nu = \lambda/(\lambda+\mu)$. Then, (1) and Gauss' theorem give rise to the boundary integral equation

$$\frac{1}{2}u_\alpha(\boldsymbol{x}) + \sum_\beta\int_\Gamma p^*_{\alpha\beta}(\boldsymbol{y};\boldsymbol{x})u_\beta(\boldsymbol{y})\mathrm{d}\Gamma(\boldsymbol{y}) = \sum_\beta\int_\Gamma u^*_{\alpha\beta}(\boldsymbol{y};\boldsymbol{x})p_\beta(\boldsymbol{y})\mathrm{d}\Gamma(\boldsymbol{y}), \quad \boldsymbol{x} \in \Gamma. \tag{5}$$

Next, the discretization of (5) is considered. First, the boundary Γ is approximated by N line segments, i.e.

$$\Gamma = \sum_{i=1}^{N}\Gamma_i,$$

where $\sum$ represents the union of sets without any intersection except for their boundaries. Each Γ_i is called a boundary element. Let $\boldsymbol{x}_i$ be the representative point of Γ_i, e.g. its mid point. Also assume that $u_\alpha(\boldsymbol{x})$ and $p_\alpha(\boldsymbol{x})$ remain constant within each boundary element Γ_i, i.e.

$$u_{i\alpha} \equiv u_\alpha(\boldsymbol{x}), \quad p_{i\alpha} \equiv p_\alpha(\boldsymbol{x}), \qquad \boldsymbol{x} \in \Gamma_i.$$

Then, (5) can be discretized as

$$\frac{1}{2}u_{i\alpha} + \sum_{j,\beta}u_{j\beta}\int_{\Gamma_j}p^*_{\alpha\beta}(\boldsymbol{y};\boldsymbol{x}_i)\mathrm{d}\Gamma(\boldsymbol{y}) = \sum_{j,\beta}p_{j\beta}\int_{\Gamma_j}u^*_{\alpha\beta}(\boldsymbol{y};\boldsymbol{x}_i)\mathrm{d}\Gamma(\boldsymbol{y}). \tag{6}$$

Let

$$g_{i\alpha,j\beta} \equiv \int_{\Gamma_j}u^*_{\alpha\beta}(\boldsymbol{y};\boldsymbol{x}_i)\mathrm{d}\Gamma(\boldsymbol{y}),$$

$$h_{i\alpha,j\beta} \equiv \int_{\Gamma_j}p^*_{\alpha\beta}(\boldsymbol{y};\boldsymbol{x}_i)\mathrm{d}\Gamma(\boldsymbol{y}) + \frac{1}{2}\delta_{\alpha\beta}\delta_{ij},$$

$$G \equiv (g_{i\alpha,j\beta}), \qquad H \equiv (h_{i\alpha,j\beta}),$$

$$\boldsymbol{U} \equiv (u_{11}\ u_{12}\ u_{21}\ \cdots u_{N2})^{\mathrm{T}}, \quad \boldsymbol{P} \equiv (p_{11}\ p_{12}\ p_{21}\ \cdots p_{N2})^{\mathrm{T}}.$$

Using these notations, (6) can be expressed in matrix vector form as

$$HU = GP. \tag{7}$$

Let

$$\boldsymbol{U} \equiv \begin{pmatrix} \boldsymbol{U}^{(\mathrm{D})} \\ \boldsymbol{U}^{(\mathrm{N})} \end{pmatrix}, \qquad \boldsymbol{P} \equiv \begin{pmatrix} \boldsymbol{P}^{(\mathrm{D})} \\ \boldsymbol{P}^{(\mathrm{N})} \end{pmatrix},$$

where the entries of $\boldsymbol{U}^{(\mathrm{D})}$ and $\boldsymbol{P}^{(\mathrm{D})}$ are the values of $u_{i\alpha}$ and $p_{i\alpha}$, respectively, on $\Gamma^{(\mathrm{D})}$, and the entries of $\boldsymbol{U}^{(\mathrm{N})}$ and $\boldsymbol{P}^{(\mathrm{N})}$ are the values on $\Gamma^{(\mathrm{N})}$. Then $\boldsymbol{U}^{(\mathrm{D})}$ and $\boldsymbol{P}^{(\mathrm{N})}$ are known values, and $\boldsymbol{U}^{(\mathrm{N})}$ and $\boldsymbol{P}^{(\mathrm{D})}$ are unknown values. By bringing the unknowns to the left hand side, equation(7) yields a system of linear equations

$$\begin{pmatrix} -G^{(\mathrm{D})} & H^{(\mathrm{N})} \end{pmatrix} \begin{pmatrix} \boldsymbol{P}^{(\mathrm{D})} \\ \boldsymbol{U}^{(\mathrm{N})} \end{pmatrix} = \begin{pmatrix} -H^{(\mathrm{D})} & G^{(\mathrm{N})} \end{pmatrix} \begin{pmatrix} \boldsymbol{U}^{(\mathrm{D})} \\ \boldsymbol{P}^{(\mathrm{N})} \end{pmatrix}, \tag{8}$$

where $\begin{pmatrix} -G^{(\mathrm{D})} & H^{(\mathrm{N})} \end{pmatrix}$ is a dense non-symmetric matrix.

THE MULTIPOLE METHOD

THE BASIC IDEA OF THE MULTIPOLE METHOD

In the standard boundary element method, (8) is solved using LU decomposition, requiring $O(N^3)$ computation time and $O(N^2)$ memory.

If one were to apply iterative methods such as GCR [2] or Bi-CGSTAB [8] for the non-symmetric system of linear equations, computation time may be reduced provided the number of iterations is less than $O(N)$. However, memory and computation time of $O(N^2)$ is still required to perform the dense matrix vector multiplication for each iteration. Hence, one would like to reduce this bottleneck.

Let us, for the moment, consider the Dirichlet problem, where $\boldsymbol{P}$ is to be solved for given $\boldsymbol{U}$ in equation (7). Then, in order to apply an iterative solver, one must be able to evaluate the inner product between any row vector of G and an iteration vector

$$\boldsymbol{q} \equiv \begin{pmatrix} q_{11} & q_{12} & q_{21} & \cdots & q_{N2} \end{pmatrix}^{\mathrm{T}}$$

to compute $G\boldsymbol{q}$.

From (6), the inner product related to the observation point $\boldsymbol{x}_i$ is given by

$$I_{i\alpha} \equiv \sum_{j=1}^{N} \int_{\Gamma_j} \sum_{\beta} u^*_{\alpha\beta}(\boldsymbol{y}; \boldsymbol{x}_i) q_{j\beta} \mathrm{d}\Gamma(\boldsymbol{y}).$$

Define $\Gamma_i^{(\mathrm{n})}$ as the set of boundary elements Γ_j which are near $\boldsymbol{x}_i$. Assume that the size of the set $\Gamma_i^{(\mathrm{n})}$ is a constant independent of N, and let

$$\Gamma_i^{(\mathrm{f})} \equiv \Gamma - \Gamma_i^{(\mathrm{n})}.$$

To distinguish between the contributions of $\Gamma_i^{(n)}$ and $\Gamma_i^{(f)}$ to $I_{i\alpha}$, denote $I_{i\alpha}$ as

$$\begin{aligned} I_{i\alpha} &= \sum_\beta I_{i\alpha\beta}^{(n)} + \sum_\beta I_{i\alpha\beta}^{(f)}, \\ I_{i\alpha\beta}^{(n)} &\equiv \int_{\Gamma_i^{(n)}} u_{\alpha\beta}^*(\boldsymbol{y}; \boldsymbol{x}_i) q_{j\beta} \mathrm{d}\Gamma(\boldsymbol{y}), \\ I_{i\alpha\beta}^{(f)} &\equiv \int_{\Gamma_i^{(f)}} u_{\alpha\beta}^*(\boldsymbol{y}; \boldsymbol{x}_i) q_{j\beta} \mathrm{d}\Gamma(\boldsymbol{y}). \end{aligned}$$

First, the near contribution $I_{i\alpha}^{(n)}$ is calculated directly using

$$I_{i\alpha\beta}^{(n)} = \sum_{\Gamma_j \subset \Gamma_i^{(n)}} g_{i\alpha,j\beta} q_{j\beta},$$

where

$$g_{i\alpha,j\beta} = \int_{\Gamma_j} u_{\alpha\beta}^*(\boldsymbol{y}; \boldsymbol{x}_i) \mathrm{d}\Gamma(\boldsymbol{y}).$$

Next, consider the calculation of the far contribution $I_{i\alpha\beta}^{(f)}$. Construct clusters G_h by assembling boundary elements Γ_j which are included in $\Gamma_i^{(f)}$. G_h satisfy

$$G_h \equiv \sum_j \Gamma_j, \qquad \Gamma_i^{(f)} = \sum_h G_h, \tag{9}$$

i.e. G_h is a disjoint union of boundary elements Γ_j, and $\Gamma_i^{(f)}$ in turn is a disjoint union of clusters G_h. The clusters must be formed so that there exists $\boldsymbol{C}_h$ for each cluster G_h such that

$$|\boldsymbol{x}_i - \boldsymbol{C}_h| > 2\,|\boldsymbol{y} - \boldsymbol{C}_h|, \qquad \forall \boldsymbol{y} \in G_h,$$

as shown in Fig. 1. $\boldsymbol{C}_h$ is called the center of cluster G_h.

In order to reduce memory and computation time, the G_h's are formed so that they can be shared among different $\boldsymbol{x}_i$'s, and so that the total number of clusters G_h is $O(N)$. Once the clusters G_h are formed, one chooses which clusters G_h are required to form $\Gamma_i^{(f)}$ before calculating each $I_{i\alpha\beta}$. Let H_i be the set of indices h of G_h which are required to form $\Gamma_i^{(f)}$. We will refer to H_i as the cluster list for $\boldsymbol{x}_i$. Then, (9) can be written more precisely as

$$\Gamma_i^{(f)} = \sum_{h \in H_i} G_h.$$

More specifically, the clusters G_h are formed by recursively halving the rectangular region containing Ω in alternate directions and assigning a cluster G_h to the set of elements contained in the sub-rectangle, as shown in Fig. 2.

Let $\boldsymbol{r} \equiv \boldsymbol{y} - \boldsymbol{x}_i$, $r \equiv |\boldsymbol{r}|$. If we can expand each term of equation

$$\begin{aligned} I_{i\alpha\beta}^{(f)} &= \sum_h \int_{G_h} u_{\alpha\beta}^*(\boldsymbol{y}; \boldsymbol{x}_i) q_{j\beta} \mathrm{d}\Gamma(\boldsymbol{y}) \\ &= \sum_h \frac{1}{8\pi\mu(1-\nu)} \int_{G_h} \left\{ (3-4\nu) \log\left(\frac{1}{r}\right) \delta_{\alpha\beta} + \frac{r_\alpha r_\beta}{r^2} \right\} q_{j\beta} \mathrm{d}\Gamma(\boldsymbol{y}) \end{aligned} \tag{10}$$

into Taylor series around $\boldsymbol{C}_h$, and truncate them at the p-th term to obtain

$$\sum_h \sum_{k=0}^{p} F(\boldsymbol{x}_i - \boldsymbol{C}_h; k)\, c(G_h, k), \tag{11}$$

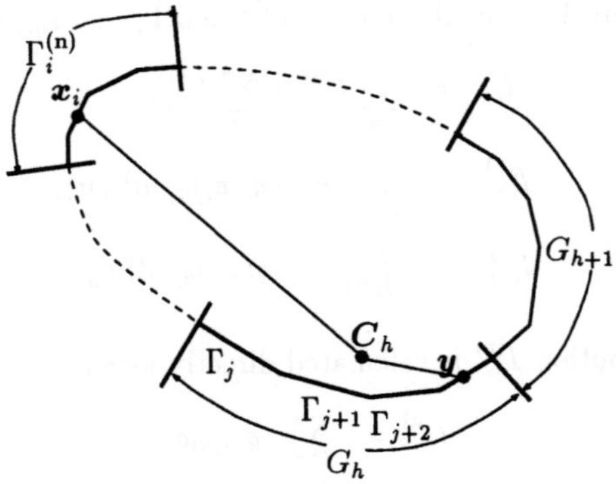

Figure 1: Observation point, source point and cluster.

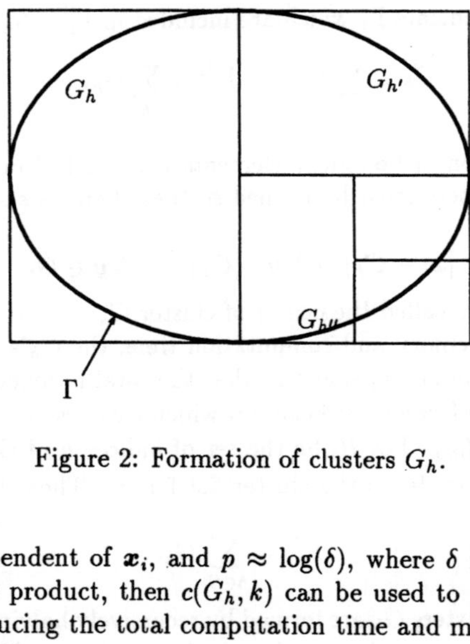

Figure 2: Formation of clusters G_h.

where $c(G_h, k)$ is independent of x_i, and $p \approx \log(\delta)$, where δ is the required accuracy in computing the inner product, then $c(G_h, k)$ can be used to calculate inner products for many x_i's, thus reducing the total computation time and memory in computing the matrix vector product $G\boldsymbol{q}$.

MULTIPOLE EXPANSION OF THE TWO DIMENSIONAL ELASTOSTATIC KERNEL

First, consider the expansion of the term

$$\int_{G_h} \log(r) q_{j\beta} \mathrm{d}\Gamma(\boldsymbol{y}),$$

in the first term of (10). For convenience, identify $\mathbf{R}^2$ with the complex plain $\mathbf{C}$ and regard $\boldsymbol{x}_i, \boldsymbol{y}, \boldsymbol{C}_h \in \mathbf{C}$. In other words, regard $\boldsymbol{x} \equiv (x_1\ x_2) \in \mathbf{R}^2$ as $\boldsymbol{x} = x_1 + \mathrm{i}x_2 \in \mathbf{C}$ and so on. Then,

$$\begin{aligned}\int_{G_h} \log(r) q_{j\beta} \mathrm{d}\Gamma(\boldsymbol{y}) &= \mathrm{Re} \int_{G_h} \log(\boldsymbol{x}_i - \boldsymbol{y}) q_{j\beta} \mathrm{d}\Gamma(\boldsymbol{y}) \\ &= \mathrm{Re} \int_{G_h} \left\{ \log(\boldsymbol{x}_i - \boldsymbol{C}_h) + \sum_{k>0} \frac{-1}{k} \frac{(\boldsymbol{y} - \boldsymbol{C}_h)^k}{(\boldsymbol{x}_i - \boldsymbol{C}_h)^k} \right\} q_{j\beta} \mathrm{d}\Gamma(\boldsymbol{y}),\end{aligned}$$

so that

$$\int_{G_h} \log(r) q_{j\beta} \mathrm{d}\Gamma(\boldsymbol{y}) = \mathrm{Re} \left[\sum_k f(\boldsymbol{x}_i - \boldsymbol{C}_h; k)\, A_\beta(G_h, k) \right], \tag{12}$$

where

$$a_\beta(G_h, k) \equiv \int_{G_h} (\boldsymbol{y} - \boldsymbol{C}_h)^k q_{j\beta} \mathrm{d}\Gamma(\boldsymbol{y}), \tag{13}$$

$$f(\boldsymbol{x}; k) \equiv \begin{cases} \log(\boldsymbol{x}), & k = 0, \\ \dfrac{1}{\boldsymbol{x}^k}, & k = 1, 2, \ldots, \end{cases}$$

and

$$A_\beta(G_h, k) \equiv \begin{cases} a_\beta(G_h, 0), & k = 0, \\ -\frac{1}{k} a_\beta(G_h, k), & k = 1, 2, \ldots \end{cases}.$$

Hence, the form of (11) is obtained by truncating at the p-th term.

Next, consider the second term of (10) with $\alpha = \beta = 1$. Still regarding $\boldsymbol{r} = r_1 + \mathrm{i}r_2$, the equation

$$\frac{r_1}{r^2} = \mathrm{Re}\frac{1}{\boldsymbol{r}} = -\mathrm{Re}\frac{1}{(\boldsymbol{x}_i - \boldsymbol{C}_h) - (\boldsymbol{y} - \boldsymbol{C}_h)} = -\mathrm{Re}\sum_{k\geq 0} \frac{(\boldsymbol{y} - \boldsymbol{C}_h)^k}{(\boldsymbol{x}_i - \boldsymbol{C}_h)^{k+1}}$$

leads to the expansion

$$\begin{aligned}\frac{r_1^2}{r^2} &= -\mathrm{Re}\sum_{k\geq 0} \frac{(\boldsymbol{y} - \boldsymbol{C}_h)^k}{(\boldsymbol{x}_i - \boldsymbol{C}_h)^{k+1}} \mathrm{Re}\boldsymbol{r} \\ &= \mathrm{Re}\left\{ \sum_{k\geq 0} \frac{(\boldsymbol{y} - \boldsymbol{C}_h)^k}{(\boldsymbol{x}_i - \boldsymbol{C}_h)^{k+1}} \right\} \mathrm{Re}(\boldsymbol{x}_i - \boldsymbol{C}_h) \\ &\quad -\mathrm{Re}\left\{ \sum_{k\geq 0} \frac{(\boldsymbol{y} - \boldsymbol{C}_h)^k}{(\boldsymbol{x}_i - \boldsymbol{C}_h)^{k+1}} \right\} \mathrm{Re}(\boldsymbol{y} - \boldsymbol{C}_h).\end{aligned}$$

Besides, since $\mathrm{Re}\, z_1 \mathrm{Re}\, z_2 = \mathrm{Re}\,(z_1 \mathrm{Re}\, z_2)$ for arbitrary complex numbers z_1 and z_2,

$$\begin{aligned}\int_{G_h} \frac{r_1^2}{r^2} q_{j\beta} \mathrm{d}\Gamma(\boldsymbol{y}) &= \mathrm{Re}\left[\int_{G_h} \sum_{k\geq 0} \frac{(\boldsymbol{y} - \boldsymbol{C}_h)^k}{(\boldsymbol{x}_i - \boldsymbol{C}_h)^{k+1}} \mathrm{Re}(\boldsymbol{x}_i - \boldsymbol{C}_h) q_{j\beta} \mathrm{d}\Gamma(\boldsymbol{y}) \right] \\ &\quad -\mathrm{Re}\left[\int_{G_h} \sum_{k\geq 0} \frac{(\boldsymbol{y} - \boldsymbol{C}_h)^k}{(\boldsymbol{x}_i - \boldsymbol{C}_h)^{k+1}} \mathrm{Re}(\boldsymbol{y} - \boldsymbol{C}_h) q_{j\beta} \mathrm{d}\Gamma(\boldsymbol{y}) \right].\end{aligned}$$

If we now let

$$
\begin{aligned}
a_\beta^{(\mathrm{r})}(G_h,k) &\equiv \int_{G_h} (\boldsymbol{y}-\boldsymbol{C}_h)^k \mathrm{Re}(\boldsymbol{y}-\boldsymbol{C}_h) q_{j\beta} \mathrm{d}\Gamma(\boldsymbol{y}), \qquad (14)\\
a_\beta^{(\mathrm{i})}(G_h,k) &\equiv \int_{G_h} (\boldsymbol{y}-\boldsymbol{C}_h)^k \mathrm{Im}(\boldsymbol{y}-\boldsymbol{C}_h) q_{j\beta} \mathrm{d}\Gamma(\boldsymbol{y}),\\
f^{(\mathrm{r})}(\boldsymbol{x};k) &\equiv \frac{\mathrm{Re}\,\boldsymbol{x}}{\boldsymbol{x}^k},\\
f^{(\mathrm{i})}(\boldsymbol{x};k) &\equiv \frac{\mathrm{Im}\,\boldsymbol{x}}{\boldsymbol{x}^k},
\end{aligned}
$$

we obtain

$$
\begin{aligned}
\int_{G_h} \frac{r_1^2}{r^2} q_{j\beta} \mathrm{d}\Gamma(\boldsymbol{y}) &= \mathrm{Re}\left[\sum_k f^{(\mathrm{r})}(\boldsymbol{x}_i-\boldsymbol{C}_h;k+1) a_\beta(G_h,k)\right] \qquad (15)\\
&\quad -\mathrm{Re}\left[\sum_k f(\boldsymbol{x}_i-\boldsymbol{C}_h;k+1) a_\beta^{(\mathrm{r})}(G_h,k)\right].
\end{aligned}
$$

Hence, the second term of (10) can also be expressed in the form of (11). From (10), (12) and (15), we have

$$
\begin{aligned}
8\pi\mu(1-\nu) I_{i11}^{(\mathrm{f})} &= (4\nu-3)\mathrm{Re}\left[\sum_{h,k} f(\boldsymbol{x}_i-\boldsymbol{C}_h;k) A_1(G_h,k)\right] \qquad (16)\\
&\quad +\mathrm{Re}\left[\sum_{h,k} f^{(\mathrm{r})}(\boldsymbol{x}_i-\boldsymbol{C}_h;k+1) a_1(G_h,k)\right]\\
&\quad -\mathrm{Re}\left[\sum_{h,k} f(\boldsymbol{x}_i-\boldsymbol{C}_h;k+1) a_1^{(\mathrm{r})}(G_h,k)\right],
\end{aligned}
$$

where $a_\beta(G_h,k)$, $a_\beta^{(\mathrm{r})}(G_h,k)$ can be shared by many $\boldsymbol{x}_i$'s. The integrals over each boundary element Γ_j constituting G_h for $a_\beta(G_h,k)$ and $a_\beta^{(\mathrm{r})}(G_h,k)$ in (13) and (14) are available in closed form [9].

Similar results follow for $\alpha = 1$, $\beta = 2$ and $\alpha = 2$ by replacing Re by Im, and for the integrals involving $p_{\alpha\beta}^*$ [9].

THE PROCEDURE FOR COMPUTING THE INNER PRODUCT

In order to evaluate the inner products which constitute the matrix vector product in the iterative solver, we take the following procedure. First, the values of $a_\beta(G_h,k)$, $a_\beta^{(\mathrm{r})}(G_h,k)$ and $a_\beta^{(\mathrm{i})}(G_h,k)$ are computed for all the clusters G_h. The amount of computation for this step is $O(N\log N)$, since the time to compute $a_\beta(G_h,k)$ $a_\beta^{(\mathrm{r})}(G_h,k)$ and $a_\beta^{(\mathrm{i})}(G_h,k)$ for all the clusters G_h at any level l is proportional to N, and the number of levels of clusters is $O(N)$. Fig. 3 shows a one dimensional schematic diagram.

Next, the inner product for each observation point $\boldsymbol{x}_i$ are computed using (16). The cost of computing an inner product for one observation point $\boldsymbol{x}_i$ is $O(\log N)$, since an

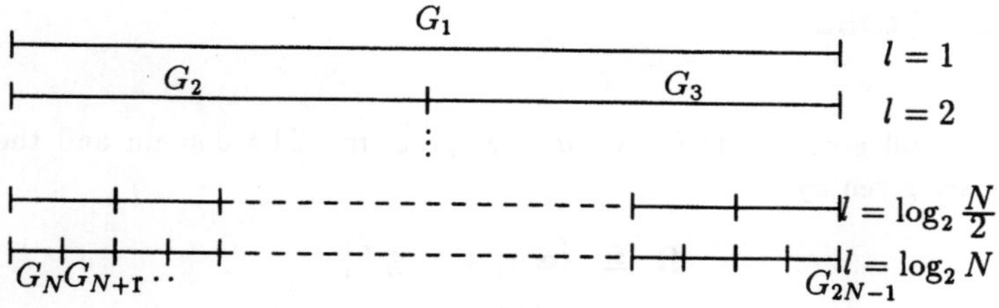

Figure 3: Clusters G_h at level l.

$O(1)$ number of clusters are required from each level. Hence, the total computation for the inner products for all the points $\boldsymbol{x}_i$, $i = 1, 2, \ldots N$ is also $O(N \log N)$.

Thus, we can conclude that the amount of computation to solve the original system of linear equations (6) by an iterative method is $O(N \log N)$ per one iteration. (One may improve the order to $O(N)$ using techniques similar to those in [5].) Hence, if M iterations are required to solve equation (6), the total amount of computation is $O(MN \log N)$.

If we adopt the Bi-CGSTAB method [8] as the iterative method, all we have to save are $a_\beta(G_h, k)$, $a_\beta^{(r)}(G_h, k)$, $a_\beta^{(i)}(G_h, k)$ and some $2N$-dimensional vectors, all of size $O(N)$. This reduces the memory requirement to $O(N \log N)$. Note that the memory required is $O(N)$ for storing $a_\beta(G_h, k)$, $a_\beta^{(r)}(G_h, k)$, $a_\beta^{(i)}(G_h, k)$ (floating numbers), but the memory required for storing the integer list of clusters related to each observation point $\boldsymbol{x}_i$ is $O(N \log N)$. However, the number of iterations M may become large if the problem is ill-conditioned. On the other hand, if we adopt the GCR method [2], it is guaranteed that M is not larger than the number of unknowns $2N$, in the absence of round off error. But the memory requirement is $O(MN)$, since we must save all the Krylov bases. Then, the memory requirement increases with M.

Once the unknowns on the boundary are determined, the strain and stress at an interior point can also be calculated efficiently using multipole expansions and the same clusters G_h [9].

NUMERICAL EXPERIMENT RESULTS

In this section we will give numerical experiment results for two dimensional elastostatic problems, comparing the proposed multipole method with previous methods. The computations were done on the workstation DEC 3000 model AXP (150 MHz) with 64 Mbyte main memory and 200 Mbyte swap area.

The term 'simple iteration' will indicate the results obtained by the usual iterative methods such as the Bi-CGSTAB [8] or the GCR [2] without using the multipole method.

The convergence threshold for the relative residual of the iterative methods (with or without the multipole method) was set to $R^{(l)} = 10^{-5}$. The multipole expansions were truncated at $p = 25$. The discretization of the boundary was done using constant straight line elements.

DIRICHLET PROBLEM

First, we will give results for a Dirichlet problem. The domain and the boundary conditions are given by

$$\begin{aligned} \Omega &\equiv \left\{ \boldsymbol{x} | x_1^2 + x_2^2 \leq 9 \right\}, \\ u(\boldsymbol{x}) &= -10^{-6} \times x_2 \qquad \boldsymbol{x} \in \Gamma. \end{aligned}$$

The elastic constants were set to $\nu = 0.1$ and $\mu = 9.45 \times 10^4$, where the length is measured in meters and the Lamé constants are in MPa.

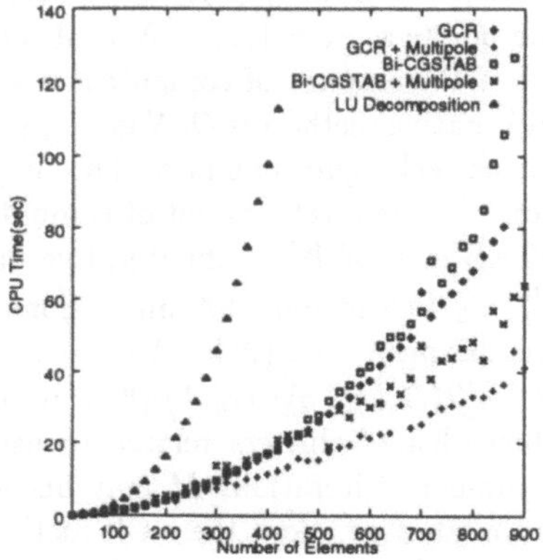

Figure 4: Comparison of the CPU time for the multipole method, the simple iteration and LU decomposition.

Fig. 4 compares the CPU time for the multipole method, simple iteration and LU decomposition, as the number of boundary elements increases. The GCR by the multipole method was the fastest, followed by the Bi-CGSTAB by the multipole method, the GCR by simple iteration, and the Bi-CGSTAB by simple iteration. The LU decomposition consumed the most CPU time, regardless of the number of elements N.

NEUMANN PROBLEM FOR A DOMAIN WITH MANY HOLES

Next, a Neumann problem defined for a domain Ω with many circular holes is considered, where

$$\begin{aligned} \Omega &\equiv \{ \boldsymbol{x} | -2 \leq x_1 \leq 2, \ -1 \leq x_2 \leq 1 \} - \sum_{m=1}^{8} \sum_{n=1}^{4} C_{lm}, \\ C_{lm} &\equiv \left\{ \boldsymbol{x} \middle| \{9x_1 - (4m - 18)\}^2 + \{9x_2 - (2n - 5)\}^2 < 0.81 \right\} \end{aligned}$$

as shown in Fig. 5. The boundary condition is given by

$$\boldsymbol{p}(\boldsymbol{x}) = \begin{cases} (1,\ 0)^{\mathrm{T}} & x_1 = \pm 2, \\ \mathbf{0} & \text{otherwise}. \end{cases}$$

The elastic constants are the same as in the previous example.

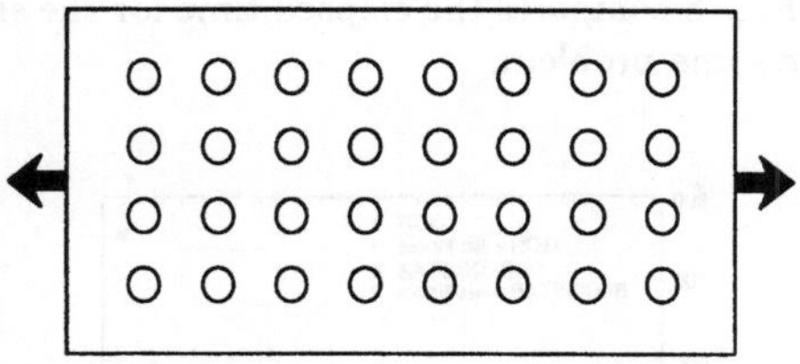

Figure 5: Domain with many holes.

This is a model problem of a porous media, and suggests that the number of boundary elements N can become quite large, even for two dimensional problems, when the geometry of the boundary Γ is complex. As shown in Fig. 6, simple iteration is faster than the multipole method when N is smaller than 900. Fig. 7 shows the number of iterations with respect to the number of elements N. Only the results for the simple iterations is shown, since the difference between the simple iteration and the multipole method is very small. This problem is very ill-conditioned and requires many iterations compared to the previous Dirichlet problem. Also the convergence behavior is oscillatory with the increase of N, depending on the discretization of the holes.

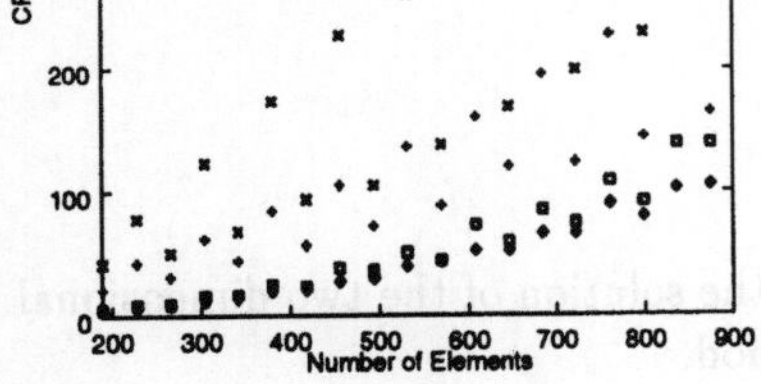

Figure 6: Comparison of CPU time between multipole method and simple iteration ($R^{(\mathrm{I})} = 10^{-5}$, $p = 25$).

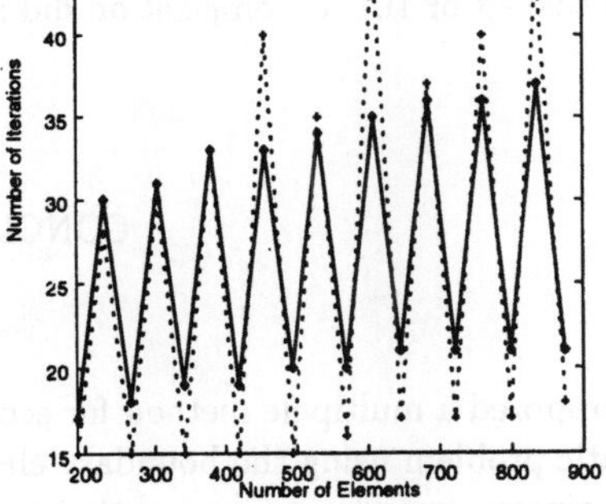

Figure 7: Number of iterations (simple iteration, $R^{(\mathrm{I})} = 10^{-5}$).

In the multipole method, the multipole expansions have to be performed for each iteration, whereas for the simple iteration, the G and H matrices are computed only once and can be shared among the iterations. Hence, the more iterations required to solve the linear systems, the lighter the weight of the CPU time for the construction of G and H matrices becomes. This is the reason why the simple iteration is relatively fast for this problem. However, when the number of elements exceeds 800, the data for the simple iterations can no longer be saved on the main memory, and frequent access to the disk makes the elapsed time far longer compared to the CPU time, making the method impractical. With the multipole method, elements up to 3000 could be treated within reasonable time. Fig. 8 compares the elapsed time for the simple iteration and the multipole method for the same problem.

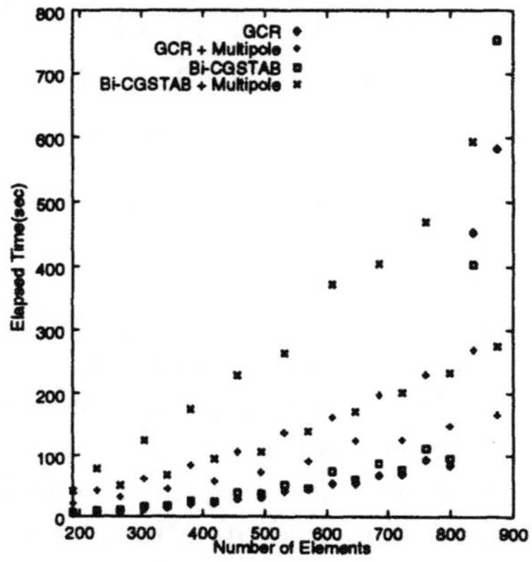

Figure 8: Comparison of the elapsed time between the multipole method and the simple iteration ($R^{(\mathrm{I})} = 10^{-5}$, $p = 25$).

In order to make the multipole method more competitive and robust, we need an efficient preconditioner to reduce the number of iterations. Preliminary attempts using diagonal scaling or ILU decomposition did not improve the situation substantially.

CONCLUSIONS

We proposed a multipole method for economizing the solution of the two dimensional elastostatic problem using the boundary element method.

Preliminary experiments showed that the method reduces the CPU time and memory remarkably. However, for ill-conditioned problems which require many iterations, the method becomes less competitive in CPU time, although it is still competitive in terms of elapsed time, which is determined by the required memory.

The search for an effective preconditioner for reducing the number of iterations, and a more efficient implementation of the multipole method are left for future work.

ACKNOWLEDGEMENTS

The authors would like to thank Professor K. Sugihara, Dr. M. Sugihara and Mr. T. Imai for valuable comments and assistance. This work was supported by the Grant-in-Aid for Scientific Research (C) from the Ministry of Education, Science and Culture, Japan.

REFERENCES

[1] BREBBIA, C.A., AND DOMINGUEZ, J.: "Boundary Elements An Introductory Course", Computational Mechanics Publications with McGraw-Hill Book Company, New York, 1992.

[2] EISENSTAT, S.C., ELMAN, C. AND SCHULTZ, M.H.: "Variational iterative methods for nonsymmetric systems of linear equations", SIAM J. Numer. Anal., 20 (1983), pp.345-357.

[3] GREENBAUM, A., GREENGARD, L., AND MAYO, A.: "On the numerical solution of the biharmonic equation in the plane", Physica D, 60 (1992), pp.216-225.

[4] GREENBAUM, A., GREENGARD, L., AND MCFADDEN, G.B.: "Laplace's equation and Dirichlet-Neumann map in multiply connected domains", J. Comput. Phys., 105 (1993), pp.267-278.

[5] GREENGARD, L.: "The Rapid Evaluation of Potential Fields in Particle Systems", MIT Press, Cambridge, 1988.

[6] ROKHLIN, V.: "Rapid solution of integral equations of classical potential theory", J. Comput. Phys., 60 (1985), pp.187-207.

[7] SAAD, Y., AND SCHULTZ, M.H.: "GMRES: A generalized minimum residual algorithm for solving nonsymmetric linear systems", SIAM J. Sci. Stat. Comput., 7 (1986), pp. 856-869.

[8] VAN DER VORST, H.A.: "Bi-CGSTAB: A fast and smoothly converging variant of Bi-CG for the solution of nonsymmetric linear systems", SIAM J. Sci. Stat. Comput., 13 (1992), pp.631-644.

[9] YAMADA, Y., AND HAYAMI, K.: "A multipole boundary element method for two dimensional elastostatics", METR 95-07, Technical Reports, Department of Mathematical Engineering and Information Physics, University of Tokyo, August, 1995.

List of Participants:

Andrä, Heiko Institut für Technische Mechanik, Festigkeitslehre, Universität Karlsruhe, Kaiserstr. 12, 76128 Karlsruhe,

Benecke, Marcus Lehrstuhl für Allgemeine und Theoretische Elektrotechnik, Technische Fakultät der CAU Kiel, Kaiserstr. 2, 24143 Kiel,

Bittner, Kai Fachbereich Mathematik, Universität Rostock, 18051 Rostock,

Brokate, Martin Lehrstuhl Praktische Mathematik, Mathematisches Seminar II, Christian-Albrechts-Universität Kiel, Hermann-Rodewald-Str. 3, 24098 Kiel,

Bücker, Martin Central Institute for Applied Mathematics, Research Centre Jülich (KFA), 52425 Jülich,

Burmeister, Jens Lehrstuhl Praktische Mathematik, Mathematisches Seminar II, Christian-Albrechts-Universität Kiel, Hermann-Rodewald-Str. 3, 24098 Kiel,

Carstensen, Carsten, Lehrstuhl Wissenschaftliches Rechnen, Mathematisches Seminar II, Christian-Albrechts-Universität Kiel, Ludewig-Meyn Str. 4, 24098 Kiel,

Chen, Ke Department of Statistics and Computational Mathematics, Victoria Building, Brownlow Hill, The University of Liverpool, P. O. Box 147, Liverpool L69 3BX, England,

Faermann, Birgit Lehrstuhl Praktische Mathematik, Mathematisches Seminar II, Christian-Albrechts-Universität Kiel, Hermann-Rodewald-Str. 3, 24098 Kiel,

Finck, Tilo Technische Universität Chemnitz-Zwickau, Fakultät fuer Mathematik, Reichenhainer Strasse 39, 09107 Chemnitz,

Fischer, Bernd, Institut für Mathematik, Medizinische Universität zu Lübeck, Wallstrasse 40, 23560 Lübeck,

Funken, Stefan A. Lehrstuhl Wissenschaftliches Rechnen, Mathematisches Seminar II, Christian-Albrechts-Universität Kiel, Ludewig-Meyn Str. 4, 24098 Kiel,

Gaspar, Csaba senior research associate, Szechenyi Istvan College, Dept. of Mathematics, P.O.Box 701, H-9026 Gyor, Hungary,

Giebermann, Klaus Universität Karlsruhe, Institut f. Praktische Mathematik, Englerstr. 2, 76128 Karlsruhe,

Grimm, Mathias Lehrstuhl für Allgemeine und Theoretische Elektrotechnik, Technische Fakultät der CAU Kiel, Kaiserstr. 2, 24143 Kiel,

Grooteman, Frank National Aerospace Laboratory NLR, Structures Department, Postbus 153, NL - 8300 AD Emmeloord, The Netherlands,

Grue, John University of Oslo, Dept. of Mathematics, P.b. 1053 Blindern, N-0316 Oslo, Norway,

Hackbusch, Wolfgang Lehrstuhl Praktische Mathematik, Mathematisches Seminar II, Christian-Albrechts-Universität Kiel, Hermann-Rodewald-Str. 3, 24098 Kiel,

Hämäläinen, Jyri University of Oulu, Faculty of Technology, Department of Mathematics, Linnanmaa, SF-90570 Oulu, Finland,

Hayami, Ken Department of Mathematical Engineering and Information Physics, Graduate School of Engineering, University of Tokyo, 7-3-1, Hongo, Bunkyo-ku, 113 Tokyo, Japan,

Hochmuth, Reinhard I. Mathematisches Institut, FU-Berlin, Arnimallee 3, 14195 Berlin,

Karahan, Sedat Lehrstuhl für Allgemeine und Theoretische Elektrotechnik, Technische Fakultät der CAU Kiel, Kaiserstr. 2, 24143 Kiel,

Katzer, Edgar Eppendorfer Landstraße 16, 20249 Hamburg,

Keller, Torsten Techn. Universität Hamburg-Harburg, Arbeitsbereich Meerestechnik II, Eißendorfer Str. 42, 21073 Hamburg,

Keunecke, Jan Lehrstuhl für Allgemeine und Theoretische Elektrotechnik, Technische Fakultät der CAU Kiel, Kaiserstr. 2, 24143 Kiel,

Kharitonov, Alexander A. Dniepropetrovsk State University, 72 Gagarin Avenue, 320625 Dniepropetrovsk 10, Russia,

Kleemann, Bernd H. Berliner Institut für Optik GmbH (BIFO), Rudower Chaussee 5, D-12484 Berlin,

Köhler, Olaf Lehrstuhl für Technische Mechanik, Universität Erlangen-Nürnberg, Egerlandstraße 5, 91058 Erlangen,

Kral, Roland Technische Universität Hamburg-Harburg, Arbeitsbereich Meerestechnik II, Eißendorfer Strasse 42, 21073 Hamburg,

Kuhn, Michael Institut f. Mathematik, Johannes Kepler Universität Linz, Altenberger Str. 69, A-4040 Linz, Austria,

Lage, Christian Seminar für Angewandte Mathematik, ETH Zürich, Rämistrasse 101, CH-8092 Zürich,

Langer, Ulrich Institut f. Mathematik, Johannes Kepler Universität Linz, Altenberger Str. 69, A-4040 Linz, Austria,

Lebert, Klaus Lehrstuhl für Regelungs- und Automatisierungstechnik, Technische Fakultät der CAU Kiel, Kaiserstr. 2, 24143 Kiel,

Lehmann, Rüdiger Geodetic Institute, University of Karlsruhe, Englerstr 7, D-76128 Karlsruhe,

Liebau, Frank Technische Universität Berlin, Fachbereich 3 - Mathematik, Sekr. MA 6-2, Straße des 17.Juni 136, 10623 Berlin,

Maar, Bernd Institut für Computeranwendungen III, Universität Stuttgart, Pfaffenwaldring 27, 70550 Stuttgart,

Marburg, Steffen Institut für Festkörpermechanik, Technische Universität, 01062 Dresden,

Matthies, Hermann G. Institut für Wissenschaftliches Rechnen, TU Braunschweig, Hans-Sommer-Str. 65, 38106 Braunschweig,

Mayer, Jörg-Peter Lehrstuhl Praktische Mathematik, Mathematisches Seminar II, Christian-Albrechts-Universität Kiel, Hermann-Rodewald-Str. 3, 24098 Kiel,

Mayo, Anita IBM TJ Watson Research Center, Yorktown, NY 10598, USA,

McKenzie, Rod School of Computing and Mathematics, University of Teesside, Southfield Road, Middlesbrough, Cleveland, TS1 3BA, United Kingdom,

Merkel, Markus Lehrstuhl für Technische Mechanik, Universität Erlangen-Nürnberg, Egerlandstraße 5, 91058 Erlangen,

Michelsson, Olaf TU Ilmenau, Fakultät E/I, FA Theor. Elektrotechnik, 98684 Ilmenau,

Modersitzki, Jan Institut für Mathematik, Medizinische Universität zu Lübeck, Wallstrasse 40, 23560 Lübeck,

Mordhorst, Uwe Rechenzentrum der Universität Kiel, Olshausenstr. 40, 24098 Kiel,

Nygaard, Jens Olav Matematisk Institutt, Postboks 1053, Blindern, Oslo, Norway,

Potts, Daniel Fachbereich Mathematik, Universität Rostock, 18051 Rostock,

Quatember, Ralf Mathematisches Institut A, 6. Lehrstuhl, Universität Stuttgart, Pfaffenwaldring 57, 70550 Stuttgart,

Rathsfeld, Andreas WIAS-Berlin, Mohrenstr.39, 10117 Berlin ,

Röck, Helmut Lehrstuhl für Regelungs- und Automatisierungstechnik, Technische Fakultät der CAU Kiel, Kaiserstr. 2, 24143 Kiel,

Saranen, J. University of Oulu, Faculty of Technology, Department of Mathematics, Linnanmaa, SF-90570 Oulu, Finland,

Sauren, Manfred Central Institute for Applied Mathematics, Research Centre Jülich (KFA), 52425 Jülich,

Sauter, Stefan A. Lehrstuhl Praktische Mathematik, Mathematisches Seminar II, Christian-Albrechts-Universität Kiel, Hermann-Rodewald-Str. 3, 24098 Kiel,

Schneider, Reinhold FB Mathematik, Technische Hochschule Darmstadt, Schloßgartenstr. 7, 64289 Darmstadt,

Schwab, Christoph SAM, ETH Zürich, Rämistrasse 101, CH-8092 Zürich,

Sonnad, Vijay Mail Stop 9233, IBM Corporation, 11400 Burnet Road, Austin, Texas 78758, USA,

Stanka, Rudolf Lehrstuhl für Regelungs- und Automatisierungstechnik, Technische Fakultät der CAU Kiel, Kaiserstr. 2, 24143 Kiel,

Steidl, Gabriele TH Darmstadt, FB Mathematik, Schlossgartenstrasse 7, 64289 Darmstadt,

Steinbach, Olaf Universität Stuttgart, Mathematisches Institut A, Pfaffenwaldring 57, 70569 Stuttgart,

Stephan, Ernst P. Institut f. Angew. Mathematik, Universität Hannover, Welfengarten 1, 30060 Hannover,

Stevenson, Rob P. Department of Mathematics, University of Nijmegen, Toernooiveld, 6525 ED Nijmegen, The Netherlands,

Szikrai, Szabolcs Institut für Technische Mechanik / Festigkeitslehre, Kaiserstr. 12, 76128 Karlsruhe,

Türke, K. Institut f. Technische Mechanik, Fertigkeitslehre, Universität Karlsruhe, Kaiserstr. 12, 76128 Karlsruhe,

Weinlich, Uwe Lehrstuhl für Regelungs- und Automatisierungstechnik, Technische Fakultät der CAU Kiel, Kaiserstr. 2, 24143 Kiel,

Wendland, Wolfgang Mathematisches Institut A, Universität Stuttgart, Pfaffenwaldring 57, 70550 Stuttgart,

How to contact the authors by email:

Andrä, Heiko .... heiko@imfsun2.mach.uni-karlsruhe.de

Chen, Ke .... K.Chen@liverpool.ac.uk

Faermann, Birgit .... bf@numerik.uni-kiel.de

Finck, Tilo .... tfinck@mathematik.tu-chemnitz.de

Funken, Stefan A. .... saf@numerik.uni-kiel.de

Gáspár, Csaba .... csgaspar@d7.szif.hu

Grooteman, F.P. .... grooten@nlr.nl

Hayami, Ken .... hayami@simplex.t.u-tokyo.ac.jp

Heuer, Norbert .... heuer@lrw.uni-bremen.de

Hochmuth, Reinhard .... hochmuth@math.fu-berlin.de

Kleemann, Bernd H. .... kleemann@wias-berlin.de

Klees, Roland .... klees@geo.tudelft.nl

Kuhn, Michael .... kuhn@miraculix.numa.uni-linz.ac.at

Lage, Christian .... lage@sam.math.ethz.ch

Lehmann, Rüdiger .... lehmann@gik.bau-verm.uni-karlsruhe.de

Mayo, Anita .... amayo@watson.ibm.com

Mund, Patrick .... mund@ifam.uni-hannover.de

Rathsfeld, Andreas .... rathsfeld@wias-berlin.de

Sauter, Stefan A. .... sas@numerik.uni-kiel.de

Schippers, H. .... schipiw@nlr.nl

Schneider, Reinhold .... schneider@mathematik.th-darmstadt.de

Schwab, Christoph .... schwab@sam.math.ethz.ch

Schulz, Heiko .... heiko@mathematik.uni-stuttgart.de

Sonnad, Vijay .... sonnad@vnet.ibm.com

Steinbach, Olaf .... steinbach@mathematik.uni-stuttgart.de

Stephan, Ernst .... stephan@ifam.uni-hannover.de

Türke, Karsten .... tuerke@imfserv.mach.uni-karlsruhe.de

Wendland, Wolfgang .... wendland@mathematik.uni-stuttgart.de

Yamada, Yosihiro .... yamada@shimadzu.co.jp

Addresses of the Editors of the Series "Notes on Numerical Fluid Mechanics"

Prof. Dr. Ernst Heinrich Hirschel (General Editor)
Herzog-Heinrich-Weg 6
D-85604 Zorneding
Federal Republic of Germany

Prof. Dr. Kozo Fujii
High-Speed Aerodynamics Div.
The ISAS
Yoshinodai 3-1-1, Sagamihara
Kanagawa 229
Japan

Prof. Dr. Bram van Leer
Department of Aerospace Engineering
The University of Michigan
3025 FXB Building
1320 Beal Avenue
Ann Arbor, Michigan 48109-2118
USA

Prof. Dr. Michael A. Leschziner
UMIST-Department of Mechanical Engineering
P.O. Box 88
Manchester M60 1QD

Prof. Dr. Maurizio Pandolfi
Dipartimento di Ingegneria Aeronautica e Spaziale
Politecnico di Torino
Corso Duca Degli Abruzzi, 24
I-10129 Torino
Italy

Prof. Dr. Arthur Rizzi
Royal Institute of Technology
Aeronautical Engineering
Dept. of Vehicle Engineering
S-10044 Stockholm
Sweden

Dr. Bernard Roux
Institut de Mécanique des Fluides
Laboratoire Associé au C.R.N.S. LA 03
1, Rue Honnorat
F-13003 Marseille
France

Brief Instruction for Authors

Manuscripts should have well over 100 pages. As they will be reproduced photomechanically they should be produced with utmost care according to the guidelines, which will be supplied on request. In print, the size will be reduced linearly to approximately 75 per cent. Figures and diagrams should be lettered accordingly so as to produce letters not smaller than 2 mm in print. The same is valid for handwritten formulae. Manuscripts (in English) or proposals should be sent to the general editor, Prof. Dr. E. H. Hirschel, Herzog-Heinrich-Weg 6, D-85604 Zorneding.

Notes on Numerical Fluid Mechanics (NNFM) Volume 54

Volume 35 Proceedings of the Ninth GAMM-Conference on Numerical Methods in Fluid Mechanics (J. B. Vos / A. Rizzi / I. L. Ryhming, Eds.)

Volume 34 Numerical Solutions of the Euler Equations for Steady Flow Problems (A. Eberle / A. Rizzi / E. H. Hirschel)

Volume 33 Numerical Techniques for Boundary Element Methods (W. Hackbusch, Ed.)

Volume 32 Adaptive Finite Element Solution Algorithm for the Euler Equations (R. A. Shapiro)

Volume 31 Parallel Algorithms for Partial Differential Equations (W. Hackbusch, Ed.)

Volume 30 Numerical Treatment of the Navier-Stokes Equations (W. Hackbusch / R. Rannacher, Eds.)

MIX
Papier aus verantwortungsvollen Quellen
Paper from responsible sources
FSC® C105338

If you have any concerns about our products,
you can contact us on
ProductSafety@springernature.com

In case Publisher is established outside the EU,
the EU authorized representative is:
Springer Nature Customer Service Center GmbH
Europaplatz 3, 69115 Heidelberg, Germany

Printed by Libri Plureos GmbH
in Hamburg, Germany